Nanomaterials for Biomedical and Biotechnological Applications

Nanomaterials for Biomedical and Biotechnological Applications

Editor

Angelo Ferraro

MDPI • Basel • Beijing • Wuhan • Barcelona • Belgrade • Manchester • Tokyo • Cluj • Tianjin

Editor
Angelo Ferraro
Electrical and Computer
Engeeniring
National Technical University
of Athnes
Athens
Greece

Editorial Office
MDPI
St. Alban-Anlage 66
4052 Basel, Switzerland

This is a reprint of articles from the Special Issue published online in the open access journal *Nanomaterials* (ISSN 2079-4991) (available at: www.mdpi.com/journal/nanomaterials/special_issues/nano_biomedical).

For citation purposes, cite each article independently as indicated on the article page online and as indicated below:

LastName, A.A.; LastName, B.B.; LastName, C.C. Article Title. *Journal Name* **Year**, *Volume Number*, Page Range.

ISBN 978-3-0365-4700-8 (Hbk)
ISBN 978-3-0365-4699-5 (PDF)

Contents

About the Editor

Angelo Ferraro

Dr. Angelo Ferraro was awarded by the University of Calabria (Italy) with a degree in Biological Sciences in 2002. In 2010, he finished his PhD in Molecular Medicine at the University of Naples "Federico II"(Italy). His current position, as a senior research fellow, is at the Department of Electronic and Computer Engineer, National Technical University of Athens (NTUA), Greece.

His main fields of research are cancer molecular biology, biomaterials and biotechnologies connected with the use of nanomaterials. Specifically, in the field of cancer molecular biology, his studies have focused on the mutational analysis of oncogenes, deregulation of the MAPK pathway and epigenetic factors such as histone modifier enzymes in thyroid, breast and colon cancers. In recent years, he has been investigating the use of magnetic and non-magnetic nanoparticles as tools against cancer diseases and for biotechnological applications. At NTUA laboratories, he is investigating new approaches to enhance the effectiveness of hyperthermia (over-hyperthermia); the possibility of using rare-earth fluoride-based nanoparticles and photodynamic therapy (PDT) to specifically kill cancer cells; and developing biosensors based on affinity ligands and interdigitated capacitors. Concerning the biotechnological use of nanomaterials, his studies are focused on magnetic nanoparticles and magnetic photobioreactors to cultivate and harvest microalgae biomass. Furthermore, he is developing a technology named Selective Magnetic Separation based on the use of affinity ligands and magnetic beads to isolate from complex mixture organic molecules an inorganic and ions. He has published 70+ peer-reviewed papers (citations>4000; h-index=30).

Preface to "Nanomaterials for Biomedical and Biotechnological Applications"

Nanomaterials have long been used in basic research at universities and research centres; therefore, the scientific community is well aware of their potential. What is still not well known is that almost all kinds of nanomaterials, when properly configurated, can also be used in medical applications and in industrial processing—in the last case both for catabolic and anabolic biochemical reactions. The scope of this Special Issue is to collect cutting-edge research works dealing with nanomaterials that have the potential to produce tools suitable for biomedical and biotechnological applications.

The authors of this reprint are specialized scholars from high-profile universities who have contributed with research and review papers. We called on academic, biomedical and industrial stakeholders who wished to be updated on the latest applications of a variety of nanomaterials.

Angelo Ferraro
Editor

nanomaterials

Editorial

Special Issue "Nanomaterials for Biomedical and Biotechnological Applications"

Angelo Ferraro

School of Electrical and Computer Engineering, National Technical University of Athens, Zografou Campus, Zografou, 15772 Athens, Greece; an.ferraro2@gmail.com

Citation: Ferraro, A. Special Issue "Nanomaterials for Biomedical and Biotechnological Applications". *Nanomaterials* **2022**, *12*, 1923. https://doi.org/10.3390/nano12111923

Received: 21 February 2022
Accepted: 9 March 2022
Published: 4 June 2022

In the last few decades, biomedical and biotechnological researchers have turned their interest to nanocomposite materials. One of the first applications of nanomaterials in medical fields was the use of Gadolinium and Iron Oxide nanoparticles as contrast agents in MRI imaging. Another well-established clinical application of nanomaterials is the use of gold nanoparticles to enhance the performance of rapid screen tools such as lateral flow tests, widely in use nowadays for the COVID-19 pandemic. Since these very first applications, a growing body of scientific work has been published showing the extraordinary advantages offered by nanomaterials in speeding up chemical reactions, enhancing the yield of cell-based catalysis, ameliorating the performance of medications and many other health care services.

The need for constant improvement to reach a high standard of safety and to make nanomaterials accessible for marketing has generated, as mentioned before, a considerable number of scientific papers that highlight new important aspects to be considered, such as synthesis, stability, biocompatibility, and easy manipulation. In order to provide a comprehensive update on the latest discoveries concerning the biomedical and biotechnological applications of nanomaterials, this Special Issue of *Nanomaterials* collected 14 papers: 10 as research articles and 4 as reviews.

The topics analysed in this Special Issue focus on the use of magnetic iron oxide nanoparticles, gold nanoparticles functionalized ciprofloxacin, boron nitride nanotubes, poly-ε-caprolactone composite sorbent, titanium nanotubes and chitosan–clay nanocomposites, which have been applied in both biomedical and biotechnological fields. On the other hand, the review papers focus on nanotechnology-based diagnostic and therapeutic approaches to tackle colorectal cancer and SARS-CoV-2-related pathologies; nanomedicine applications for ocular drug delivery and the different methods to synthetize cerium oxide nanoparticles meant to be used as a clinical tool.

Iron Oxide nanoparticles present low cellular toxicity, can be easily synthesised and, most importantly, are easily functionalized with a variety of chemical groups. Four papers have been published showing the versatility of iron oxide nanoparticles. The work of G. Banis et al. [1] focuses on the advantages presented by the superparamagnetic feature of iron oxide nanoparticles, highlighting that they are magnetic only when surrounded by a magnetic field. Such a property allows to remotely control and thus drive magnetic nanoparticles using a proper magnetic field, which can be generated either by permanent magnets or by electromagnets. In their paper, G. Banis and colleagues describe an ingenious electromagnet apparatus with the ability to precisely push small drops of magnetic nanoparticles. The novelty of this research is the possibility to turn, push or attract magnetic nanomaterials by controlling the intensity of the electric current feeding the electromagnet array. The paper is accompanied by a video (available on *Nanomaterials* Journal), where the authors show the path of magnetically driven nanoparticles suspended in mineral oil. The clinical implications of these achievements are several, indeed it is now possible to hypothesise tissue and organ-specific drug administration by conjugating medications to magnetic nanoparticles and subsequently using tuneable magnetic fields to reach the internal target sites.

In addition to superparamagnetism, iron oxide nanocomposites present another interesting physical property: the ability to increase their temperature up to hundreds of degrees Celsius when embedded in an alternating magnetic field; the so-called magnetic hyperthermia. Such controlled thermal energy can be used as a physical tool to specifically kill, e.g., dangerous tumour cells. Sánchez et al. [2] performed an interesting study on how to produce cubic magnetic nanoparticles with sizes lower than the monodomain critical value while maintaining the supermagnetism critical value, which guarantees no agglomeration and, thus, nanoparticles remain in a colloid state when in solution. They optimised several synthesis variables such as ramp-up temperature, stirring speed, type of solvent and proportions of raw materials in order to achieve the production goals. Another important issue examined in the study was the economic and technological viability of production in order to set the path for future industrial implementation to support clinical widespread applications of magnetic hyperthermia against tumours.

In this Special Issue, the use of nanoparticles to tackle cancer malignancies both at diagnostic and at therapeutic levels have also been addressed by an excellent review paper prepared by Al-Joufi and colleagues [3]. The review, in the first part, provides an exhaustive but simple overview of colorectal cancer (CRC) disease and the standard diagnostic tools, whereas in the second part it presents the use of advanced nanomaterials for CRC diagnosis and treatment.

Another important clinical topic treated in the Special Issue is antibiotic resistance towards drugs used to cure infectious diseases caused by a variety of microbes. This phenomenon is becoming a global concern since a growing number of bacteria strains are now resistant to treatments, presenting a significant threat for human health. Nawaz and colleges [4] present an interesting paper on how to enhance the efficacy of the antibiotic Ciprofloxacin by conjugating it to gold nanoparticles. The research focuses on the synthesis and physical characterization of ciprofloxacin-loaded gold nanoparticles (CIP-AuNPs) and their effect on the colonisation of the bacterium Enterococcus faecalis in the liver and kidneys of mice. Authors demonstrate the effectiveness of CIP-AuNPs in eradicating E. faecalis from the host tissues and, unlike CIP alone, CIP-AuNPs were non-haemolytic, thus enhancing also the drug's safety.

Matters of improving safety and biocompatibility have been tackled also for orthopaedic implants that are used to repair bone traumas and damages. Baker et al. [5] propose in their research paper that a new surface treatment with Titanium nanotubes (TiNT) of implants may promote and improve osseointegration—a phenomenon that leads to the integration of the implant in the damaged bone. In the study, local and systemic responses to aligned and trabecular TiNT (the two most common TiNT surface morphologies) are assessed through in vitro cellular, as well as in vivo, experiments.

In the contest of nanomedicine, Silva et al. [6] report how to improve the performance of boron nitride nanotubes (BNNTs), which are employed in several medical areas. They combined copper-64 with BNNTs and synthesized and characterized 64Cu-BNNTs with appreciable properties that suggest numerous multifunctional applications, with advantages for cancer diagnosis and therapy.

Suresh and colleagues [7] also focus their work on biomedical field and in particular on how to improve drug activity to tackle rheumatoid arthritis. In their research paper it is show a novel formulation of a methotrexate-loaded nanoemulsion for subcutaneous administration. Even though the paper presents initial data that must be confirmed, the authors show that methotrexate-loaded nanoemulsion, at lower concentration, is safer for systemic circulation and that anti-arthritic activity of the nanoemulsion in CFA-induced animals is superior compared to the market drug.In connection with the previous studies, the group of Veloso and colleagues [8] reported the performance improvement of drug carrier-like peptide-based hydrogels by using two different functionalized magnetic particles. The paper showed how to use two different functionalized nanoparticles, citrate-stabilized and lipid-coated magnetic nanoparticles, in the formation of dehydropeptide-based supramolecular magnetogels. These nanocomposites have been tested for the encapsulation and release of Doxorubicin—one of the most commonly used chemotherapeutic drugs in a wide variety of cancers.

An interesting review is also presented by Khiev and colleagues [9], who analysed the topic of nanotechnology-based drug delivery systems in treating ocular diseases. The review highlights the potential of nanomaterial tools to overcome several limitations of conventional therapies during drug delivery across the blood–retinal barrier, making it a major clinical challenge. Therefore, the review summarizes the development of organic and inorganic nanoparticles for ophthalmic applications.

The use of nanomaterials to tackle other human health threats such as the SARS-CoV-2 pandemic is similarly reviewed by Asdaq and colleagues [10] and published in the Special Issue. The review, in the first part, focuses on the virus's basic structure, pathogenesis, and the current treatment options for COVID-19. In the second part, the study addresses nanotechnology and its applications in diagnosis, prevention, treatment, and targeted vaccine delivery, laying the groundwork for a successful pandemic fight.

Another critical review study is presented by Nyoka et al. [11], where the authors focus on the synthesis of cerium oxide nanoparticles by analysing various methods and explaining how the synthesis method affects physicochemical properties, behaviour in biological environments and catalytic abilities, as well as toxicity.

Concerning the chemistry application of nanomaterials, the theme of complex biological mixture separation is discussed by Raabová and colleagues [12]. In their research paper, the authors describe the advantages of combining poly-ε-caprolactone micro- and nanofibers for the online extraction of analytes from milk and serum matrixes, and confirm the protein removal capability, good mechanical stability and re-usability of this sorbent in more than 300 analyses.

Finally, the following two papers of this Special Issue address biotechnological applications.

M. Savvidou et al. [13] illustrate in their work a biotechnological application of iron oxide nanoparticles. Additionally, in this case they take advantage of the superparamagnetic property of iron oxide nanoparticles. They report how to harvest microalgae cells from the culture medium. Microalgae are photosynthetic single cell organisms that grow both in sea and fresh water. These marvellous forms of life are able to synthetise good quality proteins, lipids, antioxidant, food dye and other metabolites used for food and cosmetic applications. They require water medium to grow and, in order to collect their valuable products, must be separated from their culture liquid. Several de-watering technologies are available, such as filtration, centrifugation, precipitation, decantation, but all of them are time-consuming, high energy-demanding and expensive. M. Savvidou and her group report how the use of magnetic nanoparticles improved de-watering yield and at the same time lowered time and cost of microalgae harvesting.

The work of Benucci et al. [14] focuses on the development of new nanocomposite materials for food application. In their research, the authors synthetised chitosan–clay nanocomposite films to be used as carriers for the covalent immobilisation of the proteolytic enzyme, Papain. This enzyme is used to stabilise various types of food and drinks. The paper demonstrates how the use of chitosan–clay nanocomposite enhances the ability of Papain to stabilise white wine.

Funding: This research received no external funding.

Institutional Review Board Statement: Not Applicable.

Informed Consent Statement: Not Applicable.

Conflicts of Interest: The author declares no conflict of interest.

References

1. Banis, G.; Tyrovolas, K.; Angelopoulos, S.; Ferraro, A.; Hristoforou, E. Pushing of Magnetic Microdroplet Using Electromagnetic Actuation System. *Nanomaterials* **2020**, *10*, 371. [CrossRef] [PubMed]
2. Sánchez, O.S.; Castelo-Grande, T.; Augusto, P.A.; Compaña, J.M.; Barbosa, D. Cubic Nanoparticles for Magnetic Hyperthermia: Process Optimization and Potential Industrial Implementation. *Nanomaterials* **2021**, *11*, 1652. [CrossRef] [PubMed]
3. Al-Joufi, F.A.; Setia, A.; Salem-Bekhit, M.M.; Sahu, R.K.; Alqahtani, F.Y.; Widyowati, R.; Aleanizy, F.S. Molecular Pathogenesis of Colorectal Cancer with an Emphasis on Recent Advances in Biomarkers, as Well as Nanotechnology-Based Diagnostic and Therapeutic Approaches. *Nanomaterials* **2022**, *12*, 169. [CrossRef]
4. Nawaz, A.; Ali, S.M.; Rana, N.F.; Tanweer, T.; Batool, A.; Webster, T.J.; Menaa, F.; Riaz, S.; Rehman, Z.; Batool, F.; et al. Ciprofloxacin-Loaded Gold Nanoparticles against Antimicrobial Resistance: An In Vivo Assessment. *Nanomaterials* **2021**, *11*, 3152. [CrossRef]
5. Baker, E.; Fleischer, M.; Vara, A.; Salisbury, M.; Baker, K.; Fortin, P.; Friedrich, C. Local and Systemic In Vivo Responses to Osseointegrative Titanium Nanotube Surfaces. *Nanomaterials* **2021**, *11*, 583. [CrossRef]
6. Silva, W.M.; Ribeiro, H.; Taha-Tijerina, J.J. Potential Production of Theranostic Boron Nitride Nanotubes (^{64}Cu-BNNTs) Radiolabeled by Neutron Capture. *Nanomaterials* **2021**, *11*, 2907. [CrossRef] [PubMed]
7. Suresh, P.; Salem-Bekhit, M.M.; Veedu, H.P.; Alshehri, S.; Nair, S.C.; Bukhari, S.I.; Viswanad, V.; Taha, E.I.; Sahu, R.K.; Ghoneim, M.M.; et al. Development of a Novel Methotrexate-Loaded Nanoemulsion for Rheumatoid Arthritis Treatment with Site-Specific Targeting Subcutaneous Delivery. *Nanomaterials* **2022**, *12*, 1299. [CrossRef] [PubMed]
8. Veloso, S.; Silva, J.; Hilliou, L.; Moura, C.; Coutinho, P.; Martins, J.; Testa-Anta, M.; Salgueiriño, V.; Correa-Duarte, M.; Ferreira, P.; et al. Impact of Citrate and Lipid-Functionalized Magnetic Nanoparticles in Dehydropeptide Supramolecular Magnetogels: Properties, Design and Drug Release. *Nanomaterials* **2020**, *11*, 16. [CrossRef] [PubMed]
9. Khiev, D.; Mohamed, Z.A.; Vichare, R.; Paulson, R.; Bhatia, S.; Mohapatra, S.; Lobo, G.P.; Valapala, M.; Kerur, N.; Passaglia, C.L.; et al. Emerging Nano-Formulations and Nanomedicines Applications for Ocular Drug Delivery. *Nanomaterials* **2021**, *11*, 173. [CrossRef] [PubMed]
10. Asdaq, S.; Ikbal, A.; Sahu, R.; Bhattacharjee, B.; Paul, T.; Deka, B.; Fattepur, S.; Widyowati, R.; Vijaya, J.; Al Mohaini, M.; et al. Nanotechnology Integration for SARS-CoV-2 Diagnosis and Treatment: An Approach to Preventing Pandemic. *Nanomaterials* **2021**, *11*, 1841. [CrossRef] [PubMed]
11. Nyoka, M.; Choonara, Y.E.; Kumar, P.; Kondiah, P.P.D.; Pillay, V. Synthesis of Cerium Oxide Nanoparticles Using Various Methods: Implications for Biomedical Applications. *Nanomaterials* **2020**, *10*, 242. [CrossRef] [PubMed]
12. Raabová, H.; Havlíková, L.C.; Erben, J.; Chvojka, J.; Švec, F.; Šatínský, D. Polycaprolactone Composite Micro/Nanofibrous Material as an Alternative to Restricted Access Media for Direct Extraction and Separation of Non-Steroidal Anti-Inflammatory Drugs from Human Serum Using Column-Switching Chromatography. *Nanomaterials* **2021**, *11*, 2669. [CrossRef] [PubMed]
13. Savvidou, M.; Dardavila, M.; Georgiopoulou, I.; Louli, V.; Stamatis, H.; Kekos, D.; Voutsas, E. Optimization of Microalga *Chlorella vulgaris* Magnetic Harvesting. *Nanomaterials* **2021**, *11*, 1614. [CrossRef] [PubMed]
14. Benucci, I.; Lombardelli, C.; Cacciotti, I.; Esti, M. Papain Covalently Immobilized on Chitosan–Clay Nanocomposite Films: Application in Synthetic and Real White Wine. *Nanomaterials* **2020**, *10*, 1622. [CrossRef] [PubMed]

nanomaterials

MDPI

Article

Pushing of Magnetic Microdroplet Using Electromagnetic Actuation System

Georgios Banis [1,*], Konstantinos Tyrovolas [1], Spyridon Angelopoulos [1], Angelo Ferraro [1] and Evangelos Hristoforou [1,2]

1 Laboratory of Electronic Sensors, National TU of Athens, Zografou Campus, 15780 Athens, Greece; kwstastyr@gmail.com (K.T.); spyrosag@gmail.com (S.A.); ferraro@eie.gr (A.F.); hristoforou@ece.ntua.gr (E.H.)

2 Department of Metal Protection and Surface Treatment, National University of Science and Technology (MISiS), 119049 Moscow, Russia

* Correspondence: gbanis@central.ntua.gr

Received: 14 January 2020; Accepted: 18 February 2020; Published: 20 February 2020

Abstract: Treatment of certain diseases requires the administration of drugs at specific areas of tissues and/or organs to increase therapy effectiveness and avoid side effects that may harm the rest of the body. Drug targeting is a research field that uses various techniques to administrate therapies at specific areas of the body, including magnetic systems able to drive nano "vehicles", as well as magnetically labeled molecules, in human body fluids and tissues. Most available actuation systems can only attract magnetic elements in a relatively small workspace, limiting drug target applications to superficial tissues, and leaving no alternative cases where deep targeting is necessary. In this paper, we propose an electromagnetic actuation system able to push and deflect magnetic particles at distance of ~10 cm, enabling the manipulation of magnetic nano- and microparticles, as well as administration of drugs in tissues, which are not eligible for localized drug targeting with state-of-the-art systems. Laboratory experiments and modeling were conducted to prove the effectiveness of the proposed system. By further implementing our device, areas of the human body that previously were impossible to treat with magnetically labeled materials such as drugs, cells, and small molecules can now be accessible using the described system.

Keywords: magnetic driving; magnetic nanoparticles; actuation system; drug targeting

1. Introduction

The remote manipulation of nano- and microparticles attracts researchers from various scientific fields. Several technologies were adopted, whereby magnetism and composite materials seem to be the most appropriate tools for this task. The applications of magnetic particles in biomedical and clinical research fields are several. They are used for drug delivery in tumors [1–4], infective diseases [5], thrombosis [6], localized administration of drugs [7], stem-cell therapies [8–10], and gene delivery [11,12]. To manipulate magnetic particles, permanent magnets or electromagnets can be used. Permanent magnets propagate high-gradient magnetic fields able to retain magnetic particles in relatively small distances (up to 5 cm); however, the propagated field intensity cannot be adjusted. Therefore, their clinical applications are limited, e.g., for retaining magnetically responsive materials at small distances. A second alternative of magnetic manipulation relies on the use of electromagnets which enable the possibility to fine-tune or switch on/off the generated magnetic field. Many attempts were made to manipulate magnetic agents for clinical purposes such as the Octomag [13] or the Mag-μBot system [14]. Nevertheless, all these systems can only attract magnetically labeled items because of the paramagnetic nature of magnetic nano- and microparticles. Furthermore, an additional obstacle in driving pharmaceutical agents in deep tissues and organs is human body size [15]. In order to reach deep targets while preserving the ability to manipulate pharmaceutical agents, it is necessary to magnetically push nano- and/or microparticles (magnetic injection) deep into the human body. In an early study using permanent magnets with a specific configuration, Sarwar and his colleagues successfully managed to push magnetic particles into rat inner ears, keeping the apparatus at a corresponding "human head" working distance [16,17]. Nevertheless, the system proposed by Sarwar has the major drawback of moving parts. To achieve the pushing effect, all magnetic apparatus must be moved, implying a complex design and a laborious use of the entire device. Electromagnetic driving devices designed and developed by Probst et al. and Bijarchi et al. were also proposed recently [18,19]. However, these electromagnetic devices with adjustable magnetic gradients cannot manipulate magnetic microdroplets at human body working distances.

In this work, we present the design of an actuation system able to manipulate (deflect and push) magnetic particles at relatively big distances from its cores without the need for moving parts. Indeed, the pushing effect is obtained by fine-tuning the electric current feeding a specifically configurated electromagnet array. The optimum design and working conditions (angle, current for each coil, distance, etc.) were optimized by using computer simulations; subsequently, the magnetic system was constructed and tested successfully. This design paves the way for driving pharmaceutical agents in deep areas of the human body. The fact that electromagnets are used offers a promising and fully controllable tool for biomedical applications to push, deflect, or pull magnetic particles at relatively long distances from the magnetic core.

2. Materials and Methods

2.1. Magnetic System

The distance between the electromagnet and the magnetic particles to be driven is a key parameter in the equation governing any magnetic system that is intended to be used for the remote control of magnetic nano- and/or microparticles [20]. Indeed, the magnetic field intensity must be increased as the distance increases from the magnets. To design our system, we took advantage of the basic principal that, thanks to a specific electromagnet configuration, the magnetic field lines sum together at one specific point due to the linearity of Maxwell's equations. On this specific point, magnetic lines cancel each other, creating a magnetic field minimum. The magnetic field strength will increase outward from this minimum and create outward forces. The apparatus that was constructed following this reasoning and presented here consists of four coils, two steel cores, four power supplies, and a control system.

To achieve an increasing magnetic field gradient which pushes magnetic particles, the magnetic lines have to cancel themselves (meet at 180°) at a desirable point and, after that, the magnetic field vectors should meet again with an angle different from 180°. We used the finite element method through ANSYS 17 (Ansys, Canonsburg, PA, USA, licensed to National Technical University of Athens) to test different electromagnet arrangements with various angles and various current combinations to optimize the magnetic line alignments and obtain a cancellation node, where only one configuration was selected to be experimentally validated and is presented hereafter. The magnetic core dimensions were $60 \times 60 \times 170$ mm^3, made from electrical steel SA1008 (Sidenor steel industry, Maroussi, Greece), and the inner and outer coils were wound in 350 and 540 turns, respectively, with 2-mm-diameter copper wire. At each core, two coils were wound (Figure 1). Magnetic cores were surrounded by wooden holders and connected by an adjustable wooden bar to regulate the angle between electromagnetic elements.

2.2. Control System

The control system was designed and developed in order to control the characteristics of the supply signal and feed each coil separately with pulsing direct current (DC) (1 s duty cycle, 1 s pulse width). In order to accomplish this, the control system was based on the pulse-width modulation (PWM) technique, according to which the modulation of the width (duration) of a pulsed signal results in the amplitude adjustment of the output voltage. As a result, the designed control system can be used to adjust the current, as well as the pulse width (duty cycle) supplied to each coil. The schematic of the designed circuit is provided in Figure S1 (Supplementary Materials). It consists of four channels, each of which can be connected to a different coil. Each channel is controlled by a current mode controller (TS3843B). The output of each controller can be adjusted by a potentiometer, which is located at the upper panel of the control system. A fifth knob is used in order to control the pulse duration, through a selection among 12 different capacitor networks, which are connected to the controller through a Schmitt trigger inverter (74HC14). Finally, the adjusted controller's output is fed to the gate of a power metal–oxide–semiconductor field-effect transistor (MOSFET) (IRFP3306PbF), which allows the current to flow to the connected coil.

The designed circuit was fabricated on a printed circuit board (PCB), having the required terminals for the connection of the coils. Four power supplies with a maximum output power of 600 W (24 V, 25 A) each were used in order to provide sufficient power to the coils.

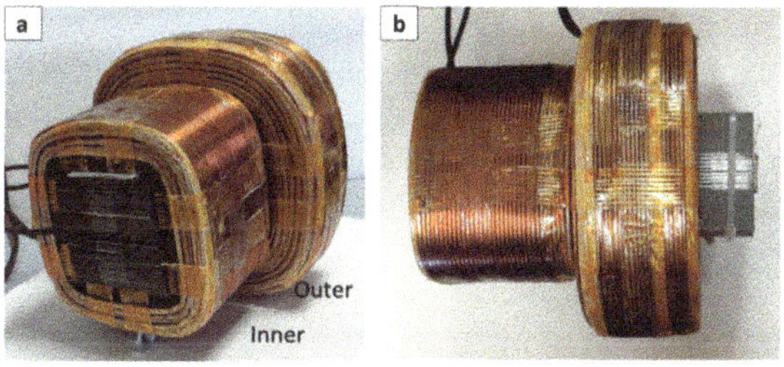

Figure 1. Electromagnet unit arrangement with the two coils (outer and inner) and the electrical steel core: (**a**) front view; (**b**) top view.

2.3. Nanoparticles

For the experiments, a microdroplet of superparamagnetic nanoparticles was used. The product (fluidMAG-Lipid, article number 4119-5) was purchased from Chemicell GmbH, Berlin, Germany. The nanoparticle core was magnetite-functionalized with a matrix phosphatidylcholine, resulting in a hydrodynamic diameter of 100 nm. The surfactant (matrix phosphatidylcholine) is hydrophilic; therefore, nanoparticles were dissolved in water. For the experiment, oleic acid was used in order to force the nanoparticles to form a microdroplet. The nanoparticle buffer was sterile water, and a micropipette with adjustable volume between 0–10 μL was used to transfer the microdroplet of nanoparticles in the oleic acid.

2.4. Microscope

Images and videos were recorded using a simple microscope made of plastic elements to avoid interference with the magnetic field (Bresser junior DM400 digital microscope, Bresser, Germany).

3. Results

3.1. Magnetic Actuator Modeling and Construction

The electromagnetic system, as well as the control system regulating the current at each coil, was constructed at our laboratory facilities. The validation of the design was done using a Hall sensor in order to compare the generated field with the values from the simulations. Measurements and simulations were taken for 18 A at the outer coils and 6 A in the inner coils. In Figure 2, the whole apparatus is shown together with the simulation results. Experimental measurements are compared and a very small deviation between them can be observed. In Figure 2b, the magnetic field intensity that propagated at a straight line from the middle of the cores is depicted for 18 A at the outer coils and

6 A in the inner coils. At a distance of 7.5 cm from the cores, for the specific values of current in the coils, the magnetic field intensity is minimized due to the opposite direction of the vectors at this point.

3.2. *Validation of the Pushing Effect*

We performed preliminary tests in order to verify the efficiency of the system and the reliability of the simulations. A paramagnetic sphere of 4-mm radius was placed on an aluminum rail at 9 cm, nearby the cancellation node (7.5 cm) but are the point where outward force was maximum (Figure 3a) according to the simulation results as seen in Figure 2. Subsequently, by feeding the coils with 18 A (outer coils) and 6 A (inner coils), the sphere received a magnetic push that moved it by about 10 cm from the initial point (Figure 3b). The results of this experiment confirmed the accuracy of our simulations, as well as the measurements with the Hall sensors. However, since the aim of our study was to demonstrate the possibility to magnetically push small particles in the human body, we set up a second round of experiments.

3.3. *Remote Control of Magnetic Microdroplet*

For this experiment we used a microdroplet with a volume of 10 μL of water-soluble ferrofluid material placed in 100% oleic acid (Figure 4). The petri disc containing the microsphere of ferrofluid nanoparticles and oleic acid was put at a specific distance (9 cm) from the cores in order to align the microdroplet with the point where outward force was maximum (Figure 4), and electromagnets were activated as described previously. Due to the high viscosity of the oleic acid and the size of the specimen, the movement of the microdroplet was restricted only to a few mm as seen in Figure 4. Nevertheless, this significant magnetic push received by the microdroplet demonstrates that it is possible to remotely control paramagnetic material.

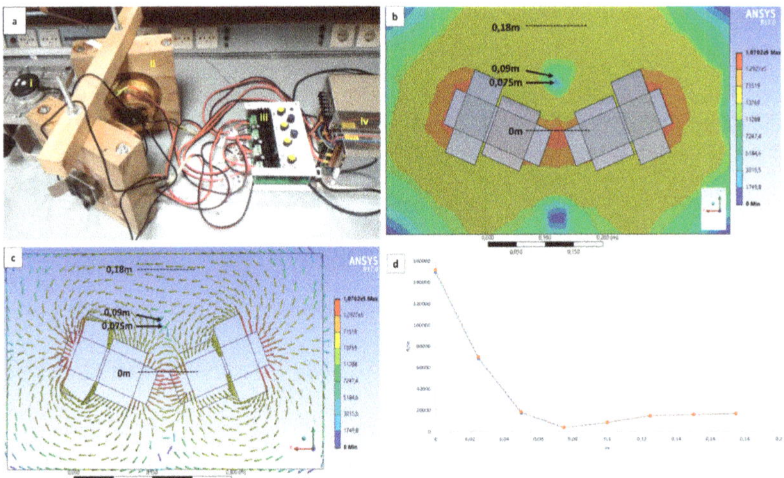

Figure 2. The whole apparatus is depicted in (**a**): (i) the microscope used to monitor the movement of the magnetic droplet; (ii) electromagnets; (iii) the control system; (iv) the power supplies. (**b**,**c**) Simulations for the magnetic field intensity propagated by the electromagnets are shown, as well as the vectors of the field. Arrows in (**b**,**c**) indicate the cancellation area created at given distance from magnet cores and the initial position of the magnetic sphere and microdroplet. (**d**) The magnetic field intensity measured by a Hall sensor (orange color) in comparison with the results of the simulation from ANSYS (blue color). The measurements are taken on the centerline starting from the last point of the cores (dashed lines). The scale bars are in meters.

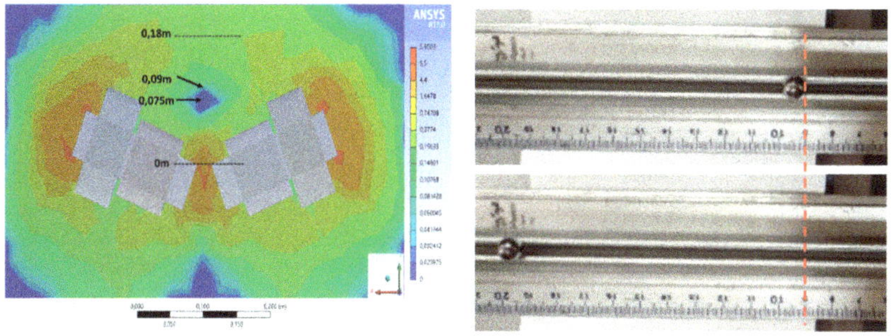

(**a**) (**b**)

Figure 3. (**a**) Map of magnetic forces generated around the electromagnets; units reported in the simulation are arbitrary and are intended to highlight areas where the force is almost zero (dark blue), while the other colors indicate forces >0. The scale bar is in meters. (**b**) A paramagnetic sphere pushed from 9 to 19 cm on an aluminum rail using the actuator system.

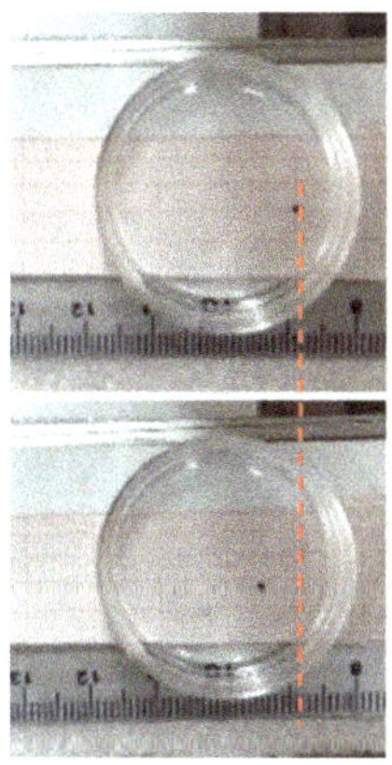

Figure 4. Ferrofluid microdroplet being pushed in oleic acid.

Having assured the reliability of the simulations, the next step was to test the deflecting efficiency. To do so, the ferrofluid microdroplet, a cluster of nanoparticles with average size of 100 nm, was put at the same starting point, and the coils were fed with specific currents (Figure 5 and Video S1, Supplementary Materials). In particular, coil 1 and coil 2 of the left array (Figure 5a,b) were fed with 15.5 A and 4 A, whereas coil 3 and coil 4 of the right array were fed with 4 A and 13 A, respectively. This differential power supplied to the four coils allowed us to create a magnetic flux that moved the droplet toward the X-axis (Figure 5c). For the deflecting step that allowed us to move the droplet toward the Y-axis (Figure 5d), coil 1 and coil 2 of the left array (Figure 5e) were fed with 17.5 A and 4 A, whereas coil 3 and coil 4 of the right array were fed with 4 A and 11 A, respectively. The differential current suppled to each coil in these experiments allowed remotely controlling a microdroplet of magnetic material; the experimental data were confirmed by the simulations that predicted a change in the direction of the magnetic field vectors (Figure 5b,f) as a consequence of power supplied to the coils. The velocity of the ferrofluid droplet was also calculated to be approximately 135 μm/s. The driving of the magnetic droplet was based on the optical feedback provided by a microscope connected to the computer. However, it is possible to predesign a route and program the actuation system accordingly, in order to drive the microdroplet at a specific route, without intervening during the procedure. The programming of the actuation system will be presented in a future work.

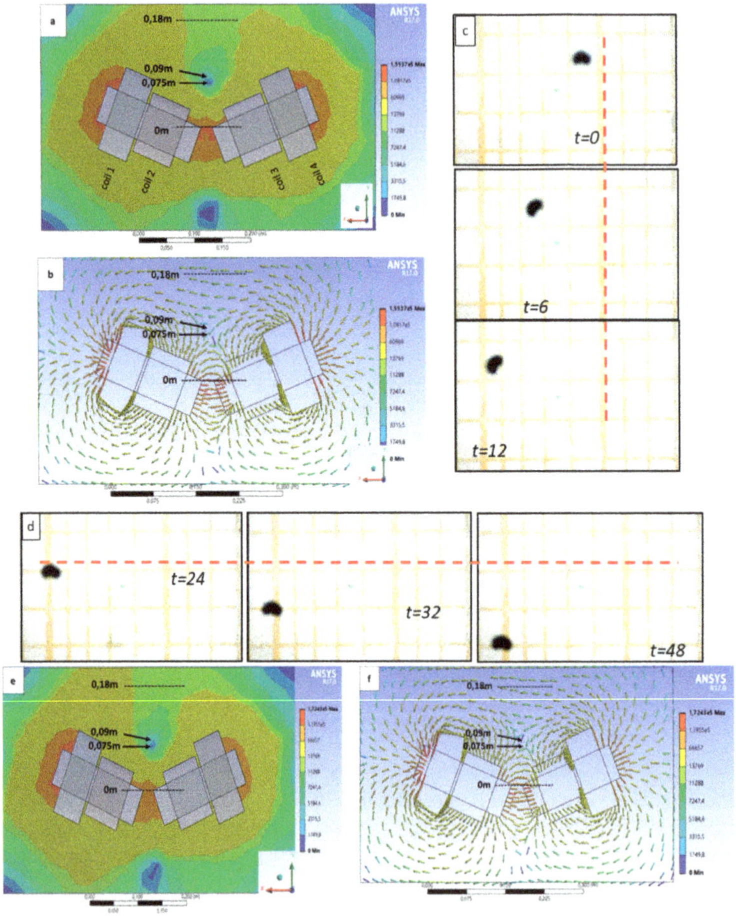

Figure 5. (**a**,**b**) Simulations of the magnetic field intensity and the vectors of the field for currents used to push forward the droplet. (**c**) The positions of the droplet at t = 0, 6, and 12 s. (**d**) The position of the droplet for t = 24, 32, and 48 s. (**e**,**f**) Simulations of the magnetic field intensity and the vectors of the field for currents used to deflect the droplet. Note the change of direction of the vector in the cancellation node (arrows) in (**b**,**f**) due to different current supply to the four coils. The measurements are taken on the centerline starting from the last point of the cores (dashed lines). The scale bars are in meters.

4. Discussion

To achieve pushing and deflecting effects of any magnetic material, a controllable cancellation node in a given bidimensional space must be realized. In fact, at this unique point, the magnetic lines at the minimum field point meet each other with opposite vector directions but with the same absolute field intensity values. Since the magnetic fields do not cancel at other points, this point is a unique location of a locally minimum (zero) magnetic field strength. The forces increase from low to high magnetic field; therefore, in the region beyond the cancelation node, the generated forces will push particles away from the magnetic system. An ideal magnetic actuator system should offer the ability to adjust the position of the local minimum point and, therefore, to remotely drive the magnetic objects by regulating the power provided to the electromagnets, and not by moving the magnetic elements.

The magnetic force applied by a non-homogeneous magnetic field on a particle suspended in a fluid is given by the following equation:

$$F_m = \mu_o\left(\left(m_p - m_f\right)\cdot\nabla\right)H_{ap}, \quad (1)$$

where μ_o is the permeability of free space, ∇ is the Hamilton operator, H_{ap} is the value of the magnetic field intensity applied at the place where the particle is located, m_p is the magnetic moment of the particle, and m_f is the magnetic moment of the liquid where the particle suspends. χ_p and χ_f are the susceptibilities of the particle and the fluid, respectively.

$$m_p - m_f = V_p\frac{3\left(\chi_p - \chi_f\right)}{\left[\left(\chi_p - \chi_f\right) + 3\left(\chi_f + 1\right)\right]}H_{ap}. \quad (2)$$

In our case, the χ_f of oleic acid is ~0; thus, the equation can be written as

$$F_m = \mu_o V_p\frac{3\chi_p}{\left(\chi_p + 3\right)}\left(H_{ap}\cdot\nabla\right)H_{ap}. \quad (3)$$

The force applied by a magnetic field with magnetic field intensity H_{ap} at a magnetic particle of radius α(m) and magnetic susceptibility χ is

$$F_m = \frac{4\pi\alpha^3}{3}\frac{\mu_o\chi}{\left(1+\frac{\chi}{3}\right)}H_{ap}\frac{dH_{ap}}{dx} = \frac{2\pi\alpha^3}{3}\frac{\mu_o\chi}{\left(1+\frac{\chi}{3}\right)}\nabla\left(\left|H_{ap}\right|^2\right). \quad (4)$$

Based on literature data, the magnetic susceptibility of the magnetic microdroplet is $\chi \approx 20$ [17] and that of the human body is $\chi \approx 10^{-6}$–10^{-4} [21]; thus, the efficiency of the magnetic field on the magnetic microdroplet is not affected remarkably by human tissue. The magnetic force applied depends strongly on the particle size. Very small objects with size on the order of nm can be manipulated by using cooling systems, which allow providing higher currents feeding the coils. Moreover, the magnetic forces can be increased by choosing particles with anisotropic shape [22] or particles with higher magnetic susceptibility. Furthermore, it must be noted that the magnetic force increases according to the increase in the magnetic field gradient. A steeper gradient results in a bigger force on a given microdroplet. Manipulation of magnetic ferrofluids attracted the interest of various scientific fields due to the multidisciplinary nature of applications. Nacev et al. and Torres-Diaz et al. [23,24] described various approaches that were done in order to manipulate ferrofluids; however, they reported the control of ferrofluid at small distances and without the ability to push outward. Other attempts were made by Zhang et al. and Hassan et al. [25,26] to manipulate ferrofluids using uniform magnetic fields which are easier to acquire, but the feasibility of this manipulation applied to human body still presents drawbacks due to system complexity and operational distances. The system presented in this work overcomes many of these issues.

A highly important field of interest is anti-tumor therapy. Indeed, magnetic drug delivery could enable a better treatment efficiency, while minimizing side effects of the chemotherapy. By functionalizing drug nanocarriers with therapeutic molecules able to interact specifically with tumor cells, it is possible to deliver drugs exclusively to the tumor tissue [27]. Another important application is targeted regenerative medicine and, more specifically, magnetically driven stem cells. Stem cells play a special role in renewing and repairing damaged tissues and organs. However, a big obstacle in such therapies is the inability to retain stem cells to the desired location. To achieve this goal, magnetic nanoparticles could be inserted in the stem cells, conferring in this way magnetic properties that allow immobilizing them via an externally applied magnetic field. Furthermore, with magnetically modified cells, it would be possible to avert the problem of the short-term cell engraftment in organs or tissue after the injection. One application of stem cells that is becoming increasingly interesting with significant opportunities in the field of neurosurgery is the treatment of spinal cord injuries. Spinal cord injuries (SCIs) are defined as complicated pathophysiological damages caused by injuries to the spinal column, and there are over 700,000 cases per year worldwide [28]. SCI pathophysiology can be divided into two distinct phases. Primary injury is the initial shearing or compression of the spinal cord tissue. The mechanical force of the primary injury can lead to hemorrhage, disruption of cell membrane integrity, and ion and neurotransmitter imbalance that immediately affects neural function. Secondary injury is related to the progressive inflammatory, ischemic, and apoptotic cascade that follows the initial mechanical assault [29]. The target of stem-cell therapies for SCI is to minimize the spread of secondary injury, improve the function of remaining cell populations, and engage regeneration of neuronal and glial populations. All these therapeutic effects could be achieved via a remote control of magnetically responsive stem cells and an apparatus like that proposed in the present work.

Another interesting field of application of magnetically responsive cells is cellular transplantation, which can be augmented with a combination of growth factors, scaffolds, or other biomaterials that improve cell survival, engraftment, and differentiation. Intraspinal application of these therapies leads to engraftment of transplanted cells, which may promote neural regeneration through several proposed mechanisms [30]. The engraftment on specific locations can be effectively increased by using magnetically modified stem cells. Furthermore, various studies were conducted on the delivery of magnetically modified stem cells at targets such as the myocardium or at the spinal canal, but the targeting in all these studies was primeval, since no study used a sophisticated method to guide and, most importantly, to inject the magnetic stem cells to the location needed [31]. Implementation of our devices could offer a solution for those pathologies that, at the moment, remain without any feasible operative option.

5. Conclusions

In this work, we presented an electromagnet system of four coils arranged in two units with a particular configuration. Laboratory experiments and modeling were conducted to prove the effectiveness of the proposed system. We showed, for the first time, an apparatus able to push and deflect magnetic particles just by acting on the currents provided to the four coils without moving any system part. The reason why two units of two coils per core were utilized is essentially due to a better control of the magnetic object during the deflection movement. Indeed, by regulating the power supplied to each coil forming a single unit, the magnetic force generated can be precisely modulated. The pushing effect could be realized with only two coils; however, in order to deflect and pull the microdroplet, the four-coil configuration is necessary. Implementation of our device could offer a solution for pathologies that can be treated with drug target therapies but, at the moment, remain without any feasible operative option.

Supplementary Materials: The following are available online at http://www.mdpi.com/2079-4991/10/2/371/s1, Figure S1: Schematic of the control system, Video S1: Magnetic microdroplet deflection and pushing.

Author Contributions: G.B. and S.A. designed and constructed the actuation system and methodized the optimization of the apparatus. K.T. conducted the experiments and calibrated the system. A.F. supervised and evaluated the experimental results and organized the structure of the draft. E.H. contributed to the conceptualization and supervised and reviewed the draft. All authors have read and agreed to the published version of the manuscript.

Funding: This project received funding from the Bio Based Industries Joint Undertaking under the European Union's Horizon 2020 research and innovation, grant agreement No. 745695. E.H. wishes to acknowledge financial support of the Ministry of Science and Higher Education of the Russian Federation in the framework of Increase Competitiveness Program of NUST "MISIS", contract No. K2-2019-012.

Conflicts of Interest: The authors declare no conflicts of interest.

References

1. Lübbe, A.S.; Alexiou, C.; Bergemann, C. Clinical Applications of Magnetic Drug Targeting. *J. Surg. Res.* **2001**, *95*, 200–206. [CrossRef] [PubMed]
2. Wu, M.; Huang, S. Magnetic Nanoparticles in Cancer Diagnosis, Drug Delivery and Treatment. *Mol. Clin. Oncol.* **2017**, *7*, 738–746. [CrossRef] [PubMed]
3. Chertok, B.; Moffat, B.A.; David, A.E.; Yu, F.; Bergemann, C.; Ross, B.D.; Yang, V.C. Iron Oxide Nanoparticles as a Drug Delivery Vehicle for MRI Monitored Magnetic Targeting of Brain Tumors. *Biomaterials* **2008**, *29*, 487–496. [CrossRef] [PubMed]
4. Toro-Cordova, A.; Flores-Cruz, M.; Santoyo-Salazar, J.; Carrillo-Nava, E.; Jurado, R.; Figueroa-Rodriguez, A.P.; Lopez-Sanchez, P.; Medina, A.L.; Garcia-Lopez, P. Liposomes Loaded with Cisplatin and Magnetic Nanoparticles: Physicochemical Characterization, Pharmacokinetics, and In-vitro Efficacy. *Molecules* **2018**, *23*, 2272. [CrossRef]
5. Carinelli, A.; Martí, M.; Alegret, S.; Pividori, M.I. Biomarker detection of global infectious diseases based on magnetic particles. *New Biotechnol.* **2015**, *32*, 521–532. [CrossRef]
6. Li, Q.; Liu, X.; Chang, M.; Lu, Z. Thrombolysis Enhancing by Magnetic Manipulation of Fe_3O_4 Nanoparticles. *Materials* **2018**, *11*, 2313. [CrossRef]
7. McBain, S.; Yiu, H.H.P.; Dobson, J. Magnetic Nanoparticles for Gene and Drug Delivery. *Int. J. Nanomed.* **2008**, *3*, 169–180.
8. Snider, E.J.; Kubelick, K.P.; Tweed, K.; Kim, R.K.; Li, Y.; Gao, K.; Read, A.T.; Emelianov, S.; Ethier, C.R. Improving Stem Cell Delivery to the Trabecular Meshwork Using Magnetic Nanoparticles. *Sci. Rep.* **2018**, *8*, 12251. [CrossRef]
9. Labusca, L.; Herea, D.D.; Mashayekhi, K. Stem Cells as Delivery Vehicles for Regenerative Medicine-Challenges and Perspectives. *World J. Stem Cells* **2018**, *10*, 43–56. [CrossRef]
10. Cores, J.; Caranasos, G.T.; Cheng, K. Magnetically Targeted Stem Cell Delivery for Regenerative Medicine. *J. Funct. Biomater.* **2015**, *6*, 526–546. [CrossRef]
11. Kami, D.; Takeda, S.; Itakura, Y.; Gojo, S.; Watanabe, M.; Toyoda, M. Application of Magnetic Nanoparticles to Gene Delivery. *Int. J. Mol. Sci.* **2011**, *12*, 3705–3722. [CrossRef] [PubMed]
12. Majidi, S.; Zeinali Sehrig, F.; Samiei, M.; Milani, M.; Abbasi, E.; Dadashzadeh, K.; Akbarzadeh, A. Magnetic Nanoparticles: Applications in Gene Delivery and Gene Therapy. *Artif. Cells Nanomed. Biotechnol.* **2016**, *44*, 1186–1193. [CrossRef] [PubMed]
13. Kummer, M.P.; Abbott, J.J.; Kratochvil, B.E.; Borer, R.; Sengul, A.; Nelson, B.J. OctoMag: An Electromagnetic System for 5-DOF Wireless Micromanipulation. In Proceedings of the 2010 IEEE International Conference on Robotics and Automation, Anchorage, AK, USA, 3–7 May 2010; pp. 1610–1616. [CrossRef]
14. Nelson, B.J. Making Microrobots Move. In Proceedings of the 2010 7th International Conference Informatics Control, Automation and Robotics (ICINCO), Funchal, Madeira, Portugal, 15–18 June 2010; Volume 1, pp. 13–18.
15. Niu, F.; Li, J.; Ma, W.; Yang, J.; Sun, D. Development of an Enhanced Electromagnetic Actuation System with Enlarged Workspace. *IEEE/ASME Trans. Mechatron.* **2017**, *22*, 2265–2276. [CrossRef]
16. Sarwar, A.; Lee, R.; Depireux, D.A.; Shapiro, B. Magnetic Injection of Nanoparticles into Rat Inner Ears at a Human Head Working Distance. *IEEE Trans. Magn.* **2013**, *49*, 440–452. [CrossRef]
17. Shapiro, B.; Dormer, K.; Rutel, I.B. A Two-Magnet System to Push Therapeutic Nanoparticles. *Aip Conf. Proc.* **2010**, *1311*, 77–88. [CrossRef]
18. Probst, R.; Lin, J.; Komaee, A.; Nacev, A.; Cummins, Z.; Shapiro, B. Planar steering of a single ferrofluid drop by optimal minimum power dynamic feedback control of four electromagnets at a distance. *J. Magn. Magn. Mater.* **2011**, *323*, 885–896. [CrossRef]
19. Bijarchi, M.A.; Favakeh, A.; Sedighi, E.; Shafii, M.B. Ferrofluid droplet manipulation using an adjustable alternating magnetic field. *Sens. Actuators A Phys.* **2020**, *301*, 1. [CrossRef]
20. Sarwar, A.; Nemirovski, A.; Shapiro, B. Optimal Halbach Permanent Magnet Designs for Maximally Pulling and Pushing Nanoparticles. *J. Magn. Magn. Mater.* **2012**, *324*, 742–754. [CrossRef]
21. Di Luzio, S.; Obletter, G.; Comani, S.; Del Gratta, C.; Romani, G.L. *Magnetic Mapping of DC Fields Related to Tissue Susceptibility in the Human Body BT—Advances in Biomagnetism*; Williamson, S.J., Hoke, M., Stroink, G., Kotani, M., Eds.; Springer: Boston, MA, USA, 1989; pp. 505–508. [CrossRef]

22. Lim, J.K.; Yeap, S.P.; Chee, H.L.; Toh, P.Y.; Low, S.C. Magnetophoresis of iron oxide nanoparticles at low field gradient: The role of shape anisotropy. *J. Colloid Interface Sci.* **2014**, 170–177. [CrossRef]
23. Nacev, A.; Komaee, A.; Sarwar, A.; Probst, R.; Kim, S.H.; Emmert-Back, M.; Shapiro, B. Towards Control of Magnetic Fluids in Patients: Directing Therapeutic Nanoparticles to Disease Locations. *IEEE Control Syst.* **2012**, *32*, 32–74. [CrossRef]
24. Torres-Diaz, I.; Rinaldi, C. Recent progress in ferrofluids research: Novel applications of magnetically controllable and tunable fluids. *Soft Matter* **2014**, *43*, 8584–8602. [CrossRef] [PubMed]
25. Zhang, J.; Hassan, M.R.; Rallabandi, B.; Wang, C. Migration of ferrofluid droplets in shear flow under a uniform magnetic field. *Soft Matter* **2019**, *15*, 2439–2446. [CrossRef] [PubMed]
26. Hassan, M.R.; Zhang, J.; Wang, C. Deformation of a ferrofluid droplet in simple shear flows under uniform magnetic fields. *Phys. Fluids* **2018**, *30*, 092002. [CrossRef]
27. Kheirkhah, P.; Denyer, S.; Bhimani, A.D.; Arnone, G.D.; Esfahani, D.R.; Aguilar, T.; Zakrzewski, J.; Venugopal, I.; Habib, N.; Gallia, G.L.; et al. Magnetic Drug Targeting: A Novel Treatment for Intramedullary Spinal Cord Tumors. *Sci. Rep.* **2018**, *8*, 11417. [CrossRef] [PubMed]
28. Kumar, R.; Lim, J.; Mekary, R.A.; Rattani, A.; Dewan, M.C.; Sharif, S.Y.; Osorio-Fonseca, E.; Park, K.B. Traumatic Spinal Injury: Global Epidemiology and Worldwide Volume. *World Neurosurg.* **2018**, *113*, e345–e363. [CrossRef] [PubMed]
29. Sekhon, L.H.S.; Fehlings, M.G. Epidemiology, Demographics, and Pathophysiology of Acute Spinal Cord Injury. *Spine* **2001**, *26*, S24. [CrossRef]
30. Jin, M.C.; Medress, Z.A.; Azad, T.D.; Doulames, V.M.; Veeravagu, A. Stem Cell Therapies for Acute Spinal Cord Injury in Humans: A Review. *Neurosurg. Focus Foc.* **2019**, *46*, E10. [CrossRef]
31. Santoso, M.R.; Yang, P.C. Magnetic Nanoparticles for Targeting and Imaging of Stem Cells in Myocardial Infarction. *Stem Cells Int.* **2016**, *2016*. [CrossRef]

nanomaterials

MDPI

Article

Cubic Nanoparticles for Magnetic Hyperthermia: Process Optimization and Potential Industrial Implementation

Omar Sánchez Sánchez [1], Teresa Castelo-Grande [2], Paulo A. Augusto [1,3,*], José M. Compaña [4] and Domingos Barbosa [2]

1 Departamento de Ingeniería Química y Textil, Facultad de Ciencias Químicas, Universidad de Salamanca, Plaza de los Caídos, 1-5, 37008 Salamanca, Spain; jovvicpl@aeiou.pt

2 LEPABE—Laboratory for Process Engineering, Environment, Biotechnology and Energy, Faculty of Engineering, University of Porto, 4200-465 Porto, Portugal; tcg@fe.up.pt (T.C.-G.); dbarbosa@fe.up.pt (D.B.)

3 Instituto de Biología Molecular y Celular del Cáncer, Campus Miguel de Unamuno, CSIC/Universidad de Salamanca (GIR Citómica), 37007 Salamanca, Spain

4 Servicio de Difracción de Rayos-X, Universidad de Salamanca, Pza. de Los Caídos s/n, 37008 Salamanca, Spain; jmcompana@usal.es

* Correspondence: pauloaugusto@usal.es

Abstract: Cubic nanoparticles are referred to as the best shaped particles for magnetic hyperthermia applications. In this work, the best set of values for obtaining optimized shape and size of magnetic particles (namely: reagents quantities and proportions, type of solvents, temperature, etc.) is determined. A full industrial implementation study is also performed, including production system design and technical and economic viability.

Keywords: magnetic hyperthermia; magnetic nanoparticles; optimization; economic analysis; plant design and process engineering; cubic particles

Citation: Sánchez, O.S.; Castelo-Grande, T.; Augusto, P.A.; Compaña, J.M.; Barbosa, D. Cubic Nanoparticles for Magnetic Hyperthermia: Process Optimization and Potential Industrial Implementation. *Nanomaterials* **2021**, *11*, 1652. https://doi.org/10.3390/nano11071652

Academic Editors: Angelo Ferraro and Claire Wilhelm

Received: 4 May 2021
Accepted: 17 June 2021
Published: 23 June 2021

1. Introduction

According to the World Health Organization (WHO), cancer is the main cause of death in the world, having been responsible for 9.6 million of deaths in 2018. In the next two decades, the increase in the number of cases is expected to be 60%, according to a new WHO report [1]. Thus, research in oncology is extremely important to develop therapies to treat this disease and increase the expectation and quality of life of patients. Currently, the main treatments are based on surgery, radiotherapy, and chemotherapy. Thus, these techniques have concentrated most of the research and development till now. Nevertheless, the known side effects of these treatments have not been completely eliminated yet. Due to all this, the search for more efficient techniques and more specific action, with fewer unwanted effects, is important in promoting a better quality of life for patients. Various types of complementary therapies exist and are continuously being developed. In particular, the so-called hyperthermia therapies, which take advantage of the fact that tumor cells have a lower thermal tolerance than healthy cells [2]. In fact, hyperthermia is a therapeutic procedure in which tissues/cells are heated above normal physiological ranges (between 41 and 46 °C) in order to kill tumor or sensitize cells to increase the efficiency of standard therapies.

A more recent modality is magnetic hyperthermia (MHT) [3–6], where the increase in temperature occurs by applying alternating magnetic fields to a magnetic material with specific characteristics. In fact, the use of Magnetic Nanoparticles (MNPs) in medicine allows to treat hard-to-reach regions of the body. Chemical manipulation at a nanoscale size has conferred the ability to conjugate biomolecules, such as antibodies, with magnetic particles, for a more effective therapy or to achieve a specific goal, making MNPs ideal vehicles for new therapies based on localized and selective heat dissipation. Thus, MNPs can simultaneously combine various therapeutic functionalities, such as drug carriers, contrast agents in magnetic resonance imaging, or magnetic heating agents. Although MHT has been incorporated into clinical practice to treat a relatively wide range of types of cancer (prostate, esophagus, brain, etc.), today, it has shown to be advantageous in the treatment of sarcomas, carcinomas, and lymphomas.

The ability of the magnetic nanoparticles to generate heat when subjected to alternating magnetic fields is the basis of MHT. In fact, when exposed to an alternating magnetic field, MNPs produce heat through two main mechanisms [7]: hysteresis loss and Néel and Brown relaxation loss. Hysteresis losses occur in large MNPs that have multiple domains, while relaxation losses occur mainly in single domain MNPs.

Magnetite (iron oxide) particles are well-known for their biocompatibility, possible monodispersity and easiness of synthesis [7–9]. Besides, mono-domain particles are usually easy to obtain. Shape and size are important characteristics and it has been proven that critical size of magnetite to form multidomain structure is between 76 and 80 nm for cubic-shaped particles and 128 nm for spherical particles [10]. Supermagnetic behavior is known to exist, at least, below 50–20 nm for magnetite particles, depending on their shape and other characteristics [11–13]. Besides, SAR values (Specific Absorption Rate) for cubic-shaped particles have been shown to be higher than the ones presented by spherical-shaped particles [14,15]. The correlation between particles microstructure and MHT response has been demonstrated (e.g., [16–18]).

The main methods to produce MNPs are coprecipitation, thermal decomposition, microemulsion, sol-gel, hydrothermal synthesis, and synthesis with polyols [8,9]. These methods have been used to prepare particles with an appropriate composition and size distribution, according to the application in mind. For example, for environmental non-specific treatments, coprecipitation is usually the preferred method [8,19–23], while for biomedical applications where monodispersed particles are usually required, less simple methods, like thermal decomposition and hydrothermal, are preferred [8]. A summary of the main methods and their characteristics is presented in Table 1.

Table 1. Comparison between selected magnetic nanoparticle manufacturing methods.

Manufacturing Method	Synthesis	Reaction Temp. (°C)	Reaction Time	Solvent	Surface-Capping Agents	Size Distribution	Shape Control	Yield
Coprecipitation	Very simple	20–90	Minutes	Water	During/after reaction	Relatively narrow	Not good	High
Microemulsion	Complicated	20–50	Hours	Organic agents	During reaction	Relatively narrow	Good	Low
Thermal decomposition	Complicated	100–320	Hours-days	Organic agents	During reaction	Very narrow	Very good	High
Hydrothermal	Simple	200–250	Hours-days	Water-ethanol	During reaction	Very narrow	Very good	Medium
Poliol	Simple	25-Boiling Point	Hours	Ethylene PEG	During reaction	Narrow	Very Good	High

In the case of magnetic hyperthermia, magnetic nanoparticles must present good size distribution, good shape control, and a good SAR generation capability.

Most of the published literature is concerned with laboratorial methods to produce magnetic nanoparticles for hyperthermia applications, but, so far, to the best of our knowledge, none has studied the economic and technological viability of the scale-up of the method and its possible application at large-scale industrial production (which is an important step towards its widespread production and availability, especially after the recent license of a real medical clinical application of magnetic hyperthermia). This will also limit the number of producing methods that may be selected, as the method must be easily scalable (a fundamental characteristic for a production plant).

The present work is concerned with the maximization of the magnetic hyperthermia capabilities of the particles, without increasing the costs to an unbearable level. Thus, the goal is to produce cubic magnetic nanoparticles with sizes lower than the monodomain critical value and, if possible, the supermagnetism critical value. The optimization of the main variables (ramp-up temperature, stirring speed, type of solvent, proportions of raw materials) for the selected method of production is one of the goals. The other goal is to study the economic and technological viability of the production, in order to set the path for future industrial implementation to support clinical widespread applications.

2. Materials and Methods

2.1. Reagents

In this study was used: iron (III) acetylacetonate (99%) from Sigma Aldrich (St. Louis, MO, USA), oleic acid (90%) from Panreac (Barcelona, Spain), benzyl ether (99%) from Sigma Aldrich (Madrid, Spain), 4-biphenylcarboxylic acid (99%) from Sigma Aldrich (Madrid, Spain), toluene (99.8%) from Panreac (Barcelona, Spain), hexane (99%) from Panreac (Barcelona, Spain), chloroform (99.9%) from Panreac (Barcelona, Spain), and nitrogen (99%) from Air Liquide (Paris, France). All solutions were made with distilled water.

2.2. General Experimental Methodology

2.2.1. Synthesis of Cubic Magnetic Nanoparticles

For the synthesis of the magnetic nanoparticles, the method developed by Kim et al. in 2008 [24] was followed, which gives rise to MNPs with cubic morphology and with sizes close to the critical size of the superparamagnetic state. In a typical synthesis of magnetite nanocubes, 2 mmol of iron acetylacetonate ($Fe(acac)_3$)—$Fe(C_5H_7O_2)_3$, 4.5 mmol of oleic acid ($C_{18}H_{34}$), and 52.5 mmol of benzyl ether $-$ $C_{14}H_{14}O$$-$(or 2 mmol of 4-biphenylcarboxylic acid $-$ $C_{13}H_{10}O_2$—depending on the size and shape) were used. The initial mixture is first degassed with nitrogen gas. Then, temperature is risen up to 290 °C at a temperature gradient ramp-up of about 20 °C/min, under energetic stirring. Once the final temperature is reached, the reaction remains at 290 °C for 30 min. At the end of the process, the sample is cooled and a mixture of toluene and hexane in a 4:1 volumetric ratio is added, and the sample centrifuged at 1700 rpm. Finally, chloroform is used to clean the particles. The main reaction that occurs at these conditions is:

$$aFe(C_5H_7O_2)_3 + bC_{18}H_{34} + cC_{13}H_{10}O_2/dC_{14}H_{14}O \rightarrow eFe_3O_4 + \text{Subproducts} \quad (1)$$

The steps of the reaction process are represented in Figure 1.

2.2.2. Particles' Characterization

Particles were characterized by determining their magnetic properties, size, and shape and percentage of magnetite. Hence, to certify the percentage of obtained magnetite, X-ray studies were made (X-ray diffraction (XRD) Bruker D8 Advance, Karlsruhe, Germany); to determine the morphology and size of the particles, Scanning Electron Microscopy (SEM) (JEOL JSM-840, Madrid, Spain) was performed; finally, to access magnetic susceptibility, samples were analyzed with a Kappabridge KLY-4 susceptometer (a semiautomatic auto balance inductivity bridge, Porto, Portugal).

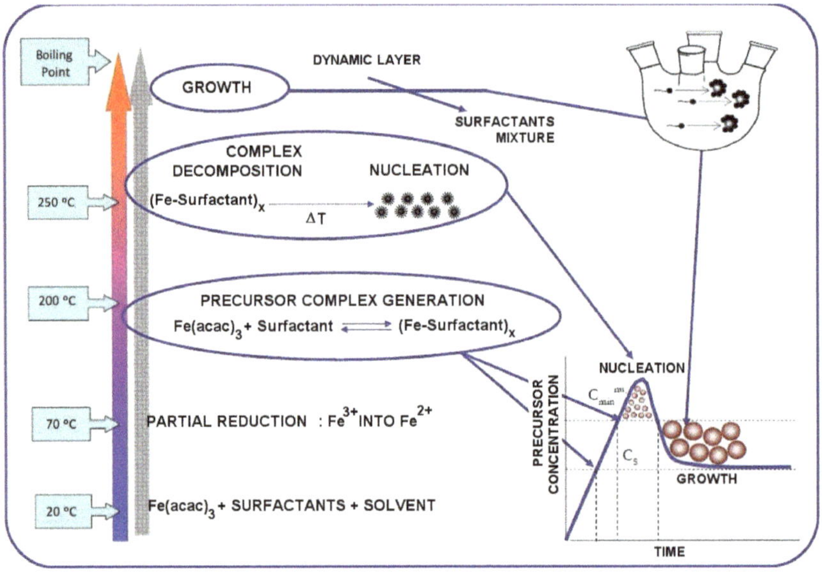

Figure 1. Reaction mechanism—based on [25].

2.2.3. Influence of the Main Variables

For all these experiments, the default options are: iron acetylacetonate (III)—2 mmol; oleic acid—4.5 mmol; 4-biphenylcarboxylic acid—2 mmol; stirring—200 rpm; reaction temperature—290 °C; temperature ramp-up: 20 °C/min; centrifugal speed—1700 rpm. The analyzed variables were:

(a) Effect of mechanical stirring (four values were used: 150, 170, 200, and 220 rpm).
(b) Effect of temperature ramp-up (the increasing temperature gradients studied were: 15, 20, 25, and 30 °C/min).
(c) Effect of the type of solvent (the solvents studied were: (a) 4-biphenylcarboxylic acid—2 mmol; (b) benzyl ether—52.5 mmol; (c) both solvents: 4-biphenylcarboxylic acid—2 mmol—and benzyl ether—52.5 mmol).
(d) Effect of reagents ratio (the ratios studied were: doubling oleic acid amount, doubling 4-biphenylcarboxylic acid amount, and doubling all reagents).
(e) Degassing time (the degassing time with nitrogen was reduced by half).

In Table 2 are shown the values used in each experiment. It is important to notice that experiment 7 corresponds to the predetermined conditions with similar results to those already presented in literature, thus no results are presented here.

Table 2. Experiments done for optimization of the synthesis process (when the value of a specific parameter differs from the default, the value of the parameter is highlighted in grey).

	Reagent/Property								
Experiment	**Iron Acetylacetonate (III) (mmol)**	**Oleic Acid (mmol)**	**4-Biphenylcarboxylic Acid (mmol)**	**Benzyl ether (mmol)**	**Reaction Temperature (°C)**	**Temperature Ramp-Up (°C/min)**	**Stirring Speed (rpm)**	**Centrifuge (rpm)**	**Degassing**
1	2	4.5	2	—	290	20	150	1700	Normal
2	2	4.5	2	—	290	20	170	1700	Normal
3	2	4.5	2	—	290	20	220	1700	Normal
4	2	4.5	2	—	290	25	220	2000	Normal
5	2	4.5	2	—	290	15	220	2000	Normal
6	2	4.5	2	—	290	30	220	2000	Normal
7	2	4.5	—	52.5	290	15	200	2500	Normal
8	2	4.5	—	52.5	290	20	200	2500	Normal
9	2	4.5	—	52.5	290	25	200	2500	Normal
10	2	4.5	2	—	290	20	220	2500	Normal
11 (*)	4	9	4	—	290	25	200	2500	Normal
12 (**)	2	4.5	2	52.5	290	25	200	2500	Normal
13	2	4.5	2	—	290	30	200	2500	Half
14	2	4.5	2	52.5	290	25	200	2500	Normal
15	2	9	2	—	290	20	200	2500	Normal
16	2	4.5	4	—	290	30	200	2500	Normal

(*): It is important to notice that experiment 11 is not an upscale version of experiment 14. In fact, the only parameters that were doubled in experiment 11 were the reagents and one of the solvents, and not all the remaining parameters (to perform an upscale all the other parameters should be also modified accordingly—as done in the upscale and economic analysis presented at the final section of the paper).
(**) Preliminary study of experiment 14 with coincident results.

2.2.4. Economic Analysis and Industrial Scale-Up Study

The following methodological steps were applied for the economic analysis and industrial scale-up [8]: (a) determine the optimized process for magnetic hyperthermia; (b) perform a market study to determine the demand of the product (magnetic nanoparticles) for hyperthermia applications; (c) determine the size of the plant; (d) analyze all the process steps, choose all the process units and perform all the mass and energy balances required in the plant; (e) detailed scale-up design of the main equipment; and (f) analyze the global economic impact and profitability.

2.3. Instrumentation

The main instrumentation used was: Three-way Reactor (Nahita, Madrid), Orbital stirrer (Nahita, Madrid), Thermal Heater (Nahita, Madrid), Centrifuge (Fischer, Barcelona), besides regular laboratory glass material.

3. Results and Discussion

In order to determine the best set of values for the main producing parameters, the obtained particles need to be analyzed and compared according to their main characteristics. In this context, it is important to analyze: magnetic properties (magnetic susceptibility), phase composition or phase distribution ratio (XRD), particle size, and morphology (SEM).

3.1. Magnetic Properties

In Table 3 are shown the results obtained for the magnetic properties of the final product corresponding to each experiment.

Table 3. Magnetic properties of the produced nanoparticles.

Experiment Number	1	2	3	4	5	6	7	8
Magnetic susceptibility (SI)	2.51×10^{-4}	5.29×10^{-4}	2.84×10^{-4}	1.56×10^{-4}	1.51×10^{-4}	2.56×10^{-4}	9.71×10^{-5}	2.00×10^{-4}
Experiment number	9	10	11	12 and 14	13	15	16	
Magnetic susceptibility (SI)	2.13×10^{-4}	9.25×10^{-5}	4.64×10^{-4}	3.12×10^{-4}	2.44×10^{-4}	—	—	

As it may be seen, the large majority of the obtained nanoparticles present the same order of magnitude for the magnetic susceptibly. However, particles obtained in experiments 2, 11, 12, and 14 present the highest values, while the lowest values were obtained for experiments 7 and 10.

3.2. Composition

Concerning their structure, X-ray analysis were made on all particles obtained from the set of the performed experiments. In Figure 2 is presented one of the diffraction patterns obtained for these. They all are almost coincident and present the same peaks, so they may be represented by the one depicted in Figure 2 (except for experiment 15 and 16 for which no analysis was possible to be done due to the nature of the obtained products). The typical magnetite peaks are shown in red, according to Powder Diffraction File PDF2 dataset #82-1533 (Powder Diffraction File, International Centre for Diffraction Data, Newtown Square, PA, USA). No additional peaks are found, so samples are essentially pure magnetite.

In addition to the qualitative analysis, an estimation of the crystallite size was calculated by using the Scherrer equation (Table 4). The main peak (at ca. 35.5°) was used for the calculations. Lanthanum hexaboride (NIST-660b) was used as crystalline standard for instrumental broadening.

Table 4. Estimation of the crystallite size (calculated by using the Scherrer equation).

Experiment Number	**1**	**2**	**3**	**4**	**5**	**6**	**7**	**8**
D (nm)	73	50	44	39	32	37	88	92
Experiment number	**9**	**10**	**11**	**12 and 14**	**13**	**15**	**16**	
D (nm)	121	36	54	32	50	—	—	

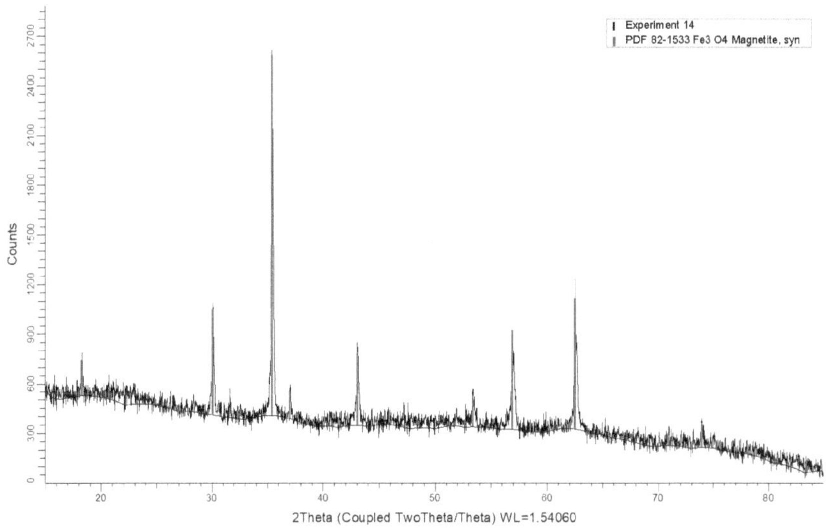

Figure 2. X-ray of experiment 14.

3.3. Particle Size and Morphology

In Figure 3 are presented the secondary electron SEM images of the particles obtained in experiments 1–14, while in Table 5 are presented their size ranges percentages, extracted from the corresponding histograms shown in Supplementary Materials. For some particles, high-quality images were not possible to obtain, mostly due to charging, even after gold sputter coating. In spite of this setback, relevant data were obtained, as summarized in Tables 5 and 6. Some general comments may be emphasized. Typically, when poorly shaped particles are obtained, some tendency to large agglomeration is present, maybe suggesting some kind of growth interference (e.g., Figure 3a,d). When moderately shaped particles are attained, there is a tendency for euhedral forms, almost cubic, though occasional octahedra may be present (e.g., some triangles in Figure 3k), but no rhombic-dodecahedra have been detected. At last, when well-shaped forms are obtained, there is an undoubted preference for cubes and parallelepipeds, in accordance to the results of Kim et al. 2008 [24]. Some stacking is observed (e.g., Figure 3g–i), but it has to be underlined that twinning is not detected in any well-shaped sample.

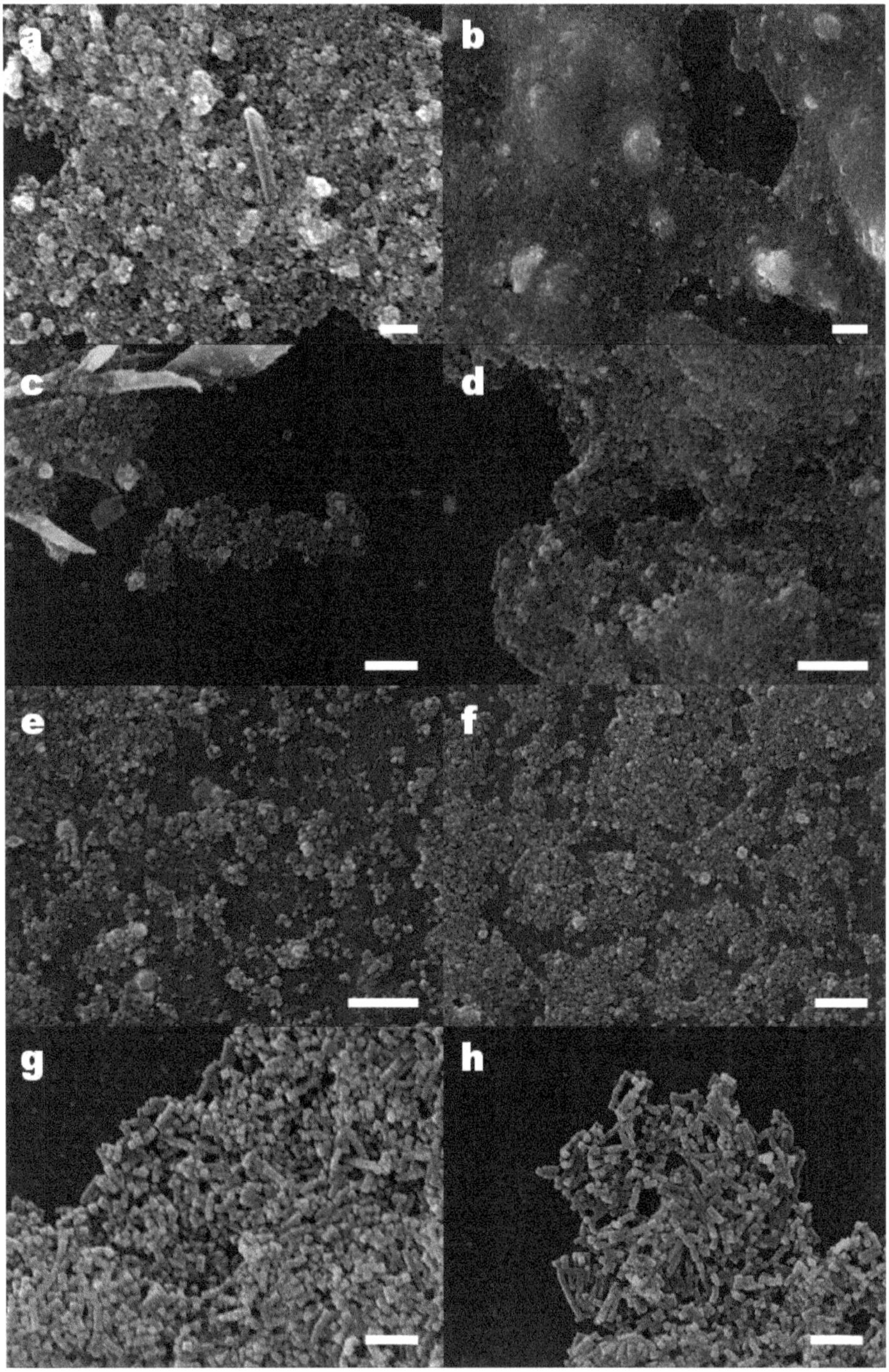

Figure 3. *Cont.*

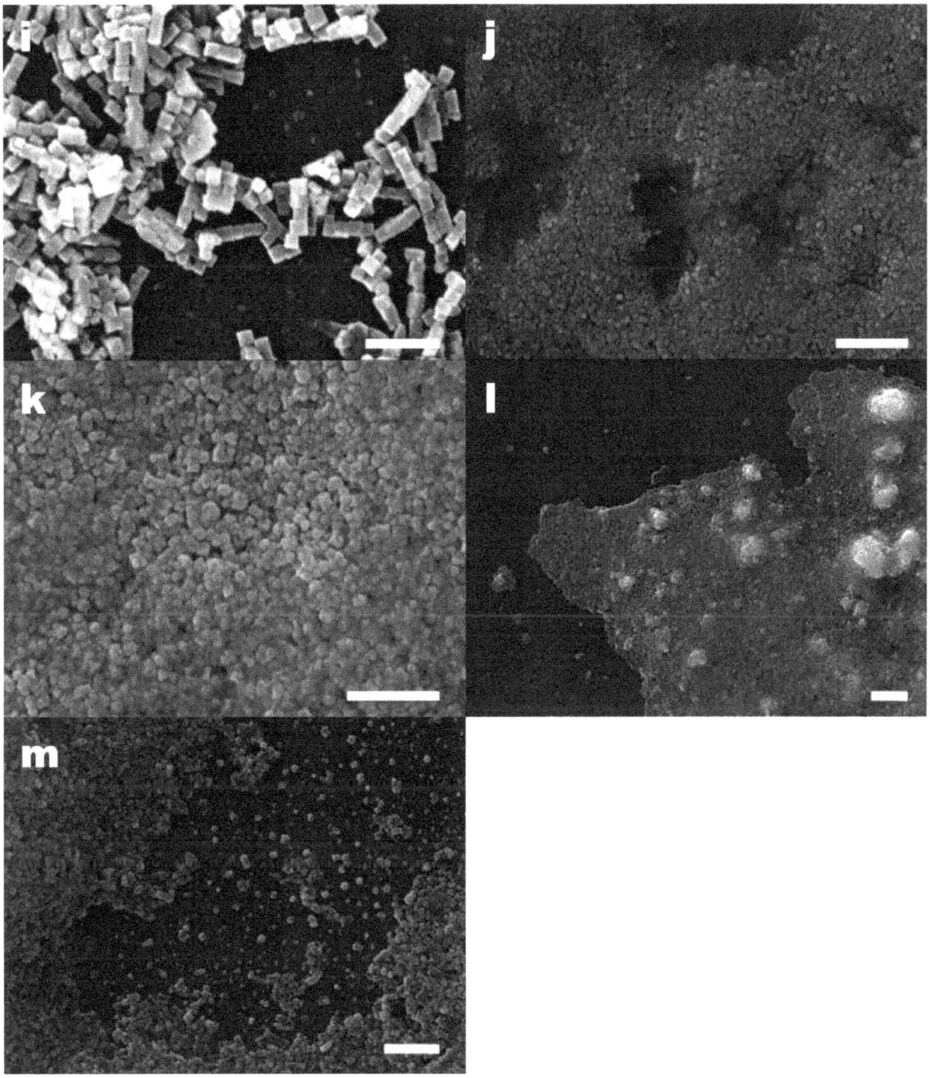

Figure 3. SEM of the nanomagnetite particles obtained in experiments: (**a**) 1; (**b**) 2; (**c**) 3; (**d**) 4; (**e**) 5; (**f**) 6; (**g**) 7; (**h**) 8; (**i**) 9; (**j**) 10; (**k**) 11; (**l**) 13; (**m**) 12 and 14. White bar length = 1 μm.

Table 5. Size ranges percentages of each experiment product. The size ranges showing highest percentage are highlighted for each case: light grey— between 10–19%; medium gray- between 20–29%; dark grey— more than 30%.

Exp. Number	10–20 (nm)	20–28 (nm)	28–36 (nm)	36–44 (nm)	44–52 (nm)	52–60 (nm)	60–68 (nm)	68–76 (nm)	76–84 (nm)	84–92 (nm)	92–100 (nm)	>100 (nm)
1	0%	0%	0%	0%	0%	0%	1%	2%	3%	2%	12%	80%
2	0%	0%	0%	0%	2%	8%	8%	9%	14%	14%	13%	32%
3	0%	0%	4%	15%	24%	27%	15%	12%	1%	0%	2%	0%
4	0%	5%	12%	28%	26%	18%	7%	2%	0%	2%	0%	0%
5	0%	3%	12%	25%	26%	14%	9%	6%	3%	1%	2%	0%
6	0%	6%	24%	25%	17%	17%	5%	2%	2%	2%	0%	0%
7	0%	0%	0%	0%	1%	4%	4%	11%	12%	12%	10%	46%
8	0%	0%	0%	0%	0%	1%	1%	3%	3%	3%	7%	82%
9	0%	0%	0%	0%	0%	0%	0%	0%	1%	2%	4%	93%
10	0%	11%	9%	19%	30%	18%	9%	4%	0%	0%	0%	0%
11	0%	4%	13%	18%	16%	20%	10%	13%	2%	3%	2%	0%
12, 14	6%	26%	32%	16%	9%	5%	4%	1%	1%	0%	0%	0%
13	0%	4%	6%	6%	15%	20%	24%	12%	5%	5%	4%	3%
15	—	—	—	—	—	—	—	—	—	—		
16	—	—	—	—	—	—	—	—	—	—		

Table 6. Main conclusions from the obtained results.

Exp. Number	Morphology	Main Size Ranges	Magnetic Character	Conclusions
1	Undefined	92–100 nm: 12% >100 nm: 80%	Regular	The low resolution does not allow to determine their precise morphology. Size much larger than desired. Presents some large agglomerations.
2	Undefined	76–84 nm: 14% 84–92 nm: 14% 92–100 nm: 13% >100 nm: 32%	Slightly Higher	The low resolution does not allow to determine their precise morphology. Size much larger than desired. Presents some large agglomerations.
3	Generally irregular. Some cubic-shaped particles are present.	36–44 nm: 15% 44–52 nm: 24% 52–60 nm: 27% 60–68 nm: 15% 68–76 nm: 12%	Regular	Acceptable morphology. Size larger than desired. Some large agglomerations are present.
4	Irregular	28–36 nm: 12% 36–44 nm: 28% 44–52 nm: 26% 52–60 nm: 18%	Regular	Poorly defined morphology. Size slightly larger than desired. Small agglomerations of medium size are seen.
5	Irregular	28–36 nm: 12% 36–44 nm: 25% 44–52 nm: 26% 52–60 nm: 14%	Regular	Poorly defined morphology. Size slightly larger than desired. Small agglomerations of medium size are seen.
6	Generally irregular. Some cubic-shaped particles are present.	28–36 nm: 24% 36–44 nm: 25% 44–52 nm: 17% 52–60 nm: 17%	Regular	Acceptable morphology. Size slightly larger than desired. Some agglomerations of medium size are seen.
7	Cubic/cobblestone particles	68–76 nm: 11% 76–84 nm: 12% 84–92 nm: 12% 92–100 nm: 10% >100 nm: 46%	Slightly Lower	Ideal morphology. Size much larger than desired. There are no agglomerations.
8	Cubic/cobblestone particles	>100 nm: 82%	Regular	Ideal morphology. Size much larger than desired. There are no agglomerations.
9	Cubic/cobblestone particles	>100 nm: 93%	Regular	Ideal morphology. Size much larger than desired. There are no agglomerations.

Table 6. *Cont.*

Exp. Number	Morphology	Main Size Ranges	Magnetic Character	Conclusions
10	Generally irregular. Some cubic-shaped particles are present.	20–28 nm: 11% 28–36 nm: 9% 36–44 nm: 19% 44–52 nm: 30% 52–60 nm: 18%	Slightly Lower	Acceptable morphology. Size slightly larger than desired. Large agglomerations are not visible.
11	Generally irregular. Some cubic-shaped particles are present.	28–36 nm: 13% 36–44 nm: 18% 44–52 nm: 16% 52–60 nm: 20% 60–68 nm: 10% 68–76 nm: 13%	Slightly Higher	Acceptable morphology. Size slightly larger than desired. There are several medium-sized agglomerations
13	Undefined	44–52 nm: 15% 52–60 nm: 20% 60–68 nm: 24% 68–76 nm: 12%	Regular	The low resolution does not allow to determine their precise morphology. Size much larger than required. Presents some large agglomerations.
12 and 14	Generally cubic-shaped particles are present. Some irregular shape also appears.	20–28 nm: 26% 28–36 nm: 32% 36–44 nm: 16%	Slightly Higher	Very good morphology. Very good size. There are no agglomerations.
15	No particles obtained	———	———	———
16	No particles obtained	———	———	———

3.4. Discussion of Results

In Table 6 is presented a summary of the main conclusions, considering all the results presented previously.

From Table 3, it may be concluded that except for the case of experiments 15 and 16 (where no particles are obtained), the particles present the same order of magnitude concerning magnetic properties (being experiments 2, 11, 12, and 14 the ones that present particles with higher magnetic properties and 7 and 10 the ones that present particles with lower magnetic properties, by a ratio of 3–4). Clearly, the process is well behaved for the majority of parameters, does not pose serious threat to the attainment of magnetite crystals, and does not have a strong influence on their anisotropy and the alignment of magnetic moments. The exceptions are experiments 15 and 16, corresponding to the increase of the surfactant (oleic acid) and one of the tested solvents (4-biphenylcarboxylic acid), respectively.

Considering the attainment of the cubic shape, only experiments 3 (highest stirring), 6 (highest ramp-up temperature scale gradient), 10 (highest centrifuging speed), and 11 (doubling of all reagents quantities) are acceptable, and experiments 7, 8, 9 (all using benzyl ether as solvent), 12, and 14 (using both solvents—benzyl ether and 4-biphenylcarboxylic acid, highest centrifuging speed and high temperature ramp-up gradient) are ideal. Comparing the products of these last experiments, considering the size of the particles, it may be concluded that particles obtained in experiments 7, 8, and 9 present a much larger size than desired (using benzyl ether as solvent alone is good for the shape, but not for obtaining small sizes); in experiments 3, 6, 10, and 11 the obtained particles present slightly larger size than desired and present agglomerations, while in experiments 12 and 14, obtained particles present the desired size and no agglomerations are observed. Therefore, steps followed to obtain particles in experiments 12 and 14 stand-out as the optimized particle production method, as these experiments present final cubic-shaped nanoparticles with high magnetic properties, with the desired size and no agglomerations. Hence, using both solvents at the same time (benzyl ether and 4-biphenylcarboxylic acid) and maintaining the selected base preferences for the other reagents and stirring speed (while increasing the centrifuging speed) seems to be the best option to obtain the desired magnetic nanoparticles.

3.4.1. Parameters Influence

By analyzing each parameter, it may be concluded that:

(a) increasing stirring speed is beneficial (experiments 1, 2, and 3),
(b) increase in temperature ramp-up does not affect the results (experiments 4, 5, 6, 7, 8, 9, and 10),
(c) substituting the solvent (4-biphenylcarboxylic acid by benzyl ether) improves the percentage of particles presenting a cubic form, but increases particle sizes (experiments 7, 8, and 9),
(d) doubling the quantities of oleic acid or 4-biphenylcarboxylic acid leads to a null production of particles (experiments 15 and 16),
(e) doubling the quantity of the two reactives and one of the tested solvents is not detrimental (experiment 11),
(f) reducing the degasification time is detrimental (experiment 13),
(g) using both solvents at the same time (4-biphenylcarboxylic acid and benzyl ether) improves the process (experiments 12 and 14).

3.4.2. Ideal Value of the Parameters

From all the above, it may be concluded that the ideal conditions are: iron acetylacetonate (III)—2 mmol; oleic acid—4.5 mmol; 4-biphenylcarboxylic acid—2 mmol; benzyl ether—52.5 mmol; stirring—200 rpm; reaction temperature—290 °C; temperature ramp-up—25 °C/min; centrifugal speed 2500 rpm.

4. Potential Industrial Implementation

To determine the potential industrial implementation of the cubic magnetic nanoparticle manufacturing process, the methodology described in [8] is followed: first, determine the potential international demand of the product (magnetic nanoparticles for magnetic hyperthermia applications), then, determine the plant size, analyze all the process steps and chose all the process units (based on the mass and energy balances), perform a detailed scale-up design of the main equipment, and finally analyze the economic impact and profitability. By following this methodology it was possible to determine the technological and economic viability of the industrial implementation of the optimized process for production of cubic-shaped magnetite nanoparticles based on the optimized values found in Section 3. In what follows, only the main results are present (in Supplementary Material are given all the details concerning these calculations).

4.1. Size of the Plant

The minimum size of the plant in order to be profitable is 3850 kg/year.

The plant will be designed for an annual income of EUR 203,200,000, at a price of EUR 30,000/kg magnetic nanoparticles. The annual production is equivalent to 6773 kg/year.

4.2. Process Engineering Design

In Figure 4 is depicted the process diagram of the proposed industrial plant and in Supplementary Material the characteristics of each stream.

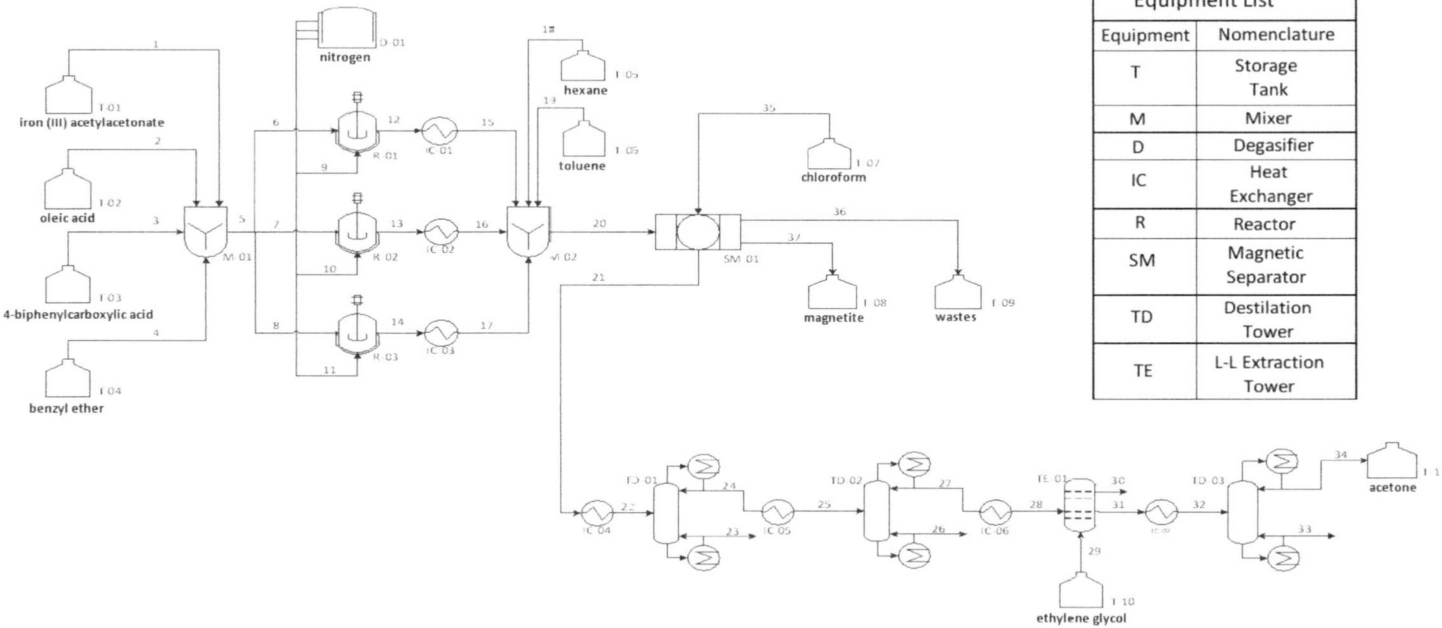

Equipment List	
Equipment	Nomenclature
T	Storage Tank
M	Mixer
D	Degasifier
IC	Heat Exchanger
R	Reactor
SM	Magnetic Separator
TD	Destilation Tower
TE	L-L Extraction Tower

Figure 4. Process Engineering Flowsheet of the Plant to Produce Cubic-Shaped Magnetite Nanoparticles.

The process begins in the M-01 mixer where the main components (stored in the T-01, T-02, T-03, and T-04 tanks) are mixed in their stoichiometric proportions. The main raw materials are: iron (III) acetylacetonate, which will act as the precursor agent, oleic acid, which will be the surfactant and intervenes in the ligand exchange reaction with iron acetylacetonate, and 4-biphenylcarboxylic acid and benzyl ether, which will act as solvents. When the raw materials have been mixed in the M-01, the mixture is taken to the R-01 reactor, as well as to reactors R-02 and R-03. In the reactors, the degassing is done with nitrogen delivered by the degasser D-01. The degassed mixture is then heated through a jacket heater up to 290 °C (563 K), with a constant temperature ramp-up rate of 25 °C/min and a stirring of 800 rpm. When the mixture reaches 290 °C, it remains constant for 30 min. Then, the sample is fed to the heat exchanger IC01, where it is cooled to room temperature. Subsequently, the mixture is introduced in the mixing container M-02 and a mixture of hexane and toluene is added, from tanks T-05 and T-06, with a ratio of 1:4, to facilitate the subsequent separation. The entire mixture is then passed through a magnetic separator SM-01, which is responsible for retaining the MNPs synthesized in the reactor. Lastly, the MNPs are cleaned with chloroform and stored in tank T-07. Regarding the remaining compounds, they are discarded by the magnetic separator through stream 21, including benzyl ether, hexane, toluene, and the other by-products (acetone), which are recycled to reduce raw material costs. For such an effect, distillation towers and a liquid-liquid extraction tower are used.

4.3. Global Economic Impact and Profitability

In Table 7 is presented a summary of the overall economic balance of the plant.

Table 7. Overall economic balance of industrial implementation of MNPs production for magnetic hyperthermia applications.

Cost	EUR	Invested Capital	EUR
1.1 Raw Materials	48,298,030	1.1 Instrumentation	827,043
1.2 Direct Human Labor	829,870	1.2 Initial Setup	473,602
1.3 Indirect Human Labor	207,358	1.3 Piping and Valves	372,270
1.4 General Services	580,909	1.4 Measuring and Control	165,409
1.5 Supplies	80,096	1.5 Heat Isolation	57,893
1.6 Maintenance	234,179	1.6 Electrical Installation	124,057
1.7 Laboratory	165,974	1.7 Land and Structures	1,122,893
1.8 Board and Technical Staff	224,564	1.8 Auxiliary Facilities	330,817
1.9 Amortization	82,704	1.9 Project and Design	361,923
1.10 Taxes and Insurances	160,192	1.10 Constructor Hiring	208,439
TOTAL COST OF FABRICATION	50,863,876	1.11 Unexpected Expenses	521,098
2.1 Commercial Expenses	10,172,775	1.12 Preliminary Studies	2,803,343
2.2 Management	131,826	1.13 Preliminary Startup	640,764
2.3 Financial Expenses	5,809,661	-	-
2.4 Research	240,286	TOTAL IMMOBILIZED	8,009,550
2.5 Technical Services	2,032,000	CIRCULATING CAPITAL	21,038,753
TOTAL COST OF MANAGEMENT	18,386,548	TOTAL INVESTED CAPITAL	29,048,303
TOTAL PRODUCTION COSTS	69,250,424	TOTAL INCOME	203,200,000

As it may be seen from Table 7 and the calculations in the Supplementary Material, a maximum net benefit of EUR 100,462,182/year and a margin of profits over costs of 145% may be attained.

4.4. Discussion of Results—Industrial Adaptation

From the obtained results, the economic viability of the industrial implementation of a plant producing magnetic nanoparticles for hyperthermia applications seems evident. Nonetheless, it is important to notice that market demand was based on the current and foreseen demand for biomedical applications. This means that the produced particles must also be able to be used in other biomedical applications in order to reach the expected selling amount and price. The characteristics of the particles seem appropriate for this goal also (although in some cases it will probably require a further process stage for functionalization, but the economic margin is enough to absorb this step and maintain economic viability).

In what concerns the technological viability, the designed process seems appropriate, also considering all the mass and heat balances that were performed and the process units applied. The recovery and recycling of the raw materials after reaction is considered, as well as an appropriate recovery and purification step concerning the produced magnetic nanoparticles.

5. Conclusions

Hyperthermia applications of magnetic nanoparticles is currently a highly intensive field of research. Cubic-shaped magnetite particles seem to be the most promising type of particles for magnetic hyperthermia. Nonetheless, only a few studies present proper ways to obtain this type of particle, no optimization of the processes has been done, and no proper study of the influence of the main operating parameters is available in the current literature. In this work, a method for the production of cubic-shaped magnetite nanoparticles was applied and the influence of the solvent, precursor and surfactant quantities and ratio, type of solvent, stirring speed, centrifuge speed, temperature ramp-up gradient, reaction temperature, and degassing time was studied. It was concluded that increasing stirring speed is beneficial, an increase in temperature ramp-up does not affect the results, substituting the solvent (4-biphenylcarboxylic acid by benzyl ether) improves cubic form presence but increases particle size, doubling the quantities of oleic acid or 4-biphenylcarboxylic acid leads to a null production of particles, doubling the quantity of the two reactives and one of the solvents is not detrimental, reducing the degasification time is detrimental, and using both solvents at the same time (4-biphenylcarboxylic acid and benzyl ether) improves the process. The best values for the studied variables are: iron acetylacetonate (III)—2 mmol; oleic acid—4.5 mmol; 4-biphenylcarboxylic acid—2 mmol; benzyl ether—52.5 mmol; stirring—200 rpm; reaction temperature—290 °C; temperature ramp-up—25 °C/min; centrifugal speed—2500 rpm.

Then, the industrial implementation of a process to produce these magnetite nanoparticles based on the optimized values reached in the research stage was fully designed and studied in order to study the potential industrial implementation viability. Full process engineering, including energy and mass balances, was also conducted for the plant. The designed plant has proven to be viable economically and technically, with a maximum net benefit of EUR 100,462,182/year. Since the profit is much larger than regular bank benefits, investors may consider attractive the construction and startup of the designed industrial plant, even if several of the parameters have to be adjusted (e.g., adjustment of the sell price of the particles, estimation of some costs, etc.). It is expected that this work will serve as a demonstration of the connection between science and practical applications, and to allow scientific improvement of other processes bearing in mind their possible industrial implementation.

Supplementary Materials: The following are available online at https://www.mdpi.com/article/10.3390/nano11071652/s1, Figure S1: Histogram and size range of the synthesized nanomagnetic particles—Sample 1.; Figure S2: Histogram and size range of the synthesized nanomagnetic particles—Sample 2.; Figure S3: Histogram and size range of the synthesized nanomagnetic particles—Sample 3.; Figure S4: Histogram and size range of the synthesized nanomagnetic particles—Sample 4.; Figure S5: Histogram and size range of the synthesized nanomagnetic particles—Sample 5.; Figure S6: Histogram and size range of the synthesized nanomagnetic particles—Sample 6.; Figure S7: Histogram and size range of the synthesized nanomagnetic particles—Sample 7.; Figure S8: Histogram and size range of the synthesized nanomagnetic particles—Sample 8.; Figure S9: Histogram and size range of the synthesized nanomagnetic particles—Sample 9.; Figure S10: Histogram and size range of the synthesized nanomagnetic particles—Sample 10.; Figure S11: Histogram and size range of the synthesized nanomagnetic particles—Sample 11.; Figure S12: Histogram and size range of the synthesized nanomagnetic particles—Sample 13.; Figure S13: Histogram and size range of the synthesized nanomagnetic particles—Sample 12 and 14.; Figure S14: Market growth for nanoparticles for biomedical applications.; Figure S15: Nanomedicine market growth rate until 2019.; Figure S16: Another study of the nanomedicine market until 2020.; Figure S17: Total costs/incomes *versus* total capacity; Table S1: Costs of Raw Materials; Table S2: Direct human labor costs.; Table S3: Labor costs for indirect human labor.; Table S4: Chief Personnel labor costs; Table S5: Management Personnel labor costs; Table S6: Streams of the Plant and their characteristics.

Author Contributions: P.A.A.: conceptualization; P.A.A. and J.M.C.: data curation; P.A.A., T.C.-G. and O.S.S.: formal analysis; P.A.A., T.C.-G. and D.B.: funding acquisition; P.A.A., T.C.-G. and O.S.S.: investigation; P.A.A., T.C.-G. and O.S.S.: methodology; P.A.A., T.C.-G. and D.B.: project administration; P.A.A.: resources; P.A.A., J.M.C. and O.S.S.: software; P.A.A., T.C.-G. and D.B.: supervision; P.A.A.: validation; P.A.A. and T.C.-G.: visualization; P.A.A. and T.C.-G.: writing—original draft; P.A.A., T.C.-G. and D.B.: writing—review and editing. All authors have read and agreed to the published version of the manuscript.

Funding: This research was funded by: (a) Project MAT2016-76824-C3-2-R, "Diseño de Nanohilos Magneticos para su uso en Tecnologias Limpias", MINECO, Spain. (b). Base Funding—UIDB/00511/2020 of the Laboratory for Process Engineering, Environment, Biotechnology and Energy—LEPABE—Funded by national funds through the FCT/MCTES (PIDDAC); (c). Project PTDC/QUIQFI/28020/2017—POCI-01-0145-FEDER-028020—Funded by FEDER funds through COMPETE2020—Programa Operacional Competitividade e Internacionalização (POCI) and by national funds (PIDDAC) through FCT/MCTES.

Institutional Review Board Statement: Not applicable.

Informed Consent Statement: Not applicable.

Data Availability Statement: Not applicable.

Acknowledgments: Helena Sant'Ovaia (Instituto de Ciências da Terra—Pólo da Faculdade de Ciências U.P.) is acknowledged for all the help concerning measuring magnetic particle properties. Joaquin Moran from the Department of Organic Chemistry, University of Salamanca, Spain, is acknowledged for all the help provided.

Conflicts of Interest: The authors declare no conflict of interest.

References

1. Wild, C.; Weiderpass, E.; Steward, B. *World Cancer Report: Cancer Research for Cancer Prevention*; International Agency for Research on Cancer, World Health Organization: Geneva, Switzerland, 2020.
2. Bruvera, J.I. *Disipación de Potencia Por Nanopartículas Magnéticas Expuestas a Campos de Radiofrecuencia Para Terapia Oncológica Por Hipertermia*; Universidad Nacional de La Plata: La Plata, Argentina, 2015.
3. Horny, M.-C.; Gamby, J.; Dupuis, V.; Siaugue, J.-M. Magnetic Hyperthermia on γ-Fe_2O_3@SiO_2 Core-Shell Nanoparticles for mi-RNA 122 Detection. *Nanomaterials* **2021**, *11*, 149. [CrossRef]
4. Caizer, C. Optimization Study on Specific Loss Power in Superparamagnetic Hyperthermia with Magnetite Nanoparticles for High Efficiency in Alternative Cancer Therapy. *Nanomaterials* **2021**, *11*, 40. [CrossRef]
5. Veloso, S.R.S.; Silva, J.F.G.; Hilliou, L.; Moura, C.; Coutinho, P.J.G.; Martins, J.A.; Testa-Anta, M.; Salgueiriño, V.; Correa-Duarte, M.A.; Ferreira, P.M.T.; et al. Impact of Citrate and Lipid-Functionalized Magnetic Nanoparticles in Dehydropeptide Supramolecular Magnetogels: Properties, Design and Drug Release. *Nanomaterials* **2021**, *11*, 16. [CrossRef] [PubMed]
6. Strbak, O.; Antal, I.; Khmara, I.; Koneracka, M.; Kubovcikova, M.; Zavisova, V.; Molcan, M.; Jurikova, A.; Hnilicova, P.; Gombos, J.; et al. Influence of Dextran Molecular Weight on the Physical Properties of Magnetic Nanoparticles for Hyperthermia and MRI Applications. *Nanomaterials* **2020**, *10*, 2468. [CrossRef]
7. Wu, K.; Wanga, J.-P. Magnetic hyperthermia performance of magnetite nanoparticle assemblies under different driving fields. *AIP Adv.* **2017**, *7*, 056327. [CrossRef]
8. Augusto, P.A.; Castelo-Grande, T.; Vargas, D.; Pascual, A.; Hernández, L.; Estevez, A.M.; Barbosa, D. Upscale Design, Process Development, and Economic Analysis of Industrial Plants for Nanomagnetic Particle Production for Environmental and Biomedical Use. *Materials* **2020**, *13*, 2477. [CrossRef] [PubMed]
9. Augusto, P.A.; Castelo-Grande, T.; Vargas, D.; Hernández, L.; Merchán, L.; Estevez, A.M.; Gómez, J.; Compaña, J.M.; Barbosa, D. Water Decontamination with Magnetic Particles by Adsorption and Chemical Degradation. Influence of the Manufacturing Parameters. *Materials* **2020**, *13*, 2219. [CrossRef] [PubMed]
10. Li, Q.; Kartikowati, C.W.; Horie, S.; Ogi, T.; Iwaki, T.; Okuyama, K. Correlation between particle size/domain structure and magnetic properties of highly crystalline Fe3O4 nanoparticles. *Sci. Rep.* **2017**, *7*, 9894. [CrossRef]
11. Butler, R.F.; Banerjee, S.K. Theoretical single-domain grain size range in magnetite and titanomagnetite. *J. Geophys. Res.* **1975**, *80*, 4049–4058. [CrossRef]

12. Muxworthy, A.R.; Williams, W. Critical superparamagnetic/single-domain grain sizes in interacting magnetite particles: Implications for magnetosome crystals. *J. R. Soc. Interface* **2009**, *6*, 1207–1212. [CrossRef]
13. Shah, R.R.; Davis, T.P.; Glover, A.L.; Nikles, D.E.; Brazel, C.S. Impact of magnetic field parameters and iron oxide nanoparticle properties on heat generation for use in magnetic hyperthermia. *J. Magn. Magn. Mater.* **2015**, *387*, 96–106. [CrossRef]
14. Khurshid, H.; Alonso, J.; Nemati, Z.; Phan, M.H.; Mukherjee, P.; Fdez-Gubieda, M.L.; Barandiarán, J.M.; Srikanth, H. Anisotropy effects in magnetic hyperthermia: A comparison between spherical and cubic exchange-coupled FeO/Fe3O4 nanoparticles. *J. Appl. Phys.* **2015**, *117*, 17A337. [CrossRef]
15. Avugadda, S.K.; Wickramasinghe, S.; Niculaes, D.; Ju, M.; Lak, A.; Silvestri, N.; Nitti, S.; Roy, I.; Samia, A.C.S.; Pellegrino, T. Uncovering the Magnetic Particle Imaging and Magnetic Resonance Imaging Features of Iron Oxide Nanocube Clusters. *Nanomaterials* **2021**, *11*, 62. [CrossRef] [PubMed]
16. Abenojar, E.C.; Wickramasinghe, S.; Bas-Concepcion, J.; Samia, A.C.S. Structural effects on the magnetic hyperthermia properties of iron oxide nanoparticles. *Prog. Nat. Sci. Mater. Int.* **2016**, *26*, 440–448. [CrossRef]
17. Martinez-Boubeta, C.; Simeonidis, K.; Makridis, A.; Angelakeris, M.; Iglesias, O.; Guardia, P.; Cabot, A.; Yedra, L.; Estradé, S.; Peiró, F.; et al. Learning from Nature to Improve the Heat Generation of Iron-Oxide Nanoparticles for Magnetic Hyperthermia Applications. *Sci. Rep.* **2013**, *3*, 1652. [CrossRef] [PubMed]
18. Reyes-Ortega, F.; Delgado, Á.V.; Iglesias, G.R. Modulation of theMagnetic Hyperthermia ResponseUsing Different Superparamagnetic Iron Oxide Nanoparticle Morphologies. *Nanomaterials* **2021**, *11*, 627. [CrossRef] [PubMed]
19. Augusto, P.A.; Castelo-Grande, T.; Merchan, L.; Estevez, A.M.; Quintero, X.; Barbosa, D. Landfill leachate treatment by sorption in magnetic particles: Preliminary study. *Sci. Total Environ.* **2019**, *648*, 636–668. [CrossRef] [PubMed]
20. Castelo-Grande, T.; Augusto, P.A.; Rico, J.; Marcos, J.; Iglesias, R.; Hernández, L.; Barbosa, D. Magnetic water treatment in a wastewater treatment plant: Part I—Sorption and magnetic particles. *J. Environ. Manag.* **2021**, *281*, 111872, ISSN 0301-4797. [CrossRef]
21. Castelo-Grande, T.; Augusto, P.A.; Rico, J.; Marcos, J.; Iglesias, R.; Hernández, L.; Barbosa, D. Magnetic water treatment in a wastewater treatment plant: Part II—Processing Waters and Kinetic Study. *J. Environ. Manag.* **2021**, acepted.
22. Hernández, L.; Augusto, P.A.; Castelo-Grande, T.; Barbosa, D. Regeneration and Reuse of Magnetic Particles for Contaminant Degradation in Water. *J. Environ. Manag.* **2021**, acepted. [CrossRef] [PubMed]
23. Estevez, A.M.; Rodriguez, J.M.; Alvaro, A.; Augusto, P.A.; Jimenez, O.; Castelo-Grande, T.; Barbosa, D. Preparation, Characterization, and Testing of Magnetic Carriers for Arsenic Removal From Water. *IEEE Trans. Magn.* **2008**, *44*, 4436–4439. [CrossRef]
24. Kim, D.; Lee, N.; Park, M.; Hyo, B.; Kwangjin, K.; Hyeon, T. Synthesis of Uniform Ferrimagnetic Magnetite Nanocubes. *J. Am. Chem. Soc.* **2009**, *131*, 454–455. [CrossRef] [PubMed]
25. Gómez Roca, A. *Preparación de Nanopartículas Magnéticas Uniformes y de Alta Cristalinidad para Biomedicina*; Instituto de Ciencia de Materiales de Madrid, Departamento de Química Física I, Universidad Complutense de Madrid: Madrid, Spain, 2009.

nanomaterials

MDPI

Review

Molecular Pathogenesis of Colorectal Cancer with an Emphasis on Recent Advances in Biomarkers, as Well as Nanotechnology-Based Diagnostic and Therapeutic Approaches

Fakhria A. Al-Joufi [1], Aseem Setia [2,*], Mounir M. Salem-Bekhit [3,4,*], Ram Kumar Sahu [5,*], Fulwah Y. Alqahtani [3], Retno Widyowati [6] and Fadilah Sfouq Aleanizy [3]

1 Department of Pharmacology, College of Pharmacy, Jouf University, Aljouf 72341, Saudi Arabia; faaljoufi@ju.edu.sa
2 Department of Pharmacy, Shri Rawatpura Sarkar University, Raipur 492015, India
3 Department of Pharmaceutics, College of Pharmacy, King Saud University, Riyadh 11451, Saudi Arabia; fyalqahtani@ksu.edu.sa (F.Y.A.); Faleanizy@ksu.edu.sa (F.S.A.)
4 Department of Microbiology and Immunology, Faculty of Pharmacy, Al-Azhar University, Cairo 11884, Egypt
5 Department of Pharmaceutical Science, Assam University (A Central University), Silchar 788011, India
6 Department of Pharmaceutical Science, Faculty of Pharmacy, Universitas Airlangga, Surabaya 60115, Indonesia; rr-retno-w@ff.unair.ac.id
* Correspondence: aseemsetia77@gmail.com (A.S.); mounirmsalem@yahoo.com (M.M.S.-B.); ramkumar.sahu@aus.ac.in (R.K.S.)

Citation: Al-Joufi, F.A.; Setia, A.; Salem-Bekhit, M.M.; Sahu, R.K.; Alqahtani, F.Y.; Widyowati, R.; Aleanizy, F.S. Molecular Pathogenesis of Colorectal Cancer with an Emphasis on Recent Advances in Biomarkers, as Well as Nanotechnology-Based Diagnostic and Therapeutic Approaches. *Nanomaterials* **2022**, *12*, 169. https://doi.org/10.3390/nano12010169

Academic Editor: Angelo Ferraro

Received: 2 November 2021
Accepted: 24 December 2021
Published: 4 January 2022

Abstract: Colorectal cancer (CRC) is a serious disease that affects millions of people throughout the world, despite considerable advances in therapy. The formation of colorectal adenomas and invasive adenocarcinomas is the consequence of a succession of genetic and epigenetic changes in the normal colonic epithelium. Genetic and epigenetic processes associated with the onset, development, and metastasis of sporadic CRC have been studied in depth, resulting in identifying biomarkers that might be used to predict behaviour and prognosis beyond staging and influence therapeutic options. A novel biomarker, or a group of biomarkers, must be discovered in order to build an accurate and clinically useful test that may be used as an alternative to conventional methods for the early detection of CRC and to identify prospective new therapeutic intervention targets. To minimise the mortality burden of colorectal cancer, new screening methods with higher accuracy and nano based diagnostic precision are needed. Cytotoxic medication has negative side effects and is restricted by medication resistance. One of the most promising cancer treatment techniques is the use of nano-based carrier system as a medication delivery mechanism. To deliver cytotoxic medicines, targeted nanoparticles might take advantage of differently expressed molecules on the surface of cancer cells. The use of different compounds as ligands on the surface of nanoparticles to interact with cancer cells, enabling the efficient delivery of antitumor medicines. Formulations based on nanoparticles might aid in early cancer diagnosis and help to overcome the limitations of traditional treatments, including low water solubility, nonspecific biodistribution, and restricted bioavailability. This article addresses about the molecular pathogenesis of CRC and highlights about biomarkers. It also provides conceptual knowledge of nanotechnology-based diagnostic techniques and therapeutic approaches for malignant colorectal cancer.

Keywords: colorectal cancer; biomarkers; cytotoxic medication; nanotechnology; genetic pathways; epigenetic changes; prognostic; phytomedicine

1. Introduction

Colorectal cancer (CRC) is a serious health issue in developed and developing countries alike, with the third highest incidence rate of all tumour-causing diseases [1]. This kind of cancer is a leading cause of morbidity and mortality in Western nations. Only a small percentage of CRC instances are caused by basic genetic abnormalities, with most cases occurring due to specific regular factors and growing age [2]. If left untreated, it can spread to the intestinal wall and the muscles beneath the surface of the skin. Environmental and genetic variables can also combine in many ways to accelerate cancer development and make it more difficult to treat.

In most cases, patients appear beyond the age of 60, with precursor starting adenomas that develop into cancer over for 1 to 2 decades [3]. Generalised screening of this population at age 50 years and older has been shown to reduce death from the disease, detect cancer at an earlier stage, and reduce the incidence of the disease. There have been four major hypotheses put out on the aetiology of colorectal cancer. To begin, the same genetic and epigenetic changes that cause colon cancer also increase the risk of developing colorectal cancer. Second, cancer develops in a staged manner, beginning with a precancerous condition and progressing to malignancy. Third, cancer development necessitates a lack of genetic stability. Fourth, in hereditary cancer syndromes, the germline forms of significant genetic abnormalities are generally associated with the development of rare colon cancers, which are caused by somatic occurrences of hereditary diseases [4].

Epigenetics, defined as heritable changes in gene expression without permanent changes in DNA sequence, plays an important role in the aetiology of different malignancies, including colorectal cancer. Recent epigenetic research has found a missing connection between certain gene expression patterns associated with CRC and the lack of genetic abnormalities. This missing link has now been discovered during the last two decades [5]. As an example, microsatellite in-stability (MSI), which is a hallmark of a molecular subgroup of CRC, results from an inability to repair DNA mismatches (MMRs) [6] properly. A genetic mutation can cause this defect in one of the MMR genes, the hypermethylation of the MLH1 gene's promoter, or both. Chromosome instability (CIN) in CRC has also been linked to global hypomethylation [7]. There is further evidence that cancer-related pathways at the post-transcriptional level are influenced by microRNAs (miRNAs), which play a role in nearly all phases of CRC, from initiation through progression and metastasis. Cell development is inhibited by miR-143, which downregulates in CRC because it tar-gets the KRAS mRNA transcript [8]. Aside from improving our knowledge of CRC pathogenesis, these discoveries have also provided new opportunities to find disease biomarkers and treatment targets. The various CRC biomarkers such as BRAF mutations, NRAS, and KRAS, microsatellite instability (MSI), DNA mismatch repair (MMR) status, and CpG island methylation have been studied. A KRAS or NRAS mutation is associated with a worse prognosis, and anti-epidermal growth factor receptor (anti-EGFR) antibody treatment will be ineffective [9]. Mutation in BRAF V600E is seen in 8–10% of individuals with CRC. These patients have a more aggressive illness and a worse prognosis in the adjuvant and metastatic context. People with BRAF mutation and microsatellite instability-high (MSI-H) tumours had a better overall prognosis than people with BRAF mutation and microsatellite stable (MSS) illness, and it turns out, using MSI to screen for Lynch syndrome and identify patients who could benefit from immunotherapy is becoming increasingly important in CRC [10].

A variety of drug-loaded nanoparticles in the 20–400-nm size range have made significant contributions to chemotherapeutic drug delivery in recent years (e.g., liposomes, dendrimers, polymeric nanoparticles, and micelles) [11]. Many potential nanomedicine advancements are based on these systems, which have evolved from basic drug-loaded nanoparticles to multifunctional nanoparticles that target particular cancer cells by attaching them to specific cell-surface proteins. Antigens such as integrin and folic acid receptors, which are differentially expressed on the surface of cancer cells, can be targeted using targeted nanoparticles [12]. Several of these nanoparticles are now in clinical trials. A great deal has been learned about using nanoparticles as therapeutic platforms for treating malignancies ranging from the prostate to the ovary. The clinical development of nanoparticles for cancer therapy is restricted, despite the high death and morbidity rates associated with CRC. In this review, we emphasise in detail the pathogenesis of colorectal cancer and their suitable candidate as biomarkers and nanotechnology-based diagnostic methods and therapeutic approaches for malignant colorectal cancer. Additionally, it provided an outlook on therapeutic nanoparticles for the management of CRC.

2. Pathogenesis of Colorectal Cancer

Genetic alterations in oncogenes and tumour suppressor genes occur sequentially from normal to the dysplastic epithelium in the adenoma–carcinoma process. This results in the development of CRC. Genomic changes, both genetic and epigenetic, convert normal glandular epithelium into adenocarcinoma when they accumulate [13]. While molecular mechanisms behind the cancer's unabated development and spread remain a mystery, several genetic pathways have been proposed to explain CRC aetiology (Figure 1). CRC involves numerous molecular pathways, including the three major pathways: CIN, MSI, and CIMP, all of which play a crucial role in the development of CRC due to genetic and epigenetic changes (Table 1). Three molecular pathways that indicate progression from cancer have been identified: microsatellite instability (associated with hypermutated group), chromosomal instability (completely within the non-hypermutated category), and the CpG Island Methylator Phenotype (CIMP; associated with both the hypermutated and non-hypermutated groups) [14]. These molecular pathways may dictate tumour development and metastatic time, as the epidemiology, mutational trials, and immune responsiveness information vary by route. The methods used to treat an illness might also differ. A biomarker known as "EMAST" refers to elevated microsatellite alterations at selected tetranucleotide repeats recently demonstrated to predict patient survival as a modulator of the three biological pathways. Individuals with sporadic CRC cannot predict whether they develop a metachronous adenoma or cancer based on the previously identified pathways [15,16]. Clinical monitoring is critical in detecting and eliminating metachronous lesions in these patients.

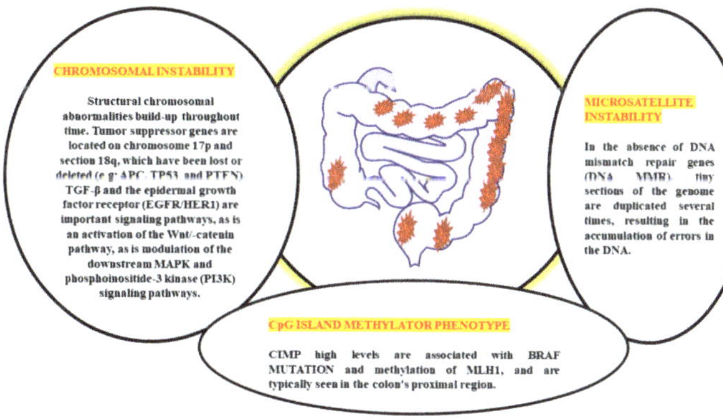

Figure 1. Multiple genetic pathways of CRC pathogenesis.

The biological behaviour, prognosis, and therapeutic response of CRC are all very variable. CRC has historically been divided into three subgroups based on the three pathophysiological routes of carcinogenesis: CIN, MSI, and CIMP from a molecular perspective [17]. CIN is the distinguishing feature for most CRCs, accounting for ~80–85% percent of the cases. CIN is characterised by the activation of growth-promoting pathways while simultaneously decreasing the activity of apoptotic pathways, with the latter being more common [18]. These tumours begin as adenomatous polyps due to the APC gene's deactivation mutation [encodes the WNT pathway effector adenomatous polyposis coli (APC)].

Table 1. An assessment of the MSI-H, EMAST, CIMP, and CIN CRC indices.

Pathophysiological Routes	Genomic Instability	Inflammation	Prognosis	Pathogenesis
CIN	Causes to the mutation and copy number variation; MSS; aneuploid	Tumour margin, lamina propria, and intraepithelial sites all have varying degrees of differentiation	Referent	Genetic changes that lead to heterozygosity loss
CIMP	Leading hypermethylation at DNA loci	Without hMLH1 hypermethylation: varied	Poor survival without hMLH1 hypermethylation	Without hMLH1 hypermethylation: unknown
MSI-H	Microsatellite instability (MSI) and diploid is appeared	Crohn's-like around tumour (tumour margin)	Better survival; early stage	Target gene frameshift mutation; $BRAF^{V600E}$
EMAST	Instability found in mostly at MSS and MSI-L, includes MSI-H	Tumour nests surrounding epithelial components have been linked to this condition.	Poor survival; later stage	Chromosome instability in combination with a frameshift mutation on a target gene

Further, the progress to adenocarcinomas by activating mutations in the KRAS (responsible for the receptor tyrosine kinase signalling) and deactivating mutations in SMAD4 (accounting for the TGFβ, it played an important role in the cell cycle control). Due to the initial event, this pathway is sometimes referred to as the APC pathway [19]. There are no mutations of MSI, MLH1, methylation, or BRAF in the CIN pathway. Moreover, Lynch syndrome (hereditary CRC's most frequent type) is caused by the germline mutations in the MMR genes, including MLH1, MSH2, MSH6, and PMS2. It is believed that MSI is initiated by a lack in the MMR DNA, causing mutations in the MLH1, MSH2, and MSH6 MMR genes. It is possible to have MMR deficiency due to MLH1 deactivation induced by biallelic hypermethylation of the MLH1 promoter or by dual somatic mutations in the MMR genes [20]. It is noticed that the individual with MMR-deficient and MSI-high CRC (dMMR–MSI-H) demonstrated a good prognosis but did not acquire therapeutics benefits from 5-fluorouracil (5-FU) and has already been characterised based on MSI status [21]. A subset of colorectal cancers known as CIMPs, which are defined by a substantial percentage of CpG island hypermethylation around the promoter of multiple tumour-suppressor genes, is identified by epigenetics rather than by CIN or MSI. Due to its close association with colorectal carcinogenesis's serrated route, hypermethylation of the MLH1 promoter, as well as its association with female gender, advanced age, and poor histology, CIMP has a significant risk of being found in patients with the disease [22]. However, the lack of a defined definition of CIMP-high makes it difficult to translate this pathophysiological feature into treatment. Although CIMP testing is most commonly done on MLH1, MINT1, CACNA1G, and CDKN2A (could potentially up to 16 distinct genes identified), there is currently does not agree about which cut-off values will be applied to distinguish between CIMP high and CIMP+ on certain scientific procedures must be employed to conduct assessments for CIMP tests [23].

As early as lesions in the polyp cancer development sequence, gene alterations have a role in cancer creation and progression. The abnormal crypt focus is the initial histological lesion linked to CRC development [24]. Tumour suppressor gene APC is often altered in colorectal cancers, and dysplastic aberrant crypt foci can spread mutations. The Wingless/Wnt pathway is activated by inhibiting APC, a systematic approach to start the polyp cancer development sequence. The Wnt signalling pathway harbours the clonal development of polyp cells to cancer due to subsequent mutations in genes like KRAS or TP53 (Figure 2). The transforming growth factor b (TGFB1)-mediated cell signalling pathway may also be modulated by KRAS and TP53 mutations, further accelerating CRC development [25].

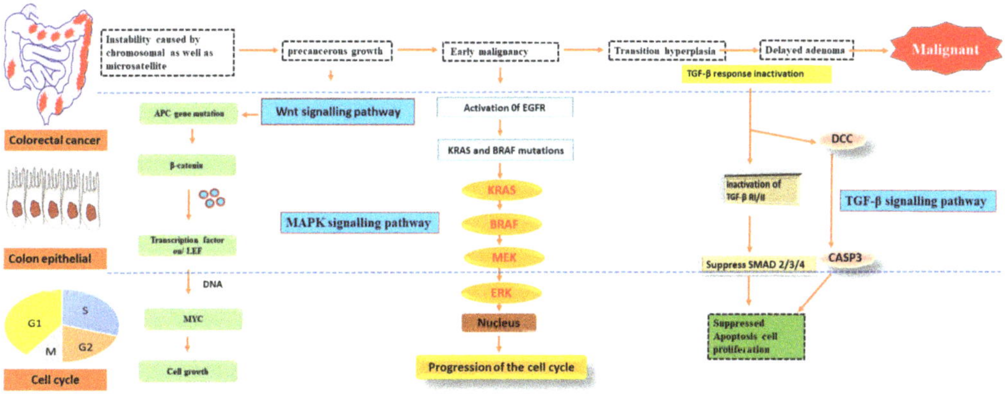

Figure 2. Pathways of signalling that have been genetically changed by CRC.

Oncogene activation and tumour suppressor gene inactivation are attributed to genetic and epigenetic alterations that occur during colon cancer development. B-Raf sends EGFR signals to KRAS, a proto-oncogene that activates the MAPK pathway [26]. Over half of all CRCs have mutations in KRAS or B-Raf, which trigger the MAPK signalling pathway and promote proliferation while also stopping cell death. The PI3 K, WNT-APC-CTNNB1, and TGFB1-SMAD signalling pathways are also part of CRC. It also includes the RAS–RAF–MAPK pathway [27].

The study of epigenetic changes in cancer has led to the identification of genes commonly altered in these specific pathways, leading to in-depth investigations into the biology of CRCs. Novel CRC diagnostic and prognostic tests and new treatments are being developed based on these findings.

3. MSI, CpG and CIMP Involvement in the Progression of CRC

CRC is a group of several illnesses, some purely molecular, while others include morphological manifestations. Colorectal carcinogenesis has been linked to three distinct molecular pathways, including chromosomal instability (CIN), microsatellite instability (MSI), and chromosomal mismatch repair (CIMP) (Figure 3). For example, chromosomal size and structure are altered in the CIN route. In contrast, changes in the number of mono and dinucleotide repeats in the coding and regulatory regions are seen in the MSI pathway. CRCs grow in two distinct routes, including the conventional adenoma–carcinoma sequence and the serrated neoplasia pathway, morphologically multistep processes [28]. An adenocarcinoma sequence that dates back to antiquity describes the progression of benign adenomas, including tubular adenomas, tubulovillous adenomas, or villous adenomas, progress and transform into invasive adenocarcinomas. The phrase "serrated neoplasia pathway" refers to the concept that sessile or conventional serrated adenomas can develop into invasive adenocarcinomas [29].

Even though Lynch syndrome is associated with inherited MSI-H tumours, the CIN pathway is the underlying biological mechanism for the typical adenoma–carcinoma sequence. Tubular adenomas with variable villosity are Lynch syndrome premalignant lesions. The great majority of spontaneous MSI-H CRCs, in contrast to Lynch syndrome tumours, are considered to develop in CIMP-H sessile serrated adenomas with BRAF mutation through methylation-associated inactivation MLH1. Hereditary and spontaneous MSI-H CRCs take the conventional adenoma–carcinoma sequence and the serrated neoplasia pathway, respectively, whereas MSI-H tumours follow a distinct morphological multistep approach. CIMP-H/non-MSI-H CRCs, in contrast to CIMP-H/MSI-H CRCs, originate from pre-existing conventional serrated adenomas [30].

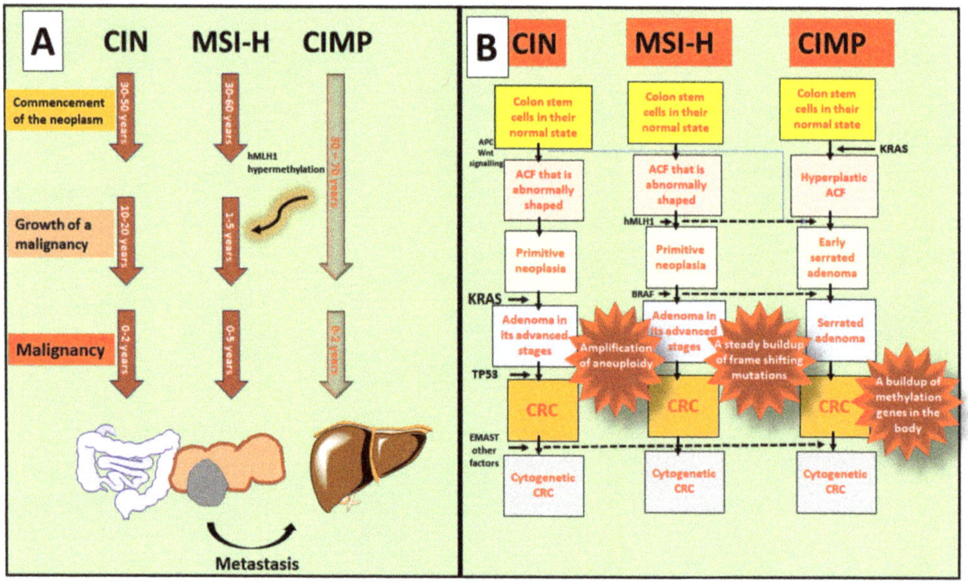

Figure 3. Involvement of MSI, CpG, and CIMP in the progression of CRC. (**A**) Every age groups are susceptible to malignancy, and progression of colorectal cancer shows after the occurrence of metastasis; (**B**) In the case of CIN, various molecular pathways play crucial roles such as Wnt, KRAS, TP53, and EMAST and forms CRC progression. While MSI-H shows progression via hMLH1 and BRAF. Whereas, CIMP is interconnected with KRAS mutation and causes cytogenetic to normal cells.

On the whole, 65–75 percent of CRCs are caused by the CIN route, whereas 20–30 percent are caused by the CIMP pathway and 5 percent by the hereditary MSI pathway. Females have a larger percentage of the CIMP route than men. Older persons have a higher percentage of the CIMP pathway than younger people. The CIMP pathway is more prevalent in the proximal intestine than the distal colon or the rectum [31].

4. Epidemiology

The frequency of colorectal cancer (CRC) has risen alarmingly in recent years around the globe (Figure 4). In 2020, an estimated 1.93 million new CRC cases were identified, and 0.94 million people died as a result of CRC-related causes globally, accounting for 10% of global cancer incidence (total 19.29 million new cases) and 9.4% of deaths due to cancer (total 9.96 million deaths). It is projected that CRC will be the third-greatest cause of cancer-related deaths in men and women globally in 2020, with 515,637 deaths in men and 419,536 in women, respectively [32]. Over 5.25 million individuals globally (5-year prevalence) have CRC, making it the second-most common disease after breast cancer (7.79 million new cases per year). Approximately 1.15 million new instances of colon cancer, 0.7 million cases of rectal cancer, and 50,000 cases of anal cancer in 2020 were found, according to the database from GLOBOCAN 2020. These figures are expected to rise to 1.92 million, 1.16 million, and 78,000 by 2040. Countries have different rates of colorectal cancer (CRC) [33]. Among countries, the greatest age-standardised incidence rates were observed in Hungary (45.3 cases per 100,000 people in 2020), Slovakia (43.9%), Norway (41.9 cases per 100,000), the Netherlands (41.0 cases per 100,000), and Denmark (40.9 cases per 100,000). Guinea, Gambia, Bangladesh, Bhutan, and Burkina Faso will have the lowest age-standardised incidence rates in 2020 at 3.3, 3.7, 3.8, 3.8, and 3.8 cases per 100,000 people, respectively.

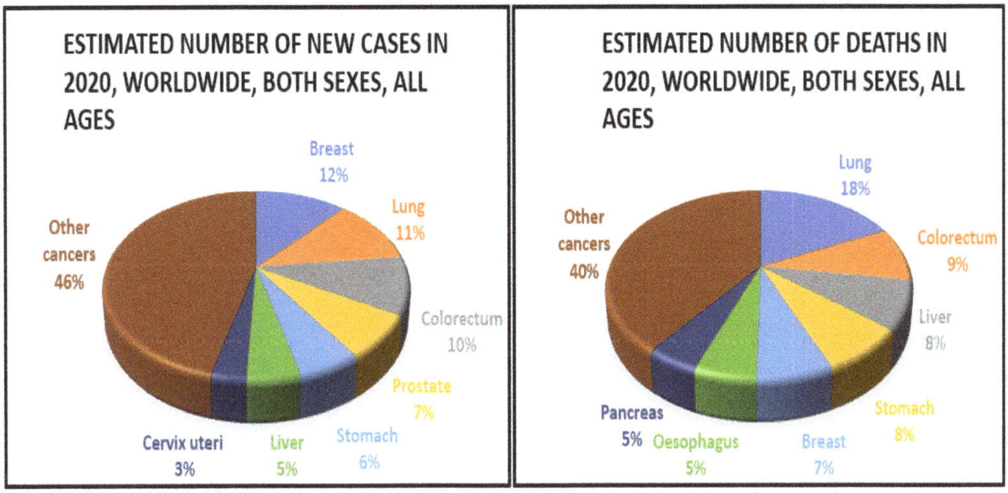

Figure 4. Epidemiology of colorectal cancer.

5. Biomarkers Based on Epigenetic Changes for CRC

When CRC progresses from an adenomas or serrated lesions stage, genetic and epigenetic changes accumulate in the lesions, giving progressively dysplastic characteristics until they become an adenocarcinoma [34]. To avoid this illness, it is critical to find precancerous lesions and early-onset colorectal cancer in people at average risk who are otherwise healthy during routine screening. As far as CRC screening is concerned, colonoscopy is the gold standard, since it can detect and eradicate precursor lesions [35]. For these reasons, many people choose not to get colonoscopies. They are painful, costly with low compliance rates, and can cause complications, including haemorrhage and perforation. For precursor lesions like adenomas, non-invasive screening procedures like the faecal occult blood test (FOBT) and faecal immunochemical test (FIT), widely used in Europe and other Western nations, have lower sensitivity and specificity than colonoscopy. As a result of these drawbacks, new non-invasive detection methods for precursor lesions and early-stage CRCs are urgently required [36].

At intermediate stages of CRC staging, the currently used tumour–node–metastasis (TNM) classification approach is inadequate for prognostication and clinical decision-making [37]. Individuals with a high risk of illness recurrence or mortality (prognostic biomarkers) and those who would benefit from chemotherapy, immunotherapy, or targeted therapy are urgently in need of biomarkers that aid identify these patients (which may be unnecessary). Many epigenetic biomarkers, including DNA methylation, histone alterations, miRNAs, and long noncoding RNA, have shown promise as clinically significant biomarkers for CRC diagnosis, prognosis, and therapy response prediction (lncRNA) [38].

Blood and other body fluids and tissues can include quantifiable indicators of pathological or physiological processes or conditions or diseases, such as molecular markers (biomarkers). Molecular markers can be used to identify conditions or diseases. Biomarkers are important diagnostic, prognostic, and therapeutic tools for cancer detection, diagnosis, and therapy selection (Figure 5). It can also be used to pinpoint the exact location of a tumour and identify any areas that are at an advanced stage of growth or are receptive to a particular therapy [39]. These molecular markers represent the mechanisms of neoplastic cell exfoliation and mucus secretion of aberrant glycoproteins in CRC by reflecting the mechanisms of cell exfoliation and mucus secretion. Shortly, researchers hope to find novel, non-invasive DNA, RNA, or protein-based molecular markers that may be used to detect colorectal cancer in the blood, faeces, and other body fluid. The prognostic, predictive, and diagnostic markers have been developed for the determination of the CRC. Diagnostic biomarkers indicate the likelihood of the illness progressing. Patient treatment measures can be predicted with the use of these molecular markers. Predictive biomarkers are utilised for forecasting the efficacy of treatment in cancer patients [40]. Colorectal polyps can be detected early with the use of the following biomarkers.

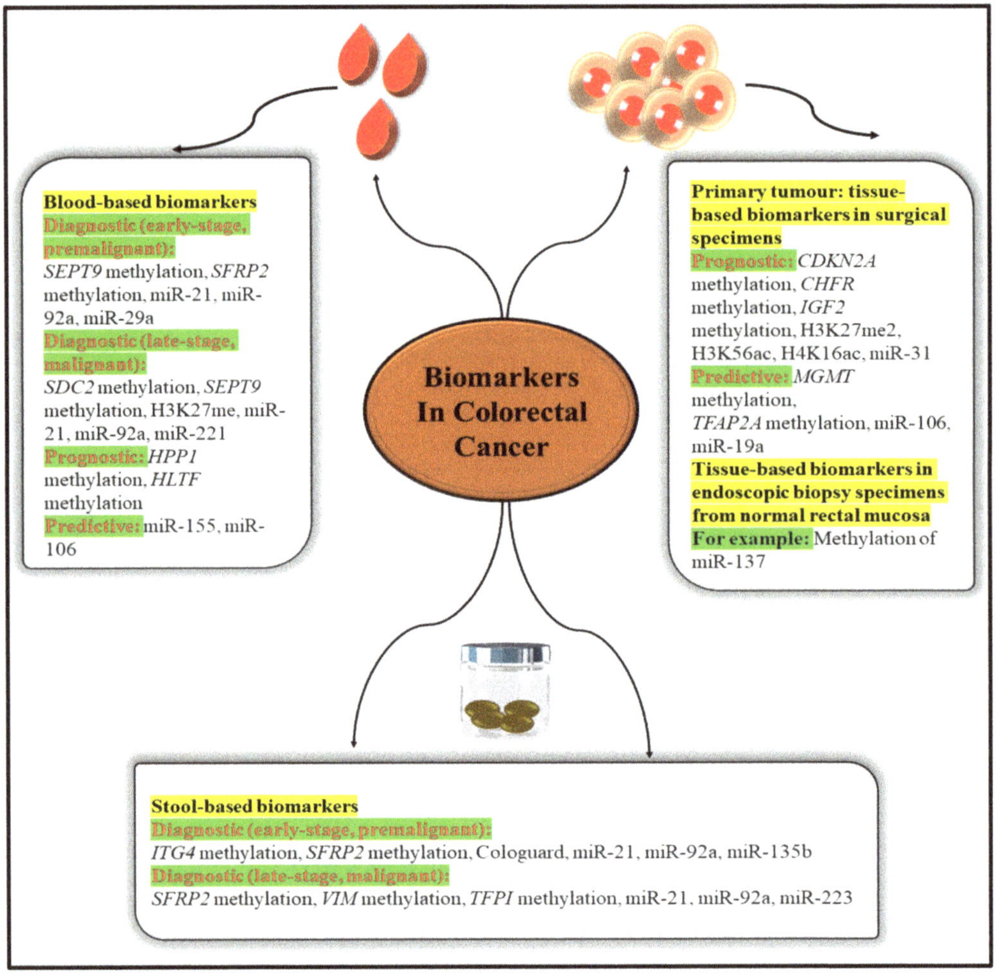

Figure 5. Epigenetic biomarkers in CRC.

5.1. Molecular Markers for Diagnosis

5.1.1. HNPCC (Hereditary Nonpolyposis CRC)

HNPCC is the most well-known autosomal dominant hereditary colon cancer condition. It is a screening test for the early identification of colorectal cancer (CRC). HNPCC is present when a germline mutation is identified in one of the four MMR genes, such as MSH6, MLH1, MSH2, and PMS2, all found in the human body. Many young patients with a new genetic condition come from families where colon cancer runs in the blood.

5.1.2. Telomerase

The telomerase is key enzyme that preserves the telomere (repeating DNA (TTAGGG/AATCCC in the human telomeres) that protects from chromosomal breakdown, elimination of operational genes, and cell death. Telomerase is necessary for the maintenance of the telomere. The activation of telomerase is an independent diagnostic and prognostic marker for malignant tumours that can be used independently.

5.1.3. Insulin Like Growth Factor Binding Protein 2 (IGFBP2)

According to research, compared to healthy individuals, CRC patients had higher serum and plasma levels of IGFBP2. It has emerged as a diagnostic device for the primary identification and development of CRC, since it is implicated in cancer cell proliferation, migration, and invasion.

5.1.4. Pyruvate Kinase M2 (PKM2)

PKM2 refers to a cytosolic enzyme expressed in both normal and malignant cells and plays an important function in energy metabolism. PKM2 expression is upregulated in CRC and other gastrointestinal malignancies. It may also be relevant for CRC inquiry and/or diagnosis based on stool or blood samples as a dismal prognostic sign.

5.2. Molecular Markers for Prognostic

5.2.1. p53

The existence of the tumour inhibitor p53 is also considered an independent prognostic factor for colon cancer. The alteration of the tumour-suppressor gene p53 is an initial and critical step in ulcerative colitis-associated carcinogenesis (sometimes referred to as TP53). The development of carcinoma in an adenoma coincides with a p53 mutation and a loss of heterozygosity (LOH) in the wild-type allele, further demonstrating its function in the regulation of malignancy. Patients with CRC who are missing this gene have a bad prognosis.

5.2.2. 18 q Loss of Heterozygosity

Patients with colorectal cancer (CRC) who have chromosomal 18q LOH have a reduced chance of survival. Patients with stage II or stage III colon cancer who lose their somatic heterozygosity have a worse prognosis than those whose tumours retain both of their parents' alleles on chromosome 18 q. Patients with stage II or stage III colon cancer with this 18qLOH have a poorer prognosis than those with normal alleles on chromosome 18q.

5.2.3. MLH1 Methylation

DNA microsatellite unstable recognition or the absence of the MLH1 protein production on immunohistochemical investigation confirm somatic MLH1 inactivation in primary colorectal malignancies, which is more common in early-stage colorectal cancers than in late illness. A more indolent illness or a better prognosis might result from this inactivation in the absence of adjuvant treatment.

5.2.4. VEGF

One of the angiogenic agents in CRC is vascular endothelial growth factor (VEGF), which is expressed in roughly 50% of CRCs but only in extremely low amounts in normal colonic mucosa and adenomas. As a result, VEGF-1 expression appears to deliver useful predictive information in CRC patients.

5.3. Predictive Molecular Markers

5.3.1. KRAS

The discovery of KRAS mutations is employed as the foremost potential prognostic marker in anti-EGFR (epidermal growth factor receptor) antibody-based therapy in CRC, such as panitumumab and cetuximab.

5.3.2. B-Raf V600E

The activation of this kinase by somatic mutation leads to an unconstrained MAPK signalling pathway in humans. The BRAF V600E activating mutation does not respond to EGFR inhibitor treatment in individuals with stage IV CRC. Although the significance of BRAF mutant level as a predicting marker gene is uncertain, BRAF mutant as a biomarker of anti-EGFR antibody resistance is perhaps the extensively researched predicting potential of BRAF alteration.

5.3.3. PIK3CA Status

The PIK3CA mutation causes an abnormal stimulation of the AKT pathway, leading to cancer cell growth. If the PI3K/AKT pathway is persistently activated, it is likely to affect CRC progression significantly. Colon cancer cells often with an induced PIK3CA mutant were more sensitive to cetuximab in preclinical studies even than PIK3CA wild-type cells.

5.3.4. ERCC—1

Repair protein for DNA excision ERCC-1 is a nucleotide excision repair gene that is involved in DNA repair. A new analysis of metastasis and stage II or III CRC and evaluating ERCC-1 as a possible predictive biomarker in direct response to oxaliplatin by studying dysregulation of ERCC-1 expression in between oxaliplatin treatment has been presented [41–44].

6. Diagnostic/Materialistic Tool for the Treatment of Colorectal Cancer

A diet rich in vegetables, fruit, and whole-grain cereal goods, as well as regular exercise, lowers the chance of developing colorectal cancer. CRC risk has been linked to environmental variables such as obesity, smoking, and high alcohol intake. Calcium and vitamin D3 supplementation may have a disease-fighting effect. It has also been shown that long-term usage of acetylsalicylic acid (NSAIDS) decreases the risk of cancer. Colorectal cancer is just one of many cancers that can be prevented by quitting smoking or abstaining from smoking. Adenomas (primary prevention) and malignancies (early detection) can be detected with screening tests (secondary prevention).

This is why regular screenings in high-risk people can help detect CRC even if symptoms are not present. After the diagnosis of colon cancer, the diagnostic tools are also useful for the TNM staging, which is used to determine the appropriate treatment. These include colonoscopy, sigmoidoscopy, magnetic resonance imaging (MRI), CTCG, transrectal ultrasonography, stool testing, double-contrast barium enema, and many others. However, all techniques have some of the other related limitations and remain inaccurate for effective diagnosis in the early phases of the disease process. There is a pressing need to develop more advanced diagnostic methods in order to increase survival rates, reduce disease-related problems due to diagnosis at later stages, and properly stage the disease so that appropriate therapy may be prescribed. Here are some of the most commonly used diagnostic tools, along with their advantages and disadvantages (Figure 6) [45,46].

6.1. CT Colonography

When referring to a computerised x-ray imaging procedure, "virtual colonoscopy" or "CT colonography", both terms refer to the same thing: a method that produces two- and three-dimensional images of the colon. It is possible that CT colonography can be used as an alternative to colonoscopy in patients with obstructive or stenotic tumours. After the colon has been completely evacuated (bowel preparation), it is insufflated with air or carbon dioxide to increase its size for better imaging during the CT scanning procedure. After insufflation, the subject is scanned in both the supine and prone positions. CT colonography is comparable to colon endoscopy in terms of patient pain because of the need for colon evacuation and insufflation, but it has the advantage of not requiring anaesthesia. For polyps larger than 10 mm in diameter (big polyps), the detection sensitivity and specificity are excellent (82 percent), but the sensitivity is poor (63 percent) for polyps between 6 and

9 mm in diameter (small polyps). Furthermore, the method cannot distinguish between flat and serrated lesions. For diagnosing the local stage of a CRC tumour, MRI is preferred over CT colonography, which performs poorly in this regard. Due to its high level of accuracy in determining the tumour's invasion and spread, preoperative MRI is useful in planning surgery. The CTCG approach has many drawbacks, including radiation exposure, a lack of sensitivity to detect small polyps, the necessity for intestinal preparation, and the need for an extra procedure to remove polyps if they are discovered. As a result, the patient's concern rises and the expense rises as a result of unneeded monitoring, which is exacerbated by extracolonic lesion identification [45,47].

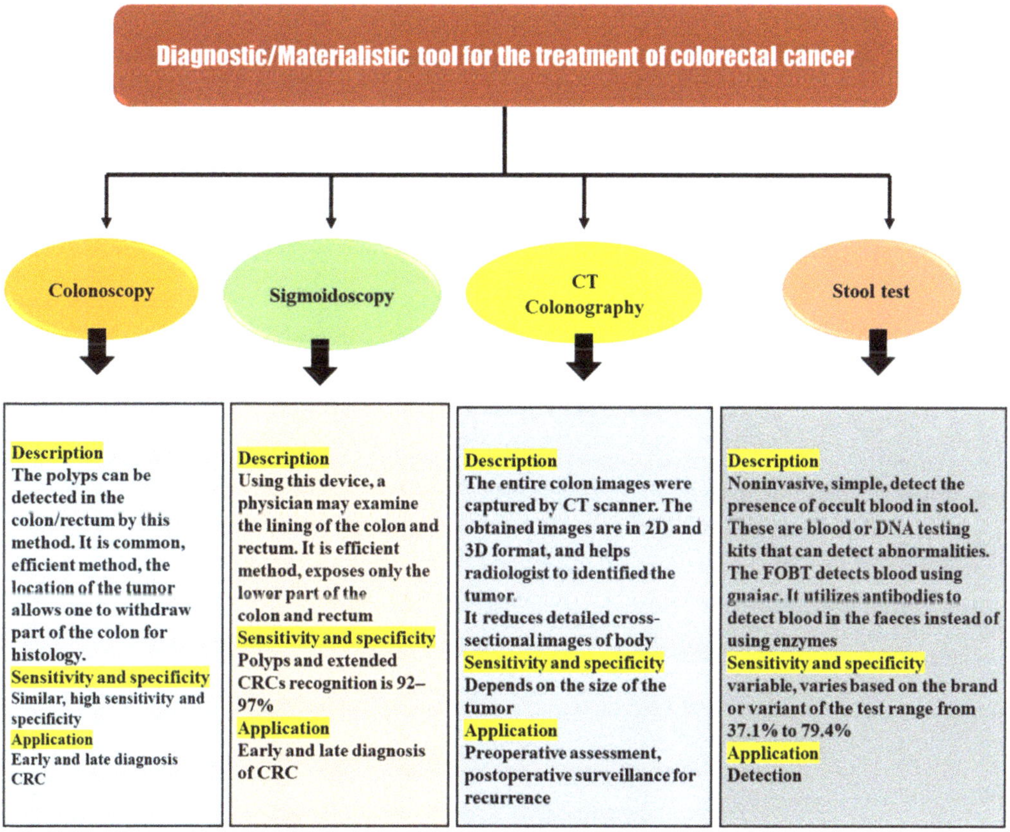

Figure 6. Diagnostic/materialistic tool for the treatment of colorectal cancer.

6.2. Colonoscopy and Sigmoidoscopy

Endoscopy is the most often used and most effective procedure for CRC diagnosis. Colonoscopy and sigmoidoscopy are included in this procedure. These tests enable the detection of a tumour and the removal of a portion of the large intestine for histological analysis to be performed. Sigmoidoscopy has a sensitivity and specificity of 92–97 percent for the detection of polyps and expanded CRCs. Sigmoidoscopy only reveals the lowest portion of the colon and the rectum when performed with a flexible endoscope. Colonoscopy provides the same level of sensitivity and specificity as a colonoscopy in terms of obtaining an image of the entire intestine. Colonoscopy has several advantages over other screening tests, including the fact that it may be completed in a shorter period and that it improves the patient's acceptance and tolerance of new sedative procedures. In comparison to the control group, the study that included individuals with an average risk of CRC following colonoscopy revealed a 67 percent of overall decrease in morbidity and a 65 percent of the overall decrease in mortality. Preliminary screening with sigmoidoscopy and colonoscopy was found in a randomised trial to reduce CRC incidence and mortality by 33% and 38–59%. As a result of this discovery, sigmoidoscopy has become increasingly popular as the main screening method in England and elsewhere. Smaller polyps (6–10 mm) are difficult to identify with this invasive method. It also restricts cancer detection in the ascending colon, the transverse colon, and the cecum. Endoscopy (sigmoidoscopy) can be used to diagnose and treat CRC in patients who are ineligible for surgery (according to the severity of the tumour or comorbidities). Methods for removing the cancer-related blockage are among these procedures. Patients typically report discomfort from endoscopy despite its many benefits. Due to the potential for consequences, people avoid or postpone the evaluation. Virtual colonoscopy has been increasingly popular in recent years. The use of computed tomography is it provides a 3D view of the large intestine. Perforation or bleeding from the large intestine can be reduced by the use of virtual colonoscopy [45,47,48].

6.3. Stool Test

GFOBT and FIT are two of the most commonly utilised main screening modalities for CRC detection. Stool testing can identify blood spilling from CRCs in the stools of asymptomatic patients. FITs are more commonly used than gFOBTs because of their ease of use and greater sensitivity. Patients who have a positive stool test must have a colonoscopy to get an accurate diagnosis and to remove any polyps that are found. The detection of DNA markers released by malignant cells in a person's stool is a promising technique for CRC screening. Stool DNA testing has a high level of sensitivity (71–91%) and specificity (93–100%) for the identification of colon cancer. False results are common because of the low sensitivity and specificity of the FOBT. Additionally, these tests fail to detect nonbleeding tumours. Faecal occult blood testing is a straightforward, inexpensive, and non-invasive diagnostic procedure. Haemoglobin in faeces implies gastrointestinal bleeding, according to the test. Since it can come from both malignant alterations and polyps larger than 1 to 2 cm, blood in faeces is an unspecific sign of colon cancer. The test's sensitivity can be increased by as much as 90% by repeating it. Heme and human globin are the two building blocks of haemoglobin, and the immunohistochemistry faecal occult blood test (FIT) confirms its presence in CRC diagnoses. Aside from sDNA, which indicates DNA changes in colorectal adenocarcinomas, molecular diagnostics, such as PCR, may be used to separate and discriminate DNA from that of bacteria in faeces. The usefulness of CRC molecular diagnostics based on genetic and epigenetic testing is restricted. Costs are expensive, because it is not widely available. Molecular diagnoses of CRC have some drawbacks, which have prompted researchers to look for biomarkers that can be found in biological samples at a cheaper cost [45,47,49,50].

7. Nanotechnology-Based CRC Diagnostic Techniques

Medicines based on nanoparticles are being developed as therapeutic methods for cancer therapy, which has led to significant improvements in pharmaceutical research [51]. These advancements have reduced the adverse effects of cytotoxic drugs while simultaneously increasing their efficacy and solubility. The encapsulation of anticancer cargo, including as siRNA, antibiotics, and chemotherapeutics has proven highly effective in the last 50 years using a variety of distinct nanoparticles with various forms, sizes, and chemical natures [52]. Using the increased permeability and retention effect afforded by tumour vascular and lymphatic drainage, these first-generation anticancer nanoparticles passively penetrate tumour tissue, allowing nanoparticle extravasation and storage within cancer cells, improving therapeutic effectiveness. Early diagnosis and therapeutic efficacy monitoring can both benefit from the use of nanoparticles [53]. Incorporating various contrast agents (such as radioactivity, superparamagnetic, or fluorescent), as well as targeting groups and biocompatible coatings, into a nanoparticle design is a feasible option [54]. Due to the disadvantages of low tissue specificity, rapid clearance, and nonspecific extracellular distribution associated with small molecular weight gadolinium and metal chelate-based contrast agents, nanotechnology may be used to modify these contrast agents to improve CRC diagnostic sensitivity and specificity [55]. Various nanostructures are being used as inputs of in vitro diagnostic assays for protein indicators or nucleic acid targets. In vivo imaging with nanostructures, similar to in vitro diagnostics, seems to be a potentially intriguing area in diagnosing colon cancer [56]. There are, however, just a few examples of how nanotechnology has been used in CRC imaging. Novel MRI contrast agents are among the most readily useful pieces of colorectal cancer technology. The introduction of nanostructures that modify standard contrast agents like gadolinium or imaging agents like iron oxide could increase the detection capability of medical image [57]. These nanostructures enhance the capabilities of traditional MRI imaging, but they also provide new possibilities for the early diagnosis and healing of the CRC. Tumour early diagnosis and healing techniques not possible with current conventional technologies have been found as beneficial nanotechnological applications in cancer biology [58]. When it comes to diagnosing, producing, and healing some malignancies, nanometre-sized particles of various structures and conformations have shown to be highly effective and promising new approaches. A patient's long-term survival can be affected by the early identification of CRC, which is critical for prevention [59]. Here are a few examples of recent Nps accomplishments that have opened up new possibilities for the early identification and successful treatment of CRC.

7.1. Quantum Dots (QDs)

QDs are semiconductor nanocrystals that generate fluorescence when excited by light and have remarkable optical properties, such as high brightness, photobleach resistance, and the ability to emit fluorescence at many wavelengths [60]. QD-based nanotechnology, with its optical and chemical benefits, is a developing platform for cancer research, particularly colorectal cancer [61]. Its bandgap energy, the degree of energy required to transport electrons from one electronic zone to the next, this time at a greater frequency, is an important characteristic of a semiconductor nanocrystal. An exciton is a pair of electrons and holes formed as a result of this excitation strategy. The unstable exciton returns its ground state and releases fluorescence photons in the form of energy [62]. When the semiconducting nanocrystal size and shape approach the bulk Bohr exciton radius, which is still generally 2–10 nm in length, electrical and optical properties are added to the particle. Scientists know that as a nanocrystal's size decreases, its bandgap increases, and its stimulation and emissions wavelength decrease because of this inverse relationship [63].

Park et al. used an enzyme-sensitive fluorescent dye in conjunction with antibody-QD conjugates to produce a multiplexed detection method for the fast and precise diagnosis of CRC. The instrument incorporated a Cresyl violet–glutamic acid (CV–Glu) derivative as the fluorescent molecular probe, as it easily transits between two fluorescent colours in reply to k-gluta myltranspeptidase enzymatic activity (GGT). A biocompatible, 690-nm emitting $AgInS_2/ZnS$ (core/shell) QD probe was created by bioconjugating an anti-matrix metalloproteinase-14 antibody (MMP14). The probe provided significant findings in ex vivo and in vitro applications by employing human CRC cell lines, ex vivo patient tumour samples, and ex vivo experimental animal model tissues. The researchers utilised two-photon microscopy to show that the co-application of both probes penetrated the tissue at depths of 10–20 mm quickly. The study's findings indicated that using both probes simultaneously allowed for quick (within 5 min) and accurate imaging of tumour lesions that would otherwise be impossible to identify using traditional colonoscopy procedures. When CRC was in its beginning phases, the dual-probe approach helped find hyperplastic polyps and adenomas [64].

Carbary-Ganz et al., created a dual-modality imaging technique that uses optical coherence tomography and laser-induced fluorescence to provide non-destructive endoscopic viewing of CRC with just a minimal amount of invasiveness. While CRC develops and progresses, this method allows for the concurrent longitudinal surveillance of biochemical and morphological changes. With great sensitivity and specificity, QDot655-VEGFR2 localised to the colon in carcinogen-treated animals and provided a significant difference between sick and healthy tissue ex vivo [65].

7.2. Iron Oxide Nanocrystals

Iron oxide NPs are incorporated in MRI as a contrast agent for human use because of their dual magnetic and photothermal properties. Apart from that, they are highly biodegradable in vivo, and then, the iron salts delivered after disintegration can be ingested into the body system through a well-coordinated biochemical mechanism [66]. There is a magnetic core to the iron oxide nanocrystals and a polymer covering, which will contain the various medicinal molecules. These nanoparticles have various biological properties due to their relatively small dimension, efficient surface region, minimal deposition rate, and ease of cellular transport [67]. The severe toxicity and oxidation sensitivity of strong magnetic metals like cobalt and nickel seem to offer limited biomedical utility. Using nanoparticles, a cancer cell is transformed into a tiny magnetic magnet that may be attached to the biopsy needle's tip. These nanoparticles are unstable in aqueous environments and easily collected and deposited [68]. It cannot detect the target tumour using nanoparticles of the iron oxide in vivo due to the diverse overexpression of the highly targeted receptor in cancer cells. Using magnetic nanoparticles, it is possible to discriminate between malignant and normal tissue. It is possible to create functional groups on these nanoparticles that crosslink with tumour-targeting molecules such as peptides or monoclonal antibodies or small molecules for the diagnostic testing or remedial agent administration due to their great surface area [69].

Kim et al. used a 3D multi-echo gradient echo to compare the particular performance of three distinct parametric techniques (normalised signal intensity, R2* (R-lymph node Region of interest), and susceptibility) to measure the number of ultra-small superparamagnetic iron oxide (USPIO) particles in lymph nodes. Before and after USPIO injection, nine rabbits with VX2 tumour implants were scanned. All of this was accomplished by combining 3D GRE amplitude and phase data to produce multi-echo combined T2* (weighted multi-echo combined magnitude image of 3D mGRE 24 h after USPIO injection)-weighted pictures and an R2* map. USPIO build-up in 18 lymph nodes (nine metastatic and nine reactive) was found to have a significant impact on signal intensity. R2* differences before and after USPIO injection were statistically significant between lymph nodes that were reactive and metastatic ($p < 0.05$), although signal intensity and vulnerability were not significantly different between the nodes. USPIO-enhanced MRI employing R2* mapping from 3D multi-echo GRE may be used to identify lymph node metastases and to perform a parametric analysis of the lymph node status in a rabbit model, as demonstrated in their work [70].

Wu et al. conducted a meta-analysis to examine how well USPIO-enhanced MRI and non-enhanced MRI, as well as USPIO-enhanced MRI in various body areas and postcontrast MRI performed in the diagnosis of lymph node metastases. As a result of this meta-analysis, USPIO-enhanced MRI is more accurate than conventional MRI in the diagnosis of lymph node metastases. For lymph node characterisation, postcontrast pictures alone are sufficient to provide the same diagnostic performance as pre-contrast MRI. There is still a lot of research to be done on the use of USPIO enhanced MRI in clinical practice [71].

Kuo et al. developed an antibody-targeting peptide AP-1(MPVA-AP1) and anticancer medicines by incorporating the magnetic nano-vehicles. The magnetic nano-vehicles disseminated in the dilute solution demonstrated great hemocompatibility and toxic-free to L929 fibroblasts, indicating that they might be used in treatments. The researchers found significant encapsulated hydrophobic and hydrophilic low-molecular-weight medicines and protein-like pharmaceuticals using a straightforward manufacturing approach. Magnetic nano-vehicles were also used to immobilise the antibody-targeting peptide AP-1, which was verified using electron spectroscopy. The results of a CRC cell (CT26-IL4Rα) assay demonstrated that the AP-1-bound nano-vehicles (MPVA-AP1) show exceptional targeting and selectivity. In the absence of a magnetic stimulation, a stable storage test revealed virtually little leakage of the encapsulated medicines. Instead of rupturing in the presence of a high-frequency magnetic field, doxorubicin-loaded nano-vehicles released the drug with pinpoint accuracy. There was further evidence from in vivo experiments that the magnetic nano-vehicles showed significant chemotherapeutic and thermotherapeutic effects. Due to this, magnetic nano-vehicles with smart properties, such as MPVA-AP1, hold great promise for anticancer applications that need precise dosing and regulated release [72].

It was found that the coprecipitation technique proposed by Syu et al., (2019) was simple and ligand-assisted to manufacture biocompatible iron oxide (IO) nanocrystals with NIR absorbance for cancer theranostics. Nanocrystals had little cytotoxicity in HT-29 colorectal cancer cells; however, the vitality of cells treated with nanocrystals dropped substantially following laser irradiation at a 808-nm wavelength. Cell death is thought to be caused by changes in protein secondary structure and membrane permeability, among other things. Nanocrystals with magnetic field (MF) application significantly enhanced tumour aggregation by fourfold in the in vivo studies, leading to a thrice-larger T2-weighted MR signal than that provided by a conventional T2-weighted MR contrast agent (Resovist®) and improved photothermal potency for the therapy of cancer. The novel iron oxide nanocrystals were shown to be highly biocompatible and show significant promise as a cancer theranostic agent [73].

8. Synthetic/Herbal Nanocrystal Used in Drug Delivery System

Biomedicine's use of nanotechnology is expanding quickly. The field of nanotechnology serves as a link between various NPs and nanophases in terms of both technology and science [74]. Particles with a size range of 1–100 nm with great specific surface area and specialised surface properties are referred to as nanoparticles (NPs) [75]. It is possible to obtain controlled drug release, site-specific drug delivery, and reduced toxicity by using nanoparticles (NPs). The NP's size significantly impacts how well it is taken in by cells and excreted from them in vivo. It has been demonstrated that greater particulates have much reduced concentrations of intracellular delivery than small nanoparticles at a similar quantity and that NPs with such a dimension of 100 nm had 2.5-fold more absorption than 1-m size nanoparticles [76]. Since the glomerular capillary sieve coefficient is 10 nm, NPs with a smaller diameter exhibit better renal clearance. Larger NPs, on the other hand, are unable to diffuse into intracellular gaps because of the increased permeability and retention (EPR) effect, lacking in a approach cancer cells [77]. The toxicology of NPs is influenced by physicochemical characteristics, including size, shape, and concentration, which affect absorption, circulation time, and total toxicity. The phospholipid bilayer structure gives the cell membrane a negative surface charge, making it even more susceptible to positively charged particles [78].

In other words, because they immediately enter the cell membrane and cause membrane rupture or damage, highly positive charged NPs have higher cytotoxicity. NP surfaces have been modified to hold a negatively charged initially to allow them to the exist specific macrophage organelles that is later converted to a positive charge by environmental cues to enhance drug administration [79]. To counteract the cationic charge, scientists used serum albumin in preincubated NPs, including shape and size, toxic effects, quantity, and charge density, which should be considered while designing nanomedicine. For example, MRI imaging, cancer diagnostics, precise cell and subcellular treatment, and the detection of genetic mutations NPs have enormous promise, with suitable surface changes in biomedical and therapeutic applications such as nanoparticle surface modification [80].

9. Phytomedicine Drug Delivery Using Nanoparticles

Developing resistance to chemotherapeutic drugs and radiation therapy is a key challenge in cancer treatment. Phytochemicals with low side effects, namely lycopene, genistein, epigallocatechin gallate, anthocyanin, curcumin, and resveratrol, are being investigated by researchers as an alternative therapy to help overcome this problem [81,82]. On the other hand, phytochemicals have several disadvantages, including poor absorption, limited water solubility, and rapid metabolism. Consequently, using NPs to formulate these phytochemicals can improve phytochemical medication delivery while also minimising adverse effects [82].

9.1. Curcumin

As a chemoprevention and anticancer agent, curcumin obtained from Indian spices turmeric is highly effective. Inhibiting different signalling pathways such as NF-κB, STAT3 and EGFR, which contribute to tumour development and metastasis, slows the advancement of many malignancies, including CRC [83]. Due to the nanoformulation, curcumin has improved water solubility and distribution while causing no harm to healthy cells. To create a nanocurcumin formulation for use in CRC treatment, NPs made of liposomes, polymers, cyclodextrin, nanogels, gold, lipids, and micelles can be used [84].

In addition to its ability to act as a free radical scavenger, curcumin also changes the expression of several stress proteins and angiogenesis-related genes. It inhibits the activity of numerous key transcription factors, namely activator protein 1 (AP-1) and nuclear factor κ-light chain enhancer of activated B cells (NF-κB) [85]. Additionally, the size of 10 μm showed significant antioxidant activity. It triggered cell death when they reached a concentration of around 50% at 50 μm, which may be due to the production of superoxide radicals [86]. Human plasma levels can be as low as 0.41–1.75 μmol/L after taking 4–8 g of turmeric orally each day. The curcumin concentration is critical for evaluating the biological consequences of frequent oral intake. Bio-transformed moieties, such as tetra hexahydro–curcumin, have drawn attention because of their poor systemic bioavailability [87].

Additionally, research shows that curcumin is an efficient chemopreventive drug in addition to its antitumour properties (Figure 7). It works by inhibiting cyclooxygenase, phospholipase A2, and phospholipase-Cr1 in tumours of the colon, stomach, and skin. It is clear that antagonists of COX-2 are effective in suppressing polyp development and cancer growth in FAP, and were it not for the negative consequences, COX-2 inhibitors would be strongly suggested for cancer polyp chemoprevention [88]. In order to avoid carcinogenesis, we require drugs that block cellular pathways that cause or promote cancer. With these characteristics, curcumin is a promising chemopreventive agent. It has recently been shown that 5-LOX plays a crucial function in controlling cell proliferation and apoptosis [89]. Some research shows 5-LOX overexpression in cancer cells and human tumours, such as those of the gastrointestinal tract and liver and those of the breast and prostate. 5-LOX has also been shown to be overexpressed in the brain and the mesothelium. 5-LOX overexpression has been connected to enhanced proliferation and tumour growth in investigations done in the lab and on mice in the wild [90].

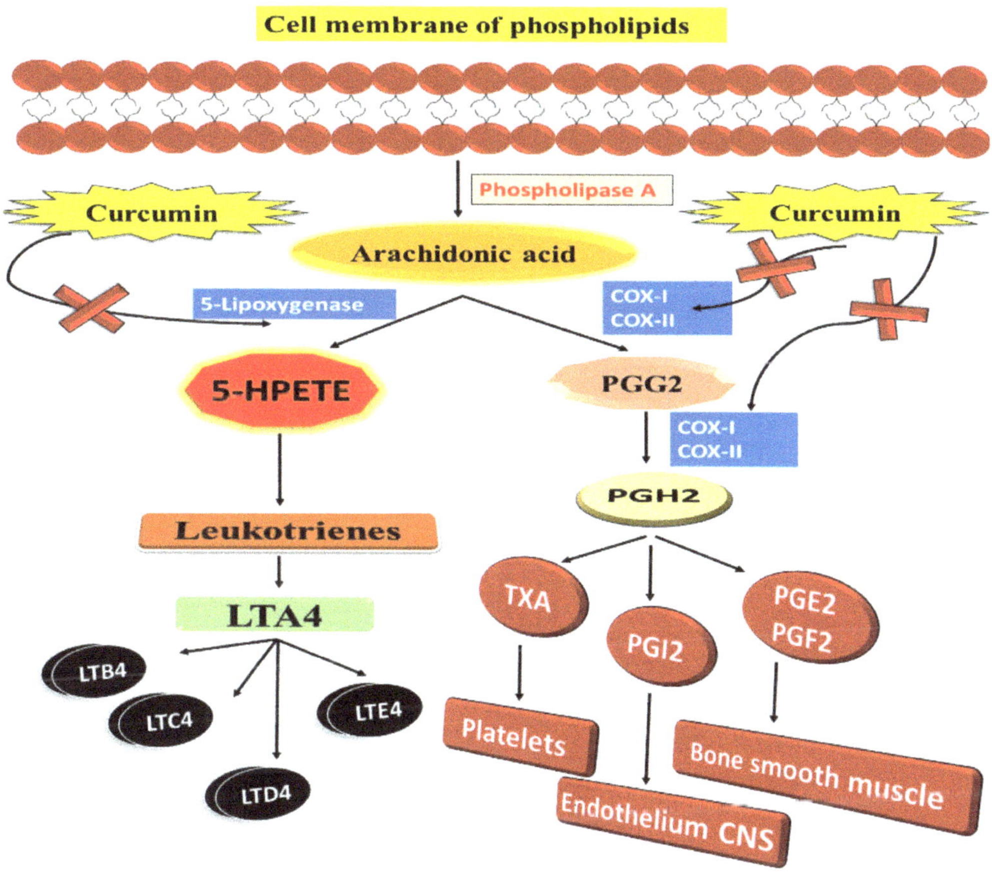

Figure 7. Curcumin and the arachidonic acid metabolism: A potential inhibitory effect.

Ohd et al. indicate that 5-LOX and cysteinyl leukotriene receptors are overexpressed in human colon adenocarcinomas and that individuals with these mutations have a poor prognosis. It indicates that increased 5-LOX expression is related to malignant transformation, although the exact processes that link 5-LOX gene expression to colon cancer development are unclear. The use of COX-2 knockout mouse models has demonstrated that COX-2 plays a function in colon carcinogenesis. However, in colon carcinogenesis, similar data from 5-LOX knockout animals are missing [91]. According to Raveendran et al., curcumin-encapsulated PCL exhibits hydrophobicity and effective NP stability due to its amphiphilic block copolymer nature. It was an effective technique, because it targeted the tumour and made tumour cells more susceptible to treatment resistance. Additionally, this formulation enhanced water solubility, resulting in increased in vitro curcumin absorption by Caco-2 CRC cells. There was significantly less inhibition in the micelles of bigger amphiphilic block copolymers than smaller ones [92].

9.2. *Resveratrol*

Natural phenols such as resveratrol are found in fruits such as red grapes, blueberries, and raspberries [93]. Resveratrol has been discovered to have anticancer, anti-cardiovascular, anti-inflammatory, and antiaging biological and pharmacological effects [94]. Further studies have shown that resveratrol has antiproliferative properties in vitro against

CRC via modulating intracellular signalling pathways such as PTEN/PI3K/Akt, Wnt/β-catenin, and others [95]. It is possible to discover therapeutic medication's biological targets and molecular processes by using network pharmacology as an approach in clinical illnesses (Figure 8). As a result, identifying biomarkers and treatment molecular pathways is crucial even before diagnosing a patient with a relapsed CRC [96]. Resveratrol belongs to the phytoalexin stilbene compound and is isolated from the various plants extracts. Chemo-sensitising, antioxidative, and anti-inflammatory characteristics are some of the things it does. Due to the poor physicochemical profile of resveratrol, it is not generally recommended therapeutically despite its positive features [97]. The iNOS, COX-2, and other inflammatory signalling pathways are all reduced by resveratrol, inhibiting proinflammatory mediators, including IL-1β and TNF-α [98]. Soo et al. created and improved an innovative drug carrier by co-encapsulating resveratrol and cyclodextrin-resveratrol inclusion complexes in liposomes' lipophilic and hydrophilic compartments. The final formulation's particle size, polydispersity index, and zeta potential were measured at 131 ± 1.30 nm, 0.089 ± 0.005 and -2.64 ± 0.51 mV, respectively. The drug release profiles of 40–60% were found to be acceptable for the standard liposomal formulations and free resveratrol, while the innovative nanoformulations demonstrated total (100%) release in 24 h. It remained steady at 4 °C for 14 days. In HT-29 colon cancer cell lines, resveratrol-encapsulated liposomes showed in vitro cytotoxicity. Liposomes were shown to have a dose-dependent cytotoxicity profile that was also boosted compared to free resveratrol (in DMSO). Using liposomal formulations to deliver hydrophobic chemotherapy agents is possible because of the study's findings, which show that the co-encapsulation of resveratrol in its purest form with its cyclodextrin complex is feasible [99]. Summerlin et al. enclosed resveratrol in colloidal MSN, and they discovered that the encapsulation efficiency and loading capacity of MCM-48–resveratrol were superior. In CRC cell lines, the formula promoted tumour cell killing via the PARP and cIAP1 pathways. Furthermore, it slowed down the production of NF-κB. As a result, this new approach to medication development may be useful in the future [100].

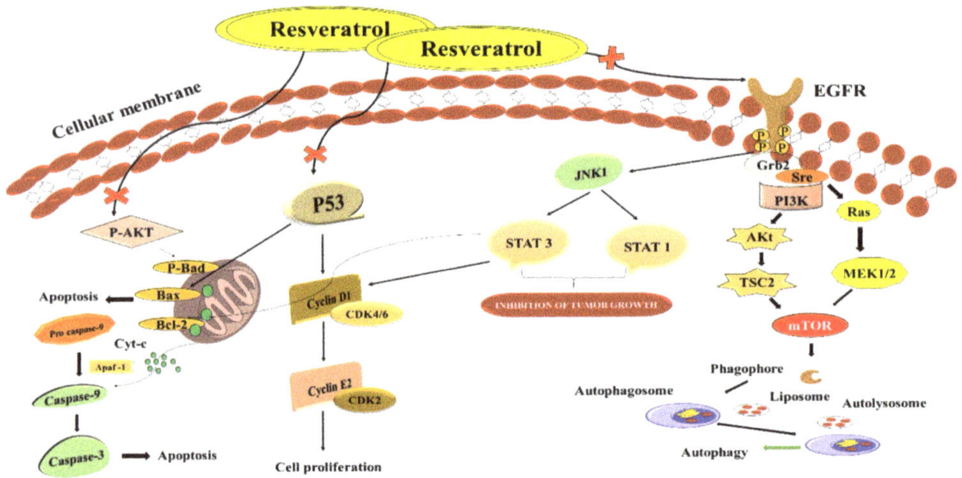

Figure 8. Schematic illustration of resveratrol as a novel healing method for treating CRC by targeting distinct signalling pathways.

10. Chemotherapeutic Drug Delivery Using Nanoparticles

Most cancers are still treated with radiotherapy and chemotherapy. It is a serious issue that a combination of high dosages and combinational therapy produces severe toxic effects to healthy colon cells. Moreover, it is also associated with a variety of cardiac and digestive side effects. Hence, it is required to reach the target location because of the short half-life of these therapies and the tendency towards resistance [101]. Since the drug's circulation time and pharmacokinetic characteristics may be improved via nanocarrier-based delivery, it may overcome these adverse effects [102].

10.1. 5-Fluorouracil (Capecitabine)

5-Fluorouracil (5-FU) was applied to heal CRC as a single agent and in combination with other drugs for more than 40 years [103]. Researchers had to develop an oral version of 5-FU because of its short half-life, the necessity of using a central line, and the demand for continual infusions [104]. When compared to conventional 5-FU, Capecitabine provides many benefits. After passing through the digestive tract, it undergoes three enzymatic processes before transforming to 5-FU (Figure 9). Thymidine phosphorylase (TP), the pathway's last enzyme, is thought to be present in tumour tissue at disproportionately high levels, which increases the agent's effectiveness and tolerance through tailored delivery [105]. Most CRC chemotherapy regimens start with 5-FU as the main active ingredient. Deoxythymidine monophosphate (dTMP) plays key role in DNA replication and repair and is inhibited by 5-FU by altering into fluorodeoxyuridine monophosphate (FdUMP) in cells [106]. FdUMP creates a stable complex mostly with the enzyme thymidylate synthase, inhibiting the production of dTMP. It often interacts mostly with the formation and activity of nucleic acids by substituting uracil or thymine in DNA and RNA as a secondary method of action. The 5-FU-related gene TYMP (formerly known as ECGF1), genes for the thymidine phosphorylase (TP) protein, varied amongst the groups. After being converted to fluorodeoxyuridine (FUDR) by the action of TP, the active metabolite FdUMP is produced. In vitro, the overexpression of TP increases the 5-FU sensitivity, which is thought to be due to the increased production of FdUMP as a result [107].

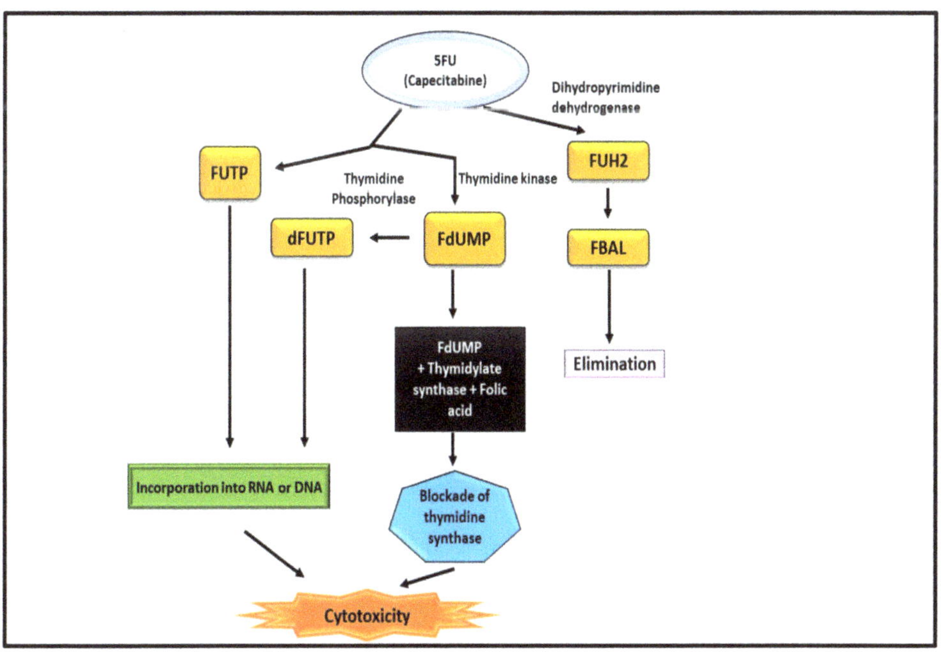

Figure 9. Mechanism of 5-FU for the treatment of CRC.

Capecitabine was used in a phase II study by Mizumoto et al. The quality of life and disease-free survival of individuals treated with Capecitabine may improve if the adverse effects are reduced throughout the drug. Capecitabine (2500 mg/m^2/day) was given to

CRC patients for five days, followed two days off (5 days on/2 days off regimen). It took three weeks to complete one course, and it took 24 weeks to complete all eight courses. Retrospectively, included patients on the traditional regimen received substantially more treatment courses than those on the 5 days on/2 days off regimen, with a p-value of 0.0438. There was a substantially greater rate of patients in the 5 days on/2 days off regimen completing their planned therapy ($p = 0.0389$). However, according to the results of this phase II research, toxic effects related to the 5 days on/2 days off regimen are less severe than those associated with the more traditional regimen, and adverse events are more common. However, there are fewer reports of high-grade toxicities. The novel regimen had a shorter time to treatment failure and showed high practicality. There was no significant difference in overall quality of life or side effects (mainly mild); therefore, the 5 days on/2 days off regimen must be a plan to be studied in future randomised controlled trials to confirm its viability [108].

10.2. Oxaliplatin

Oxaliplatin is a third-generation diamino cyclohexane (DACH) platinum compound. Oxaliplatin creates intrastrand connections between two adjacent guanine residues or between guanine and adenine, inhibiting DNA replication and transcription [109]. Oxaliplatin is the third generation of the DACH platinum compound. Oxaliplatin is successful in the treatment of colorectal cancer, but platinum compounds cisplatin and carboplatin have been shown to be ineffective [110]. It is frequently used to treat patients who have not responded to 5-fluorouracil (5FU)-based therapy. Oxaliplatin's cytotoxic effects, on the other hand, are still poorly studied in detail. Many chemotherapeutic drugs, including oxaliplatin, have been demonstrated to cause apoptosis in tumour cells [111]. Since a medication's cellular target is dependent on the accumulation in cancer cells, the clinical efficacy of oxaliplatin may be influenced by variables that regulate its accumulation due to its expression by tumour cells or by biological barriers that impact drug disposition. (e.g., enterocytes and hepatocytes) (Figure 10).

Oxaliplatin can enter cells passively or by transporters that carry it in from the outside. Its physicochemical properties are important to determine if passive diffusion is preferable to other routes [112]. Due to these characteristics, it may accumulate better in tumour cells than cisplatin, even in drug-resistant cells with a lower permeability toward the more hydrophilic drug. Efflux and influx transporters work together to determine how much medication is stored in cells. Oxaliplatin's therapeutic efficacy and side effects may be influenced by the tissue-specific expression of influx transporters or decreased efflux transporter activity when these variables are present [113].

Trifluridine/tipiracil FTD/TPI and oxaliplatin were used synergistically by Limagne et al. to enhance the immunogenic cell death (ICD). Even though all therapy groups indicated T-cell dependence, only the combination effectively induced ICD in vivo. Furthermore, neither FTD/TPI or oxaliplatin affected regulatory T cells or myeloid-derived suppressor cells. However, they did remove type-2 tumour-associated macrophages (TAM2), resulting in increased cytotoxic CD8+ T-cell infiltration and activation. Secondary T-cell exhaustion was related to both tumour cell PD-L1 expression and CD8+ T-cell PD-1 induction.

Last, but not least, even when used alone, anti–PD-1 did not boost the antitumor efficacy of FTD/TPI or oxaliplatin monotherapy, but when combined with the drugs, it did. FTD/TPI and oxaliplatin depleted TAM2 in a new immunomodulatory manner. A rationale for using oxaliplatin and FTD/TPI to remove immunosuppressive cells and boost checkpoint efficacy in patients with metastatic colorectal cancer has been provided by the induction of ICD in vivo by combining these medicines [114].

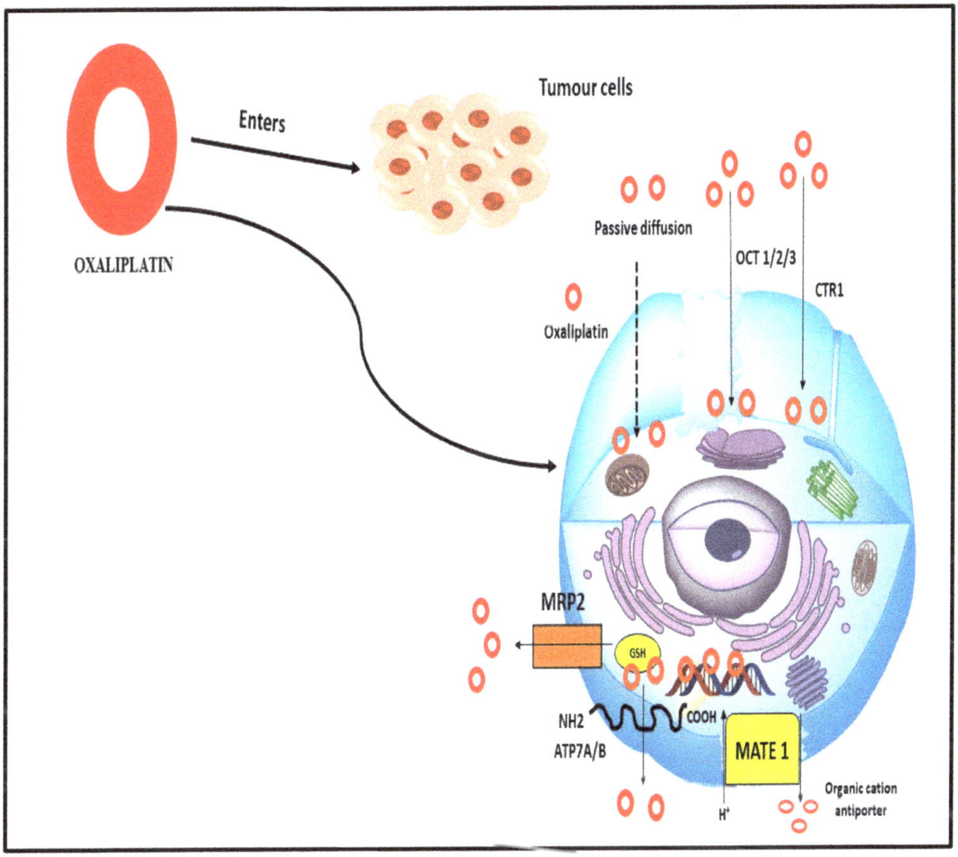

Figure 10. Numerous influx and efflux transporters that allow oxaliplatin to enter the cell passively (dashed arrow). There are several kinds of organic cation transporters, such as the OCT 1/2/3 transporters, copper transporters, MRP2 multidrug resistance-associated proteins, glutathione transporters, MATE 1 multidrug and toxin extrusion transporter, and ATP7A/B ATP-binding cassette transporters.

11. Current Colorectal Cancer Clinical Trials and State of Nanotechnology

Clinical endpoints and biomarkers are two issues that were particularly interested in when it comes to contemporary clinical studies. The alterations in OS amid two treatment groups in research evaluating a single line of therapy are becoming increasingly challenging to accomplish in advanced cancer patients with all types of clinical end goals [115]. There are a rising variety of viable therapeutic choices available in second and subsequent chemotherapy lines in addition to the regularly utilised and often ethically obligatory salvage therapies. However, the initial treatment effect may be weakened or confounded, eventually precluding the proof of a statistical significance overall survival improvement [116]. A growing number of patients will have to be included in clinical studies over an extended period to obtain significant statistical and clinically acceptable OS impacts from long-term patient survival and the ensuing mild OS risk ratio variations within the treated group [117]. Induction chemotherapy combined with a targeted drug is the standard treatment for metastatic colorectal cancer [118]. Clinical trials have evaluated various techniques for maintaining cytotoxic treatment until progression against periods of observation or using various maintenance medicines in several randomised trials. On the other hand, clinical trials have shown wildly divergent effectiveness findings, making it difficult to draw any firm conclusions about which method is best. Table 2 shows the summary of various current and finished clinical studies of CRC.

Table 2. Illustration of completed and ongoing clinical trials evidence for CRC.

Title	Location	Trial Identifier	Phase and Status
Metastatic Colorectal Cancer Database	Methodology, Biostatistics and Data Management Dijon, France	NCT04031625	Not applicable, Recruiting
Maintenance Therapy for Metastatic Colorectal Cancer After First-Line Treatment with Fruquintinib Plus Capecitabine Versus Bevacizumab Plus Capecitabine	Medical College of Zhejiang University Hangzhou, China	NCT04733963	Phase 2, Recruiting
CRC Patients having liver metastases are being treated with TAS-102 and radiation therapy	Massachusetts General Hospital, United States	NCT03223779	Phase 1, Recruiting
Anlotinib Combined With mXELIRI as Second-line Treatment of Advanced Colorectal Cancer	Guangdong Provincial Hospital of Chinese Medicine Guangzhou, China	NCT05035914	Phase 1, Recruiting
Do Colorectal Cancer Risk Estimates Affect Screening Behavior?	Stanford University School of Medicine Palo Alto, US	NCT03819920	Not Applicable, Completed
Study of Fruquintinib Efficacy and Safety in Patients with 3+ Line Colorectal Cancer (FRESCO) in Phase III	Hutchison Medi Pharma Investigational Site Hefei, Anhui, China	NCT02314819	Phase 3, Completed
Gut Microbiome Dynamics in Metastasized or Irresectable Colorectal Cancer	Wilhelmina Ziekenhuis Assen, Netherlands	NCT03941080	Not Applicable, Recruiting
Dabrafenib + Trametinib + PDR001 In Colorectal Cancer	Massachusetts General Hospital Cancer Center	NCT03668431	Phase 2, Recruiting
Colorectal Cancer Research Consortium (NCRCC) Study: National Colorectal Cancer Research Consortium	Zhejiang University College of Medicine Hangzhou, China	NCT04074538	Not Applicable, Recruiting
A Translational Study Examines the Impact of the MET Oncogene in Human Colorectal Cancer	Fondazione del Piemonte per l'Oncologia Candiolo, Italy	NCT02238821	Not Applicable, Completed
The Use of a Patient Navigator to Increase Colorectal Cancer Screening Uptake	Group Health Centre Sault Ste. Marie, Canada	NCT01506687	Phase 3, Completed
Regorafenib in the Treatment of Patients with Metastatic Colorectal Cancer: Real-World Effectiveness	Henan Cancer Hospital Zhengzhou, China	NCT05023720	Not Applicable, Recruiting
Investigating the effects on colorectal cancer patients of a walking programme	National Taiwan University Hospital Taipei, Taiwan	NCT01595256	Not Applicable, Completed
Metabolomics-Based Detection of Colorectal Cancer	Indiana University Cancer Center Indianapolis, US	NCT00507598	Not Applicable, Completed
Exploration into how often patients with colorectal cancer who are receiving irinotecan-based therapy experience nausea and vomiting	Caritas St. Elizabeth Medical Center Brighton, US	NCT00713128	Not Applicable, Completed
A Colorectal Cancer Screening Decision Aid for American Indians That Is Culturally Adapted	Robeson Health Care Corporation Lumberton, US	NCT03569761	Not Applicable, Completed
Vaccination Against MSI Colorectal Cancer	Krankenhaus Nordwest Frankfurt/Main, Germany	NCT01461148	Phase 2, Completed
Fecal Occult Blood Tests for Colorectal Cancer Screening: A Comparison of the Different Methods	Qilu Hospital Jinan, China	NCT04454099	Not Applicable, Completed
Concern for Positron Emission Tomography (PET) in Colorectal Cancer Stage II and III Follow-Up	Cancérologie et Hépato-Gastro-Entérologie Bordeaux, France	NCT00199654	Phase 3, Completed
Relapsed/refractory Colorectal Cancer Patients Receiving IMMU-130 Trial	Memorial Sloan-Kettering Cancer Center New York, US	NCT01270698	Phase 1, Completed

12. Future Prospects and Recommendation

Colorectal cancer is still the leading cause of cancer-related death in the globe. To enhance the overall survival, minimise the disease-free progress, and reduce the recurrence risk, early detection is critical. Oncogenes and tumour suppressor genes undergo successive mutations in the adenoma–carcinoma pathway, progressing from normal to dysplastic epithelium. Normal glandular epithelium becomes adenocarcinoma when genetic and epigenetic alterations to the genome accumulate. Even though the molecular processes responsible for the cancer's growth and spread remain a mystery, numerous genetic pathways have been postulated to explain CRC aetiology. A biomarker is a useful tool for detecting diseases early on and predicting how well a patient will respond to therapy. Although biomarkers employed at various stages of the illness make it easier to control CRC, each one has its limits. To make advances in biomarker creation and validation, understanding molecular interactions that aid pathogenesis is necessary. The use of nanotechnology in the treatment of colorectal cancer is critical. When it comes to the elimination or, at the very least, the chronic treatment of cancer, nanotechnology has shown to be a vital and optimistic technique. The use of nanotechnology in cancer diagnosis and therapy has been game-changing. It has been a priority in biomedicine to create drug delivery techniques that can affect biodistribution, tissue absorption, and pharmacokinetics, among other things.

Contrary to popular belief, nanomedicine can have a significant impact on human CRC diagnosis and treatment and even improve normal human physiology in some cases. We hope that this article will help scientists, research scholars, industrialists, etc. with their research work and provide ease in searching through the findings of the current research work in CRC. It also helps to identify the diagnostic, prognostic, and predictive biomarkers. Thus, it provides a conceptual knowledge of nano-based medicine for the management of CRC.

13. Conclusions

Many people around the world are still affected by CRC. mCRC management remains a challenge despite screening recommendations and advancements in treatment options. A variety of genetic and epigenetic factors contribute to the heterogeneity of CRC. There is also much variability in the tumour microenvironment in CRC, making it difficult to find therapeutic targets for people with metastatic CRC. The heterogeneity of tumours within and across primary and metastatic sites can be extremely diverse. This could be a contributing factor to a treatment failure when a patient develops a resistance to targeted medicines. Since the effectiveness of more recent focused treatments depends on a good understanding of the patient's previous response, a tumour specimen analysis is also required for understanding the underlying illness mechanism and/or tumour resistance traits. As our understanding of the colorectal carcinogenesis molecular landscape expands, new molecular biomarkers with prognostic and predictive information are being identified. For the real-time examination of tumour clonal development, medication response, the existence of minimum residual illness, and acquired resistance, liquid biopsies are a potential approach. CRC patients may benefit from real-time molecular categorisation and customised treatment using non-invasive biomarkers. With a better understanding of the epigenetic regulatory mechanisms, including cancer-specific epigenetic changes, prospective clinical implications as diagnostics or therapeutic targets in colorectal cancer can now be explored more thoroughly.

The diagnostic efficacy of these biomarkers in early-stage CRC screening, prognostication, or treatment response prediction has yet to be improved, and most of these biomarkers have not yet been validated in large independent patient cohorts. A significant step towards individualised medicine should, in our opinion, include the consideration of non-cancer-related aspects such as nutrition and lifestyle. The existing methods in the cancer management sector have been rapidly replaced by new ones that offer greater diagnostic capabilities, increased sensitivity for tumour staging, and better therapeutic approaches.

There are various advantages to using nanotechnology in colon cancer treatment. We can see early indications of nanotechnology's value in evaluating colorectal cancer imaging, medicines, and ablation methods. The first measures have already been taken to improve these areas. There should be no doubt that nanotechnology-based techniques have the potential to treat cancers other than CRC, even if this opinion has concentrated on CRC as a particularly attractive target for early application.

Author Contributions: Conceptualisation, A.S. and R.K.S.; methodology, F.A.A.-J.; validation, M.M.S.-B.; formal analysis, F.Y.A. and F.S.A.; writing—original draft preparation, A.S. and R.K.S.; writing—review and editing, R.W., M.M.S.-B., and F.A.A.-J. and supervision, R.K.S. All authors have read and agreed to the published version of the manuscript.

Funding: This paper received no external funding.

Conflicts of Interest: The authors declare no conflict of interest.

References

1. Issa, I.A.; Noureddine, M. Colorectal cancer screening: An updated review of the available options. *World J. Gastroenterol.* **2017**, *23*, 5086–5096. [CrossRef]
2. Jeon, J.; Du, M.; Schoen, R.E.; Hoffmeister, M.; Newcomb, P.A.; Berndt, S.I.; Caan, B.; Campbell, P.T.; Chan, A.T.; Chang-Claude, J.; et al. Determining risk of colorectal cancer and starting age of screening based on lifestyle, environmental, and genetic factors. *Gastroenterology* **2018**, *154*, 2152–2164. [CrossRef]
3. Carbone, M.; Arron, S.T.; Beutler, B.; Bononi, A.; Cavenee, W.; Cleaver, J.E.; Croce, C.M.; D'Andrea, A.; Foulkes, W.D.; Gaudino, G.; et al. Tumour predisposition and cancer syndromes as models to study gene–environment interactions. *Nat. Rev. Cancer* **2020**, *20*, 533–549. [CrossRef]
4. Ishaque, N.; Abba, M.L.; Hauser, C.; Patil, N.; Paramasivam, N.; Huebschmann, D.; Leupold, J.H.; Balasubramanian, G.P.; Kleinheinz, K.; Toprak, U.H.; et al. Whole genome sequencing puts forward hypotheses on metastasis evolution and therapy in colorectal cancer. *Nat. Commun.* **2018**, *9*, 1–4. [CrossRef]
5. Guo, M.; Peng, Y.; Gao, A.; Du, C.; Herman, J.G. Epigenetic heterogeneity in cancer. *Biomark. Res.* **2019**, *7*, 23. [CrossRef]
6. Ganesh, K.; Stadler, Z.K.; Cercek, A.; Mendelsohn, R.B.; Shia, J.; Segal, N.H.; Diaz, L.A. Immunotherapy in colorectal cancer: Rationale, challenges and potential. *Nat. Rev. Gastroenterol. Hepatol.* **2019**, *16*, 361–375. [CrossRef]
7. Mármol, I.; Sánchez-de-Diego, C.; Pradilla Dieste, A.; Cerrada, E.; Rodriguez Yoldi, M.J. Colorectal carcinoma: A general overview and future perspectives in colorectal cancer. *Int. J. Mol. Sci.* **2017**, *18*, 197. [CrossRef]
8. Huang, F.T.; Chen, W.Y.; Gu, Z.Q.; Zhuang, Y.Y.; Li, C.Q.; Wang, L.Y.; Peng, J.F.; Zhu, Z.; Luo, X.; Li, Y.H.; et al. The novel long intergenic noncoding RNA UCC promotes colorectal cancer progression by sponging miR-143. *Cell Death Dis.* **2017**, *8*, e2778. [CrossRef]
9. Bupathi, M.; Wu, C. Biomarkers for immune therapy in colorectal cancer: Mismatch-repair deficiency and others. *J. Gastrointest. Oncol.* **2016**, *7*, 713. [CrossRef]
10. Mauri, G.; Bonazzina, E.; Amatu, A.; Tosi, F.; Bencardino, K.; Gori, V.; Massihnia, D.; Cipani, T.; Spina, F.; Ghezzi, S.; et al. The evolutionary landscape of treatment for BRAFV600E mutant metastatic colorectal cancer. *Cancers* **2021**, *13*, 137. [CrossRef]
11. Chenthamara, D.; Subramaniam, S.; Ramakrishnan, S.G.; Krishnaswamy, S.; Essa, M.M.; Lin, F.H.; Qoronfleh, M.W. Therapeutic efficacy of nanoparticles and routes of administration. *Biomater. Res.* **2019**, *23*, 1–29. [CrossRef]
12. Kalaydina, R.V.; Bajwa, K.; Qorri, B.; Decarlo, A.; Szewczuk, M.R. Recent advances in "smart" delivery systems for extended drug release in cancer therapy. *Int. J. Nanomed.* **2018**, *13*, 4727. [CrossRef]
13. Müller, M.F.; Ibrahim, A.E.; Arends, M.J. Molecular pathological classification of colorectal cancer. *Virchows Arch.* **2016**, *469*, 125–134. [CrossRef]
14. Aghagolzadeh, P.; Radpour, R. New trends in molecular and cellular biomarker discovery for colorectal cancer. *World J. Gastroenterol.* **2016**, *22*, 5678. [CrossRef]
15. Carethers, J.M. Microsatellite instability pathway and EMAST in colorectal cancer. *Curr. Colorectal Cancer Rep.* **2017**, *13*, 73–80. [CrossRef]
16. Langner, C. Serrated and non-serrated precursor lesions of colorectal cancer. *Dig. Dis.* **2015**, *33*, 28–37. [CrossRef]
17. Fischer, J.; Walker, L.C.; Robinson, B.A.; Frizelle, F.A.; Church, J.M.; Eglinton, T.W. Clinical implications of the genetics of sporadic colorectal cancer. *ANZ J. Surg.* **2019**, *89*, 1224–1229. [CrossRef]
18. Raskov, H.; Søby, J.H.; Troelsen, J.; Bojesen, R.D.; Gögenur, I. Driver gene mutations and epigenetics in colorectal cancer. *Ann. Surg.* **2020**, *271*, 75–85. [CrossRef]
19. Koveitypour, Z.; Panahi, F.; Vakilian, M.; Peymani, M.; Forootan, F.S.; Esfahani, M.H.; Ghaedi, K. Signaling pathways involved in colorectal cancer progression. *Cell Biosci.* **2019**, *9*, 1–14. [CrossRef]
20. Boland, P.M.; Yurgelun, M.B.; Boland, C.R. Recent progress in Lynch syndrome and other familial colorectal cancer syndromes. *CA A Cancer J. Clin.* **2018**, *68*, 217–231. [CrossRef]

21. Watson, M.M. Microsatellite Instability at Tetranucleotides (EMAST) in Colorectal Cancer: Clinical Relevance, Mechanisms and Immune Markers. Ph.D. Thesis, University of Bergen, Bergen, Norway, 2020.
22. Mehta, R.S.; Song, M.; Nishihara, R.; Drew, D.A.; Wu, K.; Qian, Z.R.; Fung, T.T.; Hamada, T.; Masugi, Y.; da Silva, A.; et al. Dietary patterns and risk of colorectal cancer: Analysis by tumor location and molecular subtypes. *Gastroenterology* **2017**, *152*, 1944–1953. [CrossRef]
23. Mitsuhashi, K.; Yamamoto, I.; Kurihara, H.; Kanno, S.; Ito, M.; Igarashi, H.; Ishigami, K.; Sukawa, Y.; Tachibana, M.; Takahashi, H.; et al. Analysis of the molecular features of rectal carcinoid tumors to identify new biomarkers that predict biological malignancy. *Oncotarget* **2015**, *6*, 22114. [CrossRef]
24. Szylberg, Ł.; Janiczek, M.; Popiel, A.; Marszałek, A. Serrated polyps and their alternative pathway to the colorectal cancer: A systematic review. *Gastroenterol. Res. Pract.* **2015**, *2015*, 573814. [CrossRef]
25. Novellasdemunt, L.; Antas, P.; Li, V.S. Targeting Wnt signaling in colorectal cancer. A review in the theme: Cell signaling: Proteins, pathways and mechanisms. *Am. J. Physiol.-Cell Physiol.* **2015**, *309*, C511–C521. [CrossRef]
26. Amodio, V.; Yaeger, R.; Arcella, P.; Cancelliere, C.; Lamba, S.; Lorenzato, A.; Arena, S.; Montone, M.; Mussolin, B.; Bian, Y.; et al. EGFR blockade reverts resistance to KRASG12C inhibition in colorectal cancer. *Cancer Discov.* **2020**, *10*, 1129–1139. [CrossRef]
27. Ahronian, L.G.; Sennott, E.M.; Van Allen, E.M.; Wagle, N.; Kwak, E.L.; Faris, J.E.; Godfrey, J.T.; Nishimura, K.; Lynch, K.D.; Mermel, C.H.; et al. Clinical acquired resistance to RAF inhibitor combinations in BRAF-mutant colorectal cancer through MAPK pathway alterations. *Cancer Discov.* **2015**, *5*, 358–367. [CrossRef]
28. Ito, M.; Kanno, S.; Nosho, K.; Sukawa, Y.; Mitsuhashi, K.; Kurihara, H.; Igarashi, H.; Takahashi, T.; Tachibana, M.; Takahashi, H.; et al. Association of Fusobacterium nucleatum with clinical and molecular features in colorectal serrated pathway. *Int. J. Cancer* **2015**, *137*, 1258–1268. [CrossRef]
29. Rhee, Y.Y.; Kim, K.J.; Kang, G.H. CpG island methylator phenotype-high colorectal cancers and their prognostic implications and relationships with the serrated neoplasia pathway. *Gut Liver* **2017**, *11*, 38. [CrossRef]
30. Valo, S.; Kaur, S.; Ristimäki, A.; Renkonen-Sinisalo, L.; Järvinen, H.; Mecklin, J.P.; Nyström, M.; Peltomäki, P. DNA hypermethylation appears early and shows increased frequency with dysplasia in Lynch syndrome-associated colorectal adenomas and carcinomas. *Clin. Epigenet.* **2015**, *7*, 71. [CrossRef]
31. Suraweera, N.; Mouradov, D.; Li, S.; Jorissen, R.N.; Hampson, D.; Ghosh, A.; Sengupta, N.; Thaha, M.; Ahmed, S.; Kirwan, M.; et al. Relative telomere lengths in tumor and normal mucosa are related to disease progression and chromosome instability profiles in colorectal cancer. *Oncotarget* **2016**, *7*, 36474. [CrossRef]
32. Yu, H., Hemminki, K. Genetic epidemiology of colorectal cancer and associated cancers. *Mutagenesis* **2020**, *35*, 207–219. [CrossRef]
33. Siegel, R.L.; Jakubowski, C.D.; Fedewa, S.A.; Davis, A.; Azad, N.S. Colorectal cancer in the young: Epidemiology, prevention, management. *Am. Soc. Clin. Oncol. Educ. Book* **2020**, *40*, e75–e88. [CrossRef]
34. De Palma, F.D.; D'argenio, V.; Pol, J.; Kroemer, G.; Maiuri, M.C.; Salvatore, F. The molecular hallmarks of the serrated pathway in colorectal cancer. *Cancers* **2019**, *11*, 1017. [CrossRef]
35. Bibbins-Domingo, K.; Grossman, D.C.; Curry, S.J.; Davidson, K.W.; Epling, J.W.; García, F.A.; Gillman, M.W.; Harper, D.M.; Kemper, A.R.; Krist, A.H.; et al. Screening for colorectal cancer: US Preventive Services Task Force recommendation statement. *Jama* **2016**, *315*, 2564–2575.
36. Mousavinezhad, M.; Majdzadeh, R.; Sari, A.A.; Delavari, A.; Mohtasham, F. The effectiveness of FOBT vs. FIT: A meta-analysis on colorectal cancer screening test. *Med. J. Islam. Repub. Iran* **2016**, *30*, 366.
37. Li, M.; Zhang, J.; Dan, Y.; Yao, Y.; Dai, W.; Cai, G.; Yang, G.; Tong, T. A clinical-radiomics nomogram for the preoperative prediction of lymph node metastasis in colorectal cancer. *J. Transl. Med.* **2020**, *18*, 1–10. [CrossRef]
38. Nikolouzakis, T.K.; Vassilopoulou, L.; Fragkiadaki, P.; Mariolis Sapsakos, T.; Papadakis, G.Z.; Spandidos, D.A.; Tsatsakis, A.M.; Tsiaoussis, J. Improving diagnosis, prognosis and prediction by using biomarkers in CRC patients. *Oncol. Rep.* **2018**, *39*, 2455–2472. [CrossRef]
39. Ortiz-Quintero, B. Cell-free microRNAs in blood and other body fluids, as cancer biomarkers. *Cell Prolif.* **2016**, *49*, 281–303. [CrossRef]
40. Erstad, D.J.; Tumusiime, G.; Cusack, J.C. Prognostic and predictive biomarkers in colorectal cancer: Implications for the clinical surgeon. *Ann. Surg. Oncol.* **2015**, *22*, 3433–3450. [CrossRef]
41. Moosavi, F.; Giovannetti, E.; Saso, L.; Firuzi, O. HGF/MET pathway aberrations as diagnostic, prognostic, and predictive biomarkers in human cancers. *Crit. Rev. Clin. Lab. Sci.* **2019**, *56*, 533–566. [CrossRef]
42. Yao, Q.; Wang, W.; Jin, J.; Min, K.; Yang, J.; Zhong, Y.; Xu, C.; Deng, J.; Zhou, Y. Synergistic role of Caspase-8 and Caspase-3 expressions: Prognostic and predictive biomarkers in colorectal cancer. *Cancer Biomark.* **2018**, *21*, 899–908. [CrossRef]
43. Masuda, T.; Hayashi, N.; Kuroda, Y.; Ito, S.; Eguchi, H.; Mimori, K. MicroRNAs as biomarkers in colorectal cancer. *Cancers* **2017**, *9*, 124. [CrossRef]
44. Zarkavelis, G.; Boussios, S.; Papadaki, A.; Katsanos, K.H.; Christodoulou, D.K.; Pentheroudakis, G. Current and future biomarkers in colorectal cancer. *Ann. Gastroenterol.* **2017**, *30*, 613. [CrossRef]
45. Vinchhi, P.; Patel, M.M. Triumph against cancer: Invading colorectal cancer with nanotechnology. *Expert Opin. Drug Deliv.* **2021**, *18*, 1169–1192. [CrossRef]
46. Ladabaum, U.; Dominitz, J.A.; Kahi, C.; Schoen, R.E. Strategies for colorectal cancer screening. *Gastroenterology* **2020**, *158*, 418–432. [CrossRef]

47. Fattepur, S.; Setia, A.; Prashar, D.; Sahu, R.K.; Nilugal, K.C.; Abdullah, I.; Sreeharsah, N. Understanding diverse signaling pathways involved in colorectal carcinoma and potentials of nanotechnology in their diagnosis and therapeutics. *PharmacologyOnLine* **2021**, *2*, 1226–1243.
48. Wu, Z.; Li, Y.; Zhang, Y.; Hu, H.; Wu, T.; Liu, S.; Chen, W.; Xie, S.; Lu, Z. Colorectal Cancer Screening Methods and Molecular Markers for Early Detection. *Technol. Cancer Res. Treat.* **2020**, *19*, 1–9. [CrossRef]
49. Guimarães, D.P.; Fregnani, J.H.; Reis, R.M.; Taveira, L.N.; Scapulatempo-Neto, C.; Matsushita, M.; Silva, S.R.; Oliveira, C.Z.; Longatto-Filho, A.; Eklund, C.; et al. Comparison of a new-generation fecal immunochemical test (FIT) with guaiac fecal occult blood test (gFOBT) in detecting colorectal neoplasia among colonoscopy-referral patients. *Anticancer Res.* **2019**, *39*, 261–269. [CrossRef]
50. Segev, L.; Kalady, M.F.; Church, J.M. Left-sided dominance of early-onset colorectal cancers: A rationale for screening flexible sigmoidoscopy in the young. *Dis. Colon Rectum* **2018**, *61*, 897–902. [CrossRef]
51. Bobo, D.; Robinson, K.J.; Islam, J.; Thurecht, K.J.; Corrie, S.R. Nanoparticle-based medicines: A review of FDA-approved materials and clinical trials to date. *Pharm. Res.* **2016**, *33*, 2373–2387. [CrossRef]
52. Feng, W.; Jin, M.; Yang, K.; Pei, Y.; Pei, Z. Supramolecular delivery systems based on pillararenes. *Chem. Commun.* **2018**, *54*, 13626–13640. [CrossRef]
53. Elsabahy, M.; Heo, G.S.; Lim, S.M.; Sun, G.; Wooley, K.L. Polymeric nanostructures for imaging and therapy. *Chem. Rev.* **2015**, *115*, 10967–11011. [CrossRef]
54. Naseri, N.; Ajorlou, E.; Asghari, F.; Pilehvar-Soltanahmadi, Y. An update on nanoparticle-based contrast agents in medical imaging. *Artif. Cells Nanomed. Biotechnol.* **2018**, *46*, 1111–1121. [CrossRef]
55. Cisterna, B.A.; Kamaly, N.; Choi, W.I.; Tavakkoli, A.; Farokhzad, O.C.; Vilos, C. Targeted nanoparticles for colorectal cancer. *Nanomedicine* **2016**, *11*, 2443–2456. [CrossRef]
56. Xiao, M.; Lai, W.; Man, T.; Chang, B.; Li, L.; Chandrasekaran, A.R.; Pei, H. Rationally engineered nucleic acid architectures for biosensing applications. *Chem. Rev.* **2019**, *119*, 11631–11717. [CrossRef]
57. Siddique, S.; Chow, J.C. Application of nanomaterials in biomedical imaging and cancer therapy. *Nanomaterials* **2020**, *10*, 1700. [CrossRef]
58. Banerjee, A.; Pathak, S.; Subramanium, V.D.; Dharanivasan, G.; Murugesan, R.; Verma, R.S. Strategies for targeted drug delivery in treatment of colon cancer: Current trends and future perspectives. *Drug Discov. Today* **2017**, *22*, 1224–1232. [CrossRef]
59. Chaturvedi, V.K.; Singh, A.; Singh, V.K.; Singh, M.P. Cancer nanotechnology: A new revolution for cancer diagnosis and therapy. *Curr. Drug Metab.* **2019**, *20*, 416–429. [CrossRef]
60. Wagner, A.M.; Knipe, J.M.; Orive, G.; Peppas, N.A. Quantum dots in biomedical applications. *Acta Biomater.* **2019**, *94*, 44–63. [CrossRef]
61. Wu, X.; Xiao, T.; Luo, Z.; He, R.; Cao, Y.; Guo, Z.; Zhang, W.; Chen, Y. A micro-/nano-chip and quantum dots-based 3D cytosensor for quantitative analysis of circulating tumor cells. *J. Nanobiotechnol.* **2018**, *16*, 1–9. [CrossRef]
62. Ai, X.; Evans, E.W.; Dong, S.; Gillett, A.J.; Guo, H.; Chen, Y.; Hele, T.J.; Friend, R.H.; Li, F. Efficient radical-based light-emitting diodes with doublet emission. *Nature* **2018**, *563*, 536–540. [CrossRef]
63. Tamang, S.; Lincheneau, C.; Hermans, Y.; Jeong, S.; Reiss, P. Chemistry of InP nanocrystal syntheses. *Chem. Mater.* **2016**, *28*, 2491–2506. [CrossRef]
64. Park, Y.; Ryu, Y.M.; Wang, T.; Jung, Y.; Kim, S.; Hwang, S.; Park, J.; Bae, D.J.; Kim, J.; Moon, H.; et al. Colorectal Cancer Diagnosis Using Enzyme-Sensitive Ratiometric Fluorescence Dye and Antibody–Quantum Dot Conjugates for Multiplexed Detection. *Adv. Funct. Mater.* **2018**, *28*, 1703450. [CrossRef]
65. Carbary-Ganz, J.L.; Welge, W.A.; Barton, J.K.; Utzinger, U. In vivo molecular imaging of colorectal cancer using quantum dots targeted to vascular endothelial growth factor receptor 2 and optical coherence tomography/laser-induced fluorescence dual-modality imaging. *J. Biomed. Opt.* **2015**, *20*, 096015. [CrossRef]
66. Nosrati, H.; Salehiabar, M.; Fridoni, M.; Abdollahifar, M.A.; Manjili, H.K.; Davaran, S.; Danafar, H. New insight about biocompatibility and biodegradability of iron oxide magnetic nanoparticles: Stereological and in vivo MRI monitor. *Sci. Rep.* **2019**, *9*, 7173. [CrossRef]
67. Prasad, M.; Lambe, U.P.; Brar, B.; Shah, I.; Manimegalai, J.; Ranjan, K.; Rao, R.; Kumar, S.; Mahant, S.; Khurana, S.K.; et al. Nanotherapeutics: An insight into healthcare and multi-dimensional applications in medical sector of the modern world. *Biomed. Pharmacother.* **2018**, *97*, 1521–1537. [CrossRef]
68. Ebrahimi, N.; Bi, C.; Cappelleri, D.J.; Ciuti, G.; Conn, A.T.; Faivre, D.; Habibi, N.; Hošovský, A.; Iacovacci, V.; Khalil, I.S.; et al. Magnetic actuation methods in bio/soft robotics. *Adv. Funct. Mater.* **2021**, *31*, 2005137. [CrossRef]
69. Huang, J.; Li, Y.; Orza, A.; Lu, Q.; Guo, P.; Wang, L.; Yang, L.; Mao, H. Magnetic nanoparticle facilitated drug delivery for cancer therapy with targeted and image-guided approaches. *Adv. Funct. Mater.* **2016**, *26*, 3818–3836. [CrossRef]
70. Kim, S.H.; Oh, S.N.; Choi, H.S.; Lee, H.S.; Jun, J.; Nam, Y.; Lee, S.H.; Lee, J.K.; Lee, H.G. USPIO enhanced lymph node MRI using 3D multi-echo GRE in a rabbit model. *Contrast Media Mol. Imaging* **2016**, *11*, 544–549. [CrossRef]
71. Wu, L.; Cao, Y.; Liao, C.; Huang, J.; Gao, F. Diagnostic performance of USPIO-enhanced MRI for lymph-node metastases in different body regions: A meta-analysis. *Eur. J. Radiol.* **2011**, *80*, 582–589. [CrossRef]

72. Kuo, C.Y.; Liu, T.Y.; Chan, T.Y.; Tsai, S.C.; Hardiansyah, A.; Huang, L.Y.; Yang, M.C.; Lu, R.H.; Jiang, J.K.; Yang, C.Y.; et al. Magnetically triggered nanovehicles for controlled drug release as a colorectal cancer therapy. *Colloids Surf. B Biointerfaces* **2016**, *140*, 567–573. [CrossRef]
73. Syu, W.J.; Huang, C.C.; Hsiao, J.K.; Lee, Y.C.; Huang, Y.T.; Venkatesan, P.; Lai, P.S. Co-precipitation synthesis of near-infrared iron oxide nanocrystals on magnetically targeted imaging and photothermal cancer therapy via photoablative protein denature. *Nanotheranostics* **2019**, *3*, 236. [CrossRef]
74. Patra, J.K.; Das, G.; Fraceto, L.F.; Campos, E.V.; del Pilar Rodriguez-Torres, M.; Acosta-Torres, L.S.; Diaz-Torres, L.A.; Grillo, R.; Swamy, M.K.; Sharma, S.; et al. Nano based drug delivery systems: Recent developments and future prospects. *J. Nanobiotechnol.* **2018**, *16*, 71. [CrossRef]
75. Muhammad, Q.; Jang, Y.; Kang, S.H.; Moon, J.; Kim, W.J.; Park, H. Modulation of immune responses with nanoparticles and reduction of their immunotoxicity. *Biomater. Sci.* **2020**, *8*, 1490–1501. [CrossRef]
76. Behzadi, S.; Serpooshan, V.; Tao, W.; Hamaly, M.A.; Alkawareek, M.Y.; Dreaden, E.C.; Brown, D.; Alkilany, A.M.; Farokhzad, O.C.; Mahmoudi, M. Cellular uptake of nanoparticles: Journey inside the cell. *Chem. Soc. Rev.* **2017**, *46*, 4218–4244. [CrossRef]
77. Tee, J.K.; Yip, L.X.; Tan, E.S.; Santitewagun, S.; Prasath, A.; Ke, P.C.; Ho, H.K.; Leong, D.T. Nanoparticles' interactions with vasculature in diseases. *Chem. Soc. Rev.* **2019**, *48*, 5381–5407. [CrossRef]
78. Gupta, R.; Rai, B. Effect of size and surface charge of gold nanoparticles on their skin permeability: A molecular dynamics study. *Sci. Rep.* **2017**, *7*, 1–13. [CrossRef]
79. Fang, R.H.; Kroll, A.V.; Gao, W.; Zhang, L. Cell membrane coating nanotechnology. *Adv. Mater.* **2018**, *30*, 1706759. [CrossRef]
80. Tang, C.; He, Z.; Liu, H.; Xu, Y.; Huang, H.; Yang, G.; Xiao, Z.; Li, S.; Liu, H.; Deng, Y.; et al. Application of magnetic nanoparticles in nucleic acid detection. *J. Nanobiotechnol.* **2020**, *18*, 62. [CrossRef]
81. Singh, V.K.; Arora, D.; Ansari, M.I.; Sharma, P.K. Phytochemicals based chemopreventive and chemotherapeutic strategies and modern technologies to overcome limitations for better clinical applications. *Phytother. Res.* **2019**, *33*, 3064–3089. [CrossRef]
82. Ahmad, R.; Srivastava, S.; Ghosh, S.; Khare, S.K. Phytochemical delivery through nanocarriers: A review. *Colloids Surf. B Biointerfaces* **2021**, *197*, 111389. [CrossRef]
83. Allen, C.T.; Ricker, J.L.; Chen, Z.; Van Waes, C. Role of activated nuclear factor-κB in the pathogenesis and therapy of squamous cell carcinoma of the head and neck. *Head Neck J. Sci. Spec. Head Neck* **2007**, *29*, 959–971. [CrossRef]
84. Wong, K.E.; Ngai, S.C.; Chan, K.G.; Lee, L.H.; Goh, B.H.; Chuah, L.H. Curcumin nanoformulations for colorectal cancer: A review. *Front. Pharmacol.* **2019**, *10*, 152. [CrossRef]
85. Pistollato, F.; Giampieri, F.; Battino, M. The use of plant-derived bioactive compounds to target cancer stem cells and modulate tumor microenvironment. *Food Chem. Toxicol.* **2015**, *75*, 58–70. [CrossRef]
86. Liczbiński, P.; Michałowicz, J.; Bukowska, B. Molecular mechanism of curcumin action in signaling pathways: Review of the latest research. *Phytother. Res.* **2020**, *34*, 1992–2005. [CrossRef]
87. Canistro, D.; Chiavaroli, A.; Cicia, D.; Cimino, F.; Curro, D.; Dell'Agli, M.; Ferrante, C.; Giovannelli, L.; Leone, S.; Martinelli, G.; et al. The pharmacological basis of the curcumin nutraceutical uses: An update. *PharmAdvances* **2021**, *3*, 421–466. [CrossRef]
88. Zhang, I.; Cui, Y.; Amiri, A.; Ding, Y.; Campbell, R.E.; Maysinger, D. Pharmacological inhibition of lipid droplet formation enhances the effectiveness of curcumin in glioblastoma. *Eur. J. Pharm. Biopharm.* **2016**, *100*, 66–76. [CrossRef]
89. Che, X.H.; Chen, C.L.; Ye, X.L.; Weng, G.B.; Guo, X.Z.; Yu, W.Y.; Tao, J.; Chen, Y.C.; Chen, X. Dual inhibition of COX-2/5-LOX blocks colon cancer proliferation, migration and invasion in vitro. *Oncol. Rep.* **2016**, *35*, 1680–1688. [CrossRef]
90. Tuncer, S.; Banerjee, S. Eicosanoid pathway in colorectal cancer: Recent updates. *World J. Gastroenterol.* **2015**, *21*, 11748. [CrossRef]
91. Öhd, J.F.; Nielsen, C.K.; Campbell, J.; Landberg, G.; Löfberg, H.; Sjölander, A. Expression of the leukotriene D4 receptor CysLT1, COX-2, and other cell survival factors in colorectal adenocarcinomas. *Gastroenterology* **2003**, *124*, 57–70. [CrossRef]
92. Raveendran, R.; Bhuvaneshwar, G.S.; Sharma, C.P. In vitro cytotoxicity and cellular uptake of curcumin-loaded Pluronic/Polycaprolactone micelles in colorectal adenocarcinoma cells. *J. Biomater. Appl.* **2013**, *27*, 811–827. [CrossRef]
93. Basli, A.; Belkacem, N.; Amrani, I. Health Benefits of Phenolic Compounds Against Cancers. In *Phenolic Compounds—Biological Activity*; IntechOpen: London, UK, 2017; pp. 193–210.
94. Yang, T.; Wang, L.; Zhu, M.; Zhang, L.; Yan, L. Properties and molecular mechanisms of resveratrol: A review. *Pharm.-Int. J. Pharm. Sci.* **2015**, *70*, 501–506.
95. Pashirzad, M.; Johnston, T.P.; Sahebkar, A. Therapeutic Effects of Polyphenols on the Treatment of Colorectal Cancer by Regulating Wnt β-Catenin Signaling Pathway. *J. Oncol.* **2021**, *2021*, 3619510. [CrossRef]
96. Azer, S.A. Overview of molecular pathways in inflammatory bowel disease associated with colorectal cancer development. *Eur. J. Gastroenterol. Hepatol.* **2013**, *25*, 271–281. [CrossRef]
97. Raghav, N.; Sharma, M.R. Ginger (*Zingiber officinale* Roscoe): A mini-review of constituents and biological activities. *Chem. Biol. Interface* **2021**, *11*, 81–93.
98. Bigagli, E.; Cinci, L.; Paccosi, S.; Parenti, A.; D'Ambrosio, M.; Luceri, C. Nutritionally relevant concentrations of resveratrol and hydroxytyrosol mitigate oxidative burst of human granulocytes and monocytes and the production of pro-inflammatory mediators in LPS-stimulated RAW 264.7 macrophages. *Int. Immunopharmacol.* **2017**, *43*, 147–155. [CrossRef]
99. Soo, E.; Thakur, S.; Qu, Z.; Jambhrunkar, S.; Parekh, H.S.; Popat, A. Enhancing delivery and cytotoxicity of resveratrol through a dual nanoencapsulation approach. *J. Colloid Interface Sci.* **2016**, *462*, 368–374. [CrossRef]

100. Summerlin, N.; Qu, Z.; Pujara, N.; Sheng, Y.; Jambhrunkar, S.; McGuckin, M.; Popat, A. Colloidal mesoporous silica nanoparticles enhance the biological activity of resveratrol. *Colloids Surf. B Biointerfaces* **2016**, *144*, 1–7. [CrossRef]
101. Schirrmacher, V. From chemotherapy to biological therapy: A review of novel concepts to reduce the side effects of systemic cancer treatment. *Int. J. Oncol.* **2019**, *54*, 407–419.
102. Zhang, M.; Zang, X.; Wang, M.; Li, Z.; Qiao, M.; Hu, H.; Chen, D. Exosome-based nanocarriers as bio-inspired and versatile vehicles for drug delivery: Recent advances and challenges. *J. Mater. Chem. B* **2019**, *7*, 2421–2433. [CrossRef]
103. Vodenkova, S.; Buchler, T.; Cervena, K.; Veskrnova, V.; Vodicka, P.; Vymetalkova, V. 5-fluorouracil and other fluoropyrimidines in colorectal cancer: Past, present and future. *Pharmacol. Ther.* **2020**, *206*, 107447. [CrossRef]
104. Sara, J.D.; Kaur, J.; Khodadadi, R.; Rehman, M.; Lobo, R.; Chakrabarti, S.; Herrmann, J.; Lerman, A.; Grothey, A. 5-fluorouracil and cardiotoxicity: A review. *Ther. Adv. Med. Oncol.* **2018**, *10*, 1–18. [CrossRef]
105. Che, J.; Pan, L.; Yang, X.; Liu, Z.; Huang, L.; Wen, C.; Lin, A.; Liu, H. Thymidine phosphorylase expression and prognosis in colorectal cancer treated with 5-fluorouracil-based chemotherapy: A meta-analysis. *Mol. Clin. Oncol.* **2017**, *7*, 943–952. [CrossRef]
106. Huehls, A.M.; Huntoon, C.J.; Joshi, P.M.; Baehr, C.A.; Wagner, J.M.; Wang, X.; Lee, M.Y.; Karnitz, L.M. Genomically incorporated 5-fluorouracil that escapes UNG-initiated base excision repair blocks DNA replication and activates homologous recombination. *Mol. Pharmacol.* **2016**, *89*, 53–62. [CrossRef]
107. Reynolds, I.S.; O'Connell, E.; Fichtner, M.; McNamara, D.A.; Kay, E.W.; Prehn, J.H.; Furney, S.J.; Burke, J.P. Mucinous adenocarcinoma is a pharmacogenomically distinct subtype of colorectal cancer. *Pharm. J.* **2020**, *20*, 524–532. [CrossRef]
108. Mizumoto, Y.; Yokoyama, S.; Matsuda, K.; Iwamoto, H.; Mitani, Y.; Tamura, K.; Nakamura, Y.; Murakami, D.; Oka, M.; Kobayashi, Y.; et al. Modulation of capecitabine administration to improve continuity of adjuvant chemotherapy for patients with colorectal cancer: A phase II study. *Mol. Clin. Oncol.* **2020**, *12*, 126–133. [CrossRef]
109. Tolan, D.; Almotairy, A.R.; Howe, O.; Devereux, M.; Montagner, D.; Erxleben, A. Cytotoxicity and ROS production of novel Pt (IV) oxaliplatin derivatives with indole propionic acid. *Inorg. Chim. Acta* **2019**, *492*, 262–267. [CrossRef]
110. Cirri, D.; Bartoli, F.; Pratesi, A.; Baglini, E.; Barresi, E.; Marzo, T. Strategies for the Improvement of Metal-Based Chemotherapeutic Treatments. *Biomedicines* **2021**, *9*, 504. [CrossRef]
111. Ma, L.; Li, X.; Zhao, X.; Sun, H.; Kong, F.; Li, Y.; Sui, Y.; Xu, F. Oxaliplatin promotes siMAD2L2 induced apoptosis in colon cancer cells. *Mol. Med. Rep.* **2021**, *24*, 1–9. [CrossRef]
112. Ulbrich, K.; Hola, K.; Subr, V.; Bakandritsos, A.; Tucek, J.; Zboril, R. Targeted drug delivery with polymers and magnetic nanoparticles: Covalent and noncovalent approaches, release control, and clinical studies. *Chem. Rev.* **2016**, *116*, 5338–5431. [CrossRef]
113. Kandasamy, P.; Gyimesi, G.; Kanai, Y.; Hediger, M.A. Amino acid transporters revisited: New views in health and disease. *Trends Biochem. Sci.* **2018**, *43*, 752–789. [CrossRef]
114. Limagne, E.; Thibaudin, M.; Nuttin, L.; Spill, A.; Derangère, V.; Fumet, J.D.; Amellal, N.; Peranzoni, E.; Cattan, V.; Ghiringhelli, F. Trifluridine/tipiracil plus oxaliplatin improves PD-1 blockade in colorectal cancer by inducing immunogenic cell death and depleting macrophages. *Cancer Immunol. Res.* **2019**, *7*, 1958–1969. [CrossRef]
115. Wilson, T.A. Second Cures Draft Retains Proposals To Modernize Clinical Trials. *Inside Health Policy Com's FDA Week* **2015**, *21*, 1–9.
116. Hall, G.L. *Patient-Centered Clinical Care for African Americans: A Concise, Evidence-Based Guide to Important Differences and Better Outcomes*; Springer Nature: Cham, Switzerland, 2019.
117. Singer, P.; Blaser, A.R.; Berger, M.M.; Alhazzani, W.; Calder, P.C.; Casaer, M.P.; Hiesmayr, M.; Mayer, K.; Montejo, J.C.; Pichard, C.; et al. ESPEN guideline on clinical nutrition in the intensive care unit. *Clin. Nutr.* **2019**, *38*, 48–79. [CrossRef]
118. Sanz-Garcia, E.; Grasselli, J.; Argiles, G.; Elez, M.E.; Tabernero, J. Current and advancing treatments for metastatic colorectal cancer. *Expert Opin. Biol. Ther.* **2016**, *16*, 93–110. [CrossRef]

nanomaterials

Article

Ciprofloxacin-Loaded Gold Nanoparticles against Antimicrobial Resistance: An In Vivo Assessment

Afrah Nawaz [1], Syed Mohsin Ali [1], Nosheen Fatima Rana [1,*], Tahreem Tanweer [1], Amna Batool [1], Thomas J. Webster [2], Farid Menaa [3], Sundus Riaz [1], Zahra Rehman [1], Farhat Batool [1], Misha Fatima [1], Tuba Maryam [1], Iqra Shafique [1], Abida Saleem [1] and Arfa Iqbal [1]

1 Department of Biomedical Engineering and Sciences, School of Mechanical & Manufacturing Engineering, National University of Science & Technology, Islamabad 44000, Pakistan; anawaz.bmes16smme@student.nust.edu.pk (A.N.); sali.bmes17smme@student.nust.edu.pk (S.M.A.); ttanveer.pg@smme.edu.pk (T.T.); amnabatoolx@gmail.com (A.B.); sundusriaz@parc.gov.pk (S.R.); zrehman.bms20smme@student.nust.edu.pk (Z.R.); farhat.bms20smme@student.nust.edu.pk (F.B.); mfatima.bms20smme@student.nust.edu.pk (M.F.); tuba.bms20smme@student.nust.edu.pk (T.M.); ishafique.phd21smme@student.nust.edu.pk (I.S.); asaleem.phd21smme@student.nust.edu.pk (A.S.); aiqbal.bmes19smme@student.nust.edu.pk (A.I.)

2 Department of Chemical Engineering, Northeastern University, Boston, MA 02115, USA; websterthomas02@gmail.com

3 Departments of Internal Medicine and Nanomedicine, California Innovations Corporation, San Diego, CA 92037, USA; dr.fmenaa@gmail.com

* Correspondence: nosheen.fatima@smme.nust.edu.pk; Tel.: +92-519-085-6074; Fax: +92-908-56001

Citation: Nawaz, A.; Ali, S.M.; Rana, N.F.; Tanweer, T.; Batool, A.; Webster, T.J.; Menaa, F.; Riaz, S.; Rehman, Z.; Batool, F.; et al. Ciprofloxacin-Loaded Gold Nanoparticles against Antimicrobial Resistance: An In Vivo Assessment. *Nanomaterials* **2021**, *11*, 3152. https://doi.org/10.3390/nano11113152

Academic Editor: Angelo Ferraro

Received: 22 August 2021
Accepted: 6 November 2021
Published: 22 November 2021

Abstract: Metallic nanoparticles, such as gold nanoparticles (AuNPs), have been extensively studied as drug delivery systems for various therapeutic applications. However, drug-loaded-AuNPs have been rarely explored in vivo for their effect on bacteria residing inside tissues. Ciprofloxacin (CIP) is a second-generation fluoroquinolone with a broad-spectrum of antibiotic properties devoid of developing bacteria resistance. This research is focused on the synthesis and physical characterization of Ciprofloxacin-loaded gold nanoparticles (CIP-AuNPs) and their effect on the colonization of *Enterococcus faecalis* in the liver and kidneys of mice. The successfully prepared CIP-AuNPs were stable and exerted enhanced in vitro antibacterial activity against *E. faecalis* compared with free CIP. The optimized CIP-AuNPs were administered (500 μg/Kg) once a day via tail vein to infected mice for eight days and were found to be effective in eradicating *E. faecalis* from the host tissues. Moreover, unlike CIP, CIP-AuNPs were non-hemolytic. In summary, this study demonstrated that CIP-AuNPs are promising and biocompatible alternative therapeutics for *E.-faecalis*-induced infections resistant to conventional drugs (e.g., beta-lactams and vancomycin) and should be further investigated.

Keywords: drug delivery; antibiotics; antimicrobial resistance; gold nanoparticles; ciprofloxacin; *Enterococcus faecalis*; liver and kidney infections; nanotechnology

1. Introduction

E. faecalis has become one of the most daunting bacteria among the opportunistic pathogens [1]. This bacterium is responsible for a major percentage of nosocomial infections. It can cause infections in the gastrointestinal, respiratory, and urinary tracts; root canal; etc. Its characteristic features include extraordinary survival capacities in extreme conditions and development of resistance against antibiotics [2–4] such as β–lactam antibiotics and vancomycin [5]. Hence, innovative and effective antibiotics are required to tackle drug resistance mediated by *E. faecalis* infections.

The development of antimicrobial resistance (AMR) is one of the most pressing global issues, and great efforts have been committed to explore new and advanced antibiotics. However, extensive and impeded regulatory approval has shifted the focus to improve the efficacy of existing blockbuster drugs [6]. Advances in nanotechnology offer exciting methods to provide novel materials that can support penetration into biofilms, promote bacteria entry of the drugs, and present synergism with their own functional attributes [7–9]. Metallic nanoparticles (NPs), such as gold nanoparticles (AuNPs), are gaining recognition as improved drug delivery vehicles. They have shown effectiveness for various medical illnesses, including cancer, chronic diseases, and antibacterial, antiviral, and antifungal infections [10]. AuNPs do not exert (inherent) antibacterial activity; however, they enhance antibacterial properties of the loaded antibacterial drugs [11]. Drug-loaded AuNPs collapse bacteria membrane potential, hinder ATPase activity, and halt binding of ribosomes to tRNA leading to deteriorated bacterial cell metabolism [12]. Moreover, their site directed activity can target biofilms and quorum sensing [13,14]. The stability and the activity of such AuNPs under physiological conditions (such as pH, temperature, etc.) have also been evaluated in various studies, as such factors can modify their antibacterial activity [15,16]. Over the last years, various nanocarrier systems against *E. faecalis* have been reported such as ciprofloxacin and metronidazole encapsulated nanomatrix gels [17], bismuth nanoparticles [18], chitosan–propolis nanoparticles [19], doxycycline-functionalized polymeric nanoparticles [20], methylene blue and biogenic gold nanoparticles [21], silver nanoparticles and calcium hydroxide mixtures [22], biosynthesized silver nanoparticles [23], and graphene oxide nanoparticles [24].

Ciprofloxacin [1-cyclopropyl-6-fluoro-1,4-dihydro-4-oxo-7 piperazinylquinolone-3-carboxylic acid] (CIP) is a large spectrum fluoroquinolone antibiotic that is widely used for the treatment of numerous bacterial infections in joints, bones, skin, tooth, gastrointestinal, and urinary and respiratory tracts [25]. CIP-loaded gold nanoparticles (CIP-AuNPs) were reported by Tom et al. in 2004 [26]. Recently reported nanocarriers for CIP include CIP-loaded nanocomposites [27], PLGA–chitosan-based ciprofloxacin [28], Ciprofloxacin-containing polymer-based nanofibers (scaffolds) [29], and CIP-loaded single-walled carbon nanotubes [30]. This study aims to use ciprofloxacin protected gold nanoparticles to eradicate bacteria in the tissues of mice. The physical characterization of the nanoformulated version of this drug along with in vitro antibacterial activity, against *E. faecalis*, were carried out. For in vivo studies, an infectious mouse model was developed for *E. faecalis* and anti-colonization through the use of CIP-AuNPs was determined.

2. Materials and Methods

2.1. Bacterial Strains

The Gram-positive bacterial strain used was *E. faecalis* JH2-2 (derived from the parental strain JH2). GM17 medium was used for their growth at 37 °C [31].

2.2. Preparation of AuNPs and CIP-AuNPs

The AuNPs were prepared by a previously reported method [26]. Trisodium citrate (0.5 mM) and chloroauric acid (0.5 mM) (MERK, Munich, Germany) were fluxed together. Citrate acted as a stabilizer as well as a reducing agent. The particle formation was confirmed upon achieving a red wine color. For the loading of CIP on AuNPs, 20 mL of the above solution was mixed with 5 mL of CIP (0.5, 1.0, 1.5, 2, and 2.5 mM) at pH 6.5. Then, the solution was stirred until the red color turned to blue–purple.

2.3. Characterization of AuNPs and CIP-AuNPs

The CIP-AuNPs were characterized by UV–Vis spectrophotometry using a UV-2800 (BMS Biotechnology Medical Services, Madrid, Spain) spectrophotometer. Zeta potential was recorded using a Malvern Zeta sizer (Malvern, UK). Scanning electron microscopy (SEM) and Energy-dispersive X-ray spectroscopy (EDS) analyses were carried out using a SEM VEG 3 LMU (Tescan, Czech Republic), while Fourier-Transform Infrared Spectroscopy (FTIR) analyses were carried out using a Bruker FTIR Spectrometer ALPHA II (Westborough, MA, USA).

2.4. Drug Loading Capacity and Encapsulation Efficiency

CIP loading capacity and encapsulation efficiency by the AuNPs were evaluated using Equations (1) and (2), respectively.

$$\text{Loading capacity } (\%) = \frac{\text{Weight of CIP in CIP} - \text{AuNPs}}{\text{Weight of CIP} - \text{AuNPs}} \times 100 \tag{1}$$

$$\text{Encapsulation Efficiency } (\%) = \frac{\text{Total CIP added} - \text{Free CIP}}{\text{Total CIP added}} \times 100 \tag{2}$$

2.5. Drug Release Efficiency

The in vitro release of CIP from the CIP-AuNPs formulated with various concentrations of CIP was studied over time using a UV–Vis spectrophotometer at 280 nm from 0 to 24 h (every 2 h) by adding 20 mL of a 20-mM PBS buffer to 20 mL of the CIP-AuNPs solutions. The cumulative drug release was calculated using Equation (3).

$$\text{Cumulative drug release } (\%) = \frac{\text{CIP released from the CIP} - \text{AuNPs at t}}{\text{The total amount of CIP loaded onto the AuNPs.}} \tag{3}$$

To determine the quantity of the drug present at the absorption site, a significant predictor $T_{60\%}$, was calculated as the time taken to release 60% of the drug. For instance, Stineman interpolation using Minitab 17 software was applied. The $T_{60\%}$ drug release data were fit into the Korsmeyer-Peppas model for non-swellable matrices (Section S1.3).

2.6. Kinetic Analysis of the Drug Release

To determine the drug release from the CIP-AuNPs, numerous mathematical models (zero-order, first-order, and Higuchi's model) were employed (Section S1.3) [27,32].

2.7. In Vitro Stability of CIP-AuNPs

The effect of temperature on CIP-AuNPs was found by heating CIP-AuNPs at different temperatures (25 °C, 50 °C, 75 °C, and 100 °C) for 30 min. The effect of pH values (4, 7, and 10) and different salt concentrations (0.05, 0.1, 0.5, and 1 M) on CIP-AuNPs were also determined. For salt concentration, 10 mL of CIP-AuNPs were centrifuged at 10,000× *g* for 10 min. The resulting pellets were suspended in NaCl solutions (50 mM–1 M) at 37 °C for 24 h.

2.8. In Vitro Antibacterial Potential of CIP-AuNPs

The Minimum Inhibitory Concentration (MIC) of free CIP was calculated using different concentrations of CIP and CIP-AuNPs (0.1–10 mg/mL) against *E. faecalis.* The bacterial cultures at exponential phase of OD_{600} were harvested and counted for 10^6 CFU/mL using the standard dilution method. Using a standard inoculum of 10^6 CFUs, cultures were incubated at standard conditions for 24 h. The concentrations with a 50% reduction in bacterial count were observed as MIC. With an inoculum of approximately 10^6 CFU/mL, 50 μL of a 10-fold-diluted culture were plated on M-17 agar plates for measuring the viable cells. Colonies were counted after 24 h [33,34]. Moreover, a zone of inhibition test was also conducted by using the disc diffusion method to compare the antibacterial activity of CIP-AuNPs and free CIP. An *E. faecalis* culture was developed on the nutrient agar plate. A zone of inhibition (in mm) was then measured for *E. faecalis* using the Kirby–Bauer method. Sterile Whatman filter paper discs were impregnated with CIP, AuNPs, and CIP-AuNPs at a concentration of 10 μg/disc each.

2.9. Hemolytic Activity of CIP-AuNPs

A hemolysis test was carried out to compare the hemolytic activity of different concentrations (i.e., 10, 25, 50, and 100 μg/mL) of CIP-AuNPs (2 mM CIP), AuNPs, and CIP. Blood samples were taken from healthy female donors. Red blood cells were incubated for

4 h using the method described by Zarmina et al. [35]. As a positive control, Triton X-100 (0.5%) was used, while PBS was used as a negative control. The absorbance was measured at a wavelength of 550 nm.

2.10. Colonization of E. faecalis in BALB/c Mice

The in vivo investigation was conducted using female *BALB/c* mice (eight weeks old, weighing 25–30 g; n = 15) purchased from the National Institute of Health (NIH), Islamabad Pakistan. They were kept at 25 ± 2 °C and presented with a natural light–dark cycle (10–14 h). The mice were given autoclaved tap water and a normal diet ad libitum. For bacterial colonization in hepatic and renal tissues, a well-established intravenous (IV) infection model was used [36]. The GM17 broth at a temperature of 37 °C was used and the preculture was grown overnight. A total of 100 µL of preculture was added to brain heart infused (BHI) medium supplemented with 40% filter-sterilized serum. Phosphate buffer saline (PBS), pH 7.4, was then used to wash the subsequent pellets from the cultures, optimized by colony counting for the number of cells. The bacterial pallets were then suspended in sterile PBS. A total of 100 µL of the suspension adjusted for 1×10^9 cells/mL of bacterial suspensions were (tail vein) injected into each of the female mice (n = 15).

2.11. In Vivo Antibacterial Activity of CIP-AuNPs

To assess the in vivo antibacterial activity of CIP-AuNPs and CIP, they were suspended or dissolved in PBS buffer. The infected group (n = 15) was treated with CIP-AuNPs (500 µg/Kg, n = 5) and with free CIP (10 mg/Kg, n = 5); the remaining five mice remained untreated. CIP-AuNPs and free CIP were delivered by the tail vein once a day for eight days starting from the seventh day of infection until the day of the challenge. After a week of treatment, all mice were sacrificed, and their liver and kidneys were removed to measure the viable bacterial count through the colony-forming unit (CFU) method. For instance, organs were weighed and homogenized in 10 mL of a PBS solution. 10-fold dilutions of the homogenate were plated on the agar plate. CFUs were counted after 24 h.

2.12. Statistical Analysis

Statistical analysis (such as the mean, standard deviation, and multiple group comparison analysis by one-way ANOVA) was carried out using Origin 2021 (OriginLab Corporation, Northampton, MA, USA) and Graph Pad Prism 9.2.

3. Results

3.1. Synthesis of AuNPS and CIP-AuNPs

The synthesis of AuNPs was accomplished by a chemical reduction method. The red color confirmed the formation of AuNPs upon the addition of citrate to chloroauric acid. The UV–Vis graph shows the absorbance peak of gold nanoparticles at 521 nm (Figure 1).

Further, the addition of CIP into the AuNP solution resulted in a purple to bluish solution, indicating CIP adsorption onto the AuNPs. UV–Vis spectroscopy verified the adsorption of various CIP concentrations on AuNPs, as presented in Figure 2. The transition of the solution color took place upon adding different CIP concentrations, typically from red to purple to bluish, and finally, to dark blue (Figure 1).

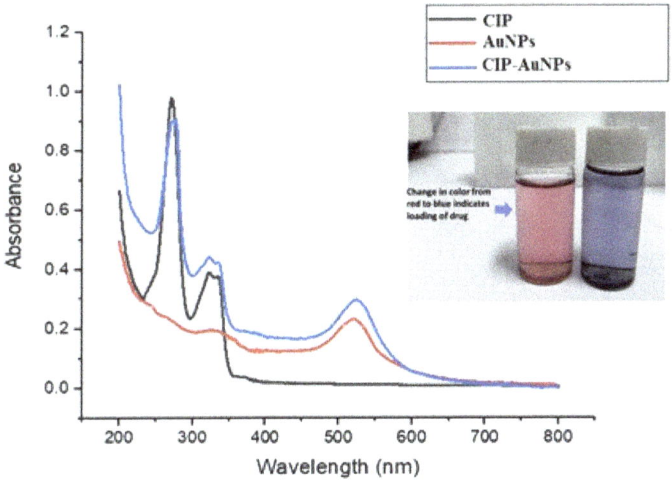

Figure 1. UV–Vis spectra of CIP, AuNPs, and CIP-AuNPs.

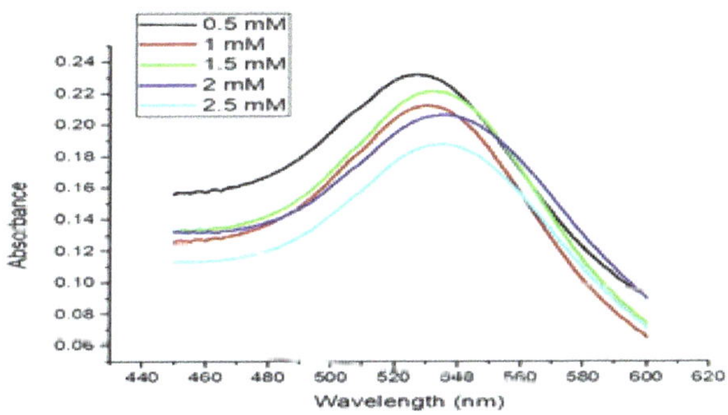

Figure 2. UV–Vis spectra of CIP-AuNPs loaded with various CIP concentrations.

3.2. *CIP Encapsulation Efficiency and CIP Loading Capacity*

CIP encapsulation efficiency was the highest (60.83%) at the highest CIP concentration used (2.5 mM); conversely, CIP encapsulation efficiency into AuNPs was the lowest (24.43%) at the lowest CIP concentration used (0.5 mM) (Table 1). Consistently, the drug loading capacity was also drug-concentration-dependent. Indeed, CIP loading capacity into AuNPs was the highest (34.54%) at the highest CIP concentration used (2.5 mM); conversely, CIP loading capacity into AuNPs was the lowest (8.85%) at the lowest CIP concentration (0.5 mM) (Table 1).

Table 1. Encapsulation efficiency (%) and loading capacity (%) of CIP onto the AuNPs at varying CIP concentrations.

CIP Concentration AuNPs	Encapsulation Efficiency (%)	Loading Capacity (%)
0.5 mM	24.43	8.85
1.0 mM	29.30	15.60
1.5 mM	30.65	28.85
2.0 mM	48.92	33.81
2.5 mM	60.83	34.54

3.3. Particle Size and Zeta Potential of AuNPs and CIP-AuNPs

The average particle size (PS) of AuNPs was found to be 23 nm. The PS for the different CIP-AuNPs are mentioned in Table 2. Upon addition of CIP (0.5 mM and 1 mM), the size of the NPs remained almost the same at 24 nm. The PS increased to 41, 88, and 128 nm using 1.5, 2.0, and 2.5 mM CIP, respectively.

Table 2. Zeta potential values for CIP-AuNPs, AuNPs, and CIP.

CIP-AuNPs	Z-Average (d. nm)	PDI	St Dev (d. nm)	Zeta Potential (mV)
0.5 mM	24.43	0.26	6.21	−32.1
1.0 mM	24.09	0.301	6.044	−33.3
1.5 mM	41	0.68	10.21	−19.7
2.0 mM	88.2	1.000	57.4	−13.4
2.5 mM	128.2	0.48	79.18	−2.12

The zeta size and charge values of the CIP-AuNPs and AuNPs are mentioned in Table 2. The AuNP had a negative charge of −32.1 mV, which remained unchanged upon the addition of 0.5 mM CIP. However, the zeta potential (ZP) and PDI values of the CIP-AuNPs at a 1.5 mM CIP concentration were -19.7 ± 6.65 mV and 0.680, respectively.

3.4. Surface Morphology and Elemental Chemical Composition of AuNPs by SEM–EDS

The surface morphology of the chemically prepared NPs (AuNPs and CIP-AuNPs) was analyzed by SEM at 10 kV. SEM images revealed spherically-shaped AuNPs (Figure 3a), 2 mM CIP-AuNPs (Figure 3b), and 2.5 mM CIP-AuNPs (Figure 3c). The SEM analysis further presented a high polydispersity of 2.5 mM CIP-AuNPs.

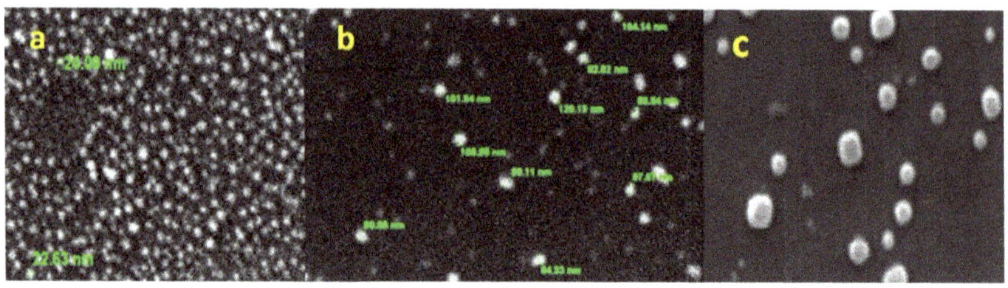

Figure 3. SEM micrographs of (**a**) AuNPs, (**b**) AuNPs (2 mM), and (**c**) CIP-AuNPs (2.5 mM).

Besides, EDX analysis was carried out for elemental mapping of CIP-AuNPs (2.5 mM) (Supplementary Materials Figure S1, Section S2.1).

3.5. Structural Analysis of CIP-AuNPs by FTIR Spectroscopy

The FTIR of CIP-AuNPs, AuNPs, and free CIP are presented in Figure 4 and Supplementary Materials Table S1. FTIR results showed the successful adsorption of CIP in AuNPs. The stretching of the N–H bond of the imino moiety on the piperazine group of the CIP was shown by the band at 3410 cm^{-1}. Absorption bands at 1655 cm^{-1} and 1070 cm^{-1} represent a primary amine (N–H) bend of the pyridone moiety and the C–F functional group of the CIP, respectively [30]. The peak at 1634.84 cm^{-1} corresponded to C = O symmetric stretching, and the band at 1381.02 cm^{-1} was assigned to the C–H stretching vibration in the AuNPs. The CIP-AuNPs exhibited bands at 3341 cm^{-1}, which corresponded to the strong stretching vibration of the O–H group of alcohols and phenols, while the band at 2062 cm^{-1} referred to strong C–H stretching [33]. The stretching bands at 3410 cm^{-1} of the N–H bond of the imino moiety on the piperazine group of the CIP moved to 3449 cm^{-1} for the CIP-AuNPs, showing adsorption of CIP on AuNPs. It has been reported that the nitrogen atom of the NH moiety of the piperazine group of CIP binds to the AuNP surface to form CIP-AuNPs. The binding of CIP to the surface of AuNPs was confirmed by the fact that the corresponding NH_2 peaks were broadened and shifted to higher wavelengths (1655 cm^{-1} to 1667 cm^{-1}). The broad stretching band of O–H was due to water traces in the CIP-AuNP and AuNP samples. The weak bands around 2900 cm^{-1} in CIP-AuNPs are explained by traces of citrate.

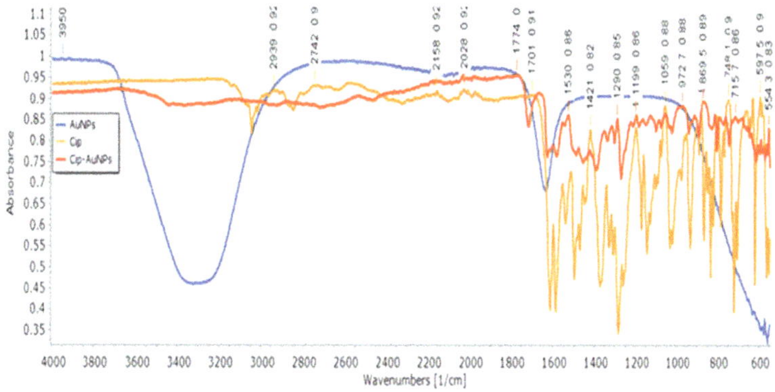

Figure 4. FTIR spectra of CIP-AuNPs, AuNPs, and free CIP.

3.6. Kinetics of CIP Drug Release from AuNPs

The drug release curve is represented by a normal biphasic drug release pattern, which can be split into two stages (Figure 5). In the initial stage (0–5 h), a rapid release of CIP occurred from all the formulations. This confirmed the adsorption of CIP on the surface of AuNPs, in which the CIP-AuNPs displayed a high surface-to-volume ratio due to their small size, which promoted burst release [34]. The second stage (5–24 h) was when the drug was released in a prolonged fashion. The T_{60} of CIP release from the AuNPs loaded with 2 mM of CIP was obtained at 27 h. The T_{60} for 0.5 mM CIP, 1.5 mM CIP, or 2.5 mM CIP was found to be around 70 at 38 h, 29 h, and 24.5 h, respectively, upon extrapolating the drug release. These results indicated that drug release was seed size-dependent i.e., the smaller the nanoparticles, the slower the drug release.

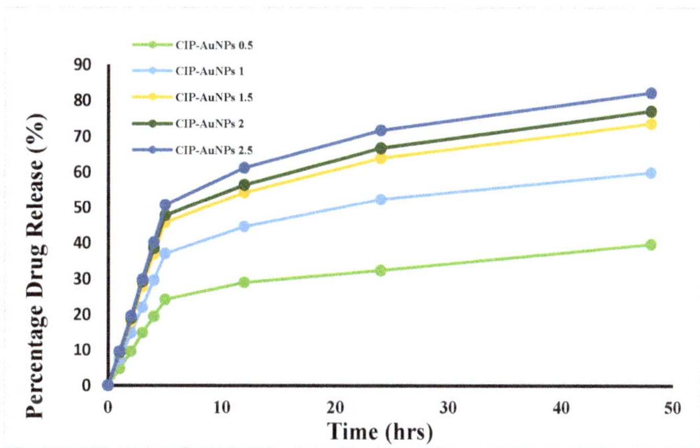

Figure 5. In vitro CIP release from AuNPs loaded with various CIP concentrations.

A first-order release was observed for CIP release from the CIP-AuNPs, which demonstrated that CIP release from CIP-AuNPs (2 mM) was concentration-dependent (Supplementary Materials Figures S2–S6). Overall, the release studies demonstrated that CIP-AuNPs loaded with >1 mM CIP are suitable nanoformulations for CIP administration in a sustained manner. The initial burst release will act as a loading dose to suppress the disease spread, while the sustained release phase will contribute towards a better therapeutic effect.

3.7. Stability Tests on CIP-AuNPs

The CIP-AuNPs with targeted drug concentration (2 mM) showed stability at room temperature and in a basic medium. They were unstable at different salt concentrations. The color change from blue to dark blue was observed with an increase in temperature for the CIP-AuNPs, and 2.0 and 2.5 mM of CIP (Supplementary Materials Figures S7–S11), which is attributed to the increase in particle size (PS) due to aggregation. CIP-AuNPs (2 mM) were stable at room temperature (25 °C), and the A_{max} increases on the order of 25 °C < 50 °C < 75 °C < 100 °C. With different pH values (Supplementary Materials Figures S12–S16) and increasing salt concentrations (Supplementary Materials Figures S17–S21), irregular trends in A_{max} (adsorption peak of CIP, 280 nm) of CIP-AuNPs in all formulations were observed. For further analysis, the CIP-AuNPs (2 mM) formulation was selected.

3.8. In Vitro Antibacterial Activity of CIP-AuNPs at the Optimized Dose

The MICs of free CIP and CIP-AuNPs were tested against *E. faecalis* JH2-2. For comparison of antibacterial activity of CIP and CIP-AuNPs (2 mM), the Minimum Inhibitory Concentration was obtained for our strain. The MIC of CIP and CIP-AuNPs were found to be 2 µg/mL and 1 µg/mL, respectively. The zone of inhibition using 10 µg/disc are presented in Figure 6. As such, a larger zone of inhibition of 23 mm was obtained for the CIP-AuNPs compared with CIP (21 mm), whereas no zone of inhibition was obtained for the bare AuNPs.

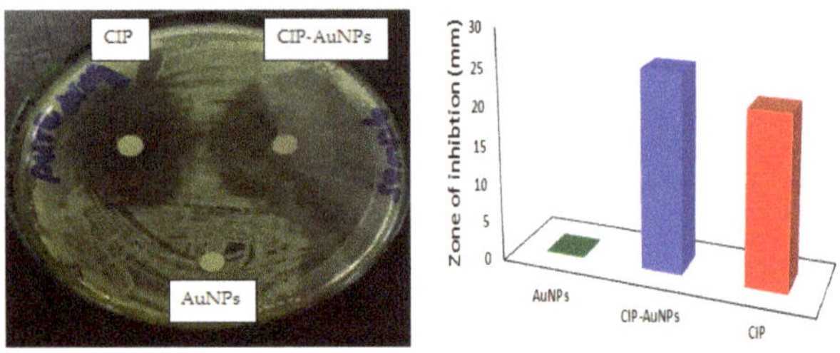

Figure 6. The zones of inhibition (ZI) displayed by CIP, AuNPs, and CIP-AuNPs for *E. faecalis*.

3.9. In Vivo Anticolonizing Potential of CIP-AuNPs in an Animal Model

The infected mouse liver and kidneys appeared to be inflamed and some lighter zones underneath the liver (Figure 7A) were observed. Since the strain was not biofilm-forming, no particular lesions were visualized in the infected liver Figure 7A(a). Mice treated with CIP and CIP-AuNPs had a normal liver appearance Figure 7A(b,c) respectively. The kidneys of infected, CIP treated and CIP-AuNPs treated mice presented no particular change in morphology or texture, Figure 7A(d–f) respectively. The average bacterial count ($\log_{10}$CFU/gm organ) in the infected liver was 39.53± 1.93 (Figure 7B(a)). CIP and CIP-AuNPs treated mice harbored 29.049 ± 1.343 $\log_{10}$CFU/mL and 28.4 ± 0.421 $\log_{10}$CFU/mL (p = 0.428) respectively, in the liver. The bacterial load in the kidneys of infected mice was 13.98 ± 0.232 $\log_{10}$CFU/mL (Figure 7B(b)) while kidneys of CIP and CIP-AuNPs treated mice presented significantly lower bacterial load, 7.91 ± 0.2132.76 ± 0.034 $\log_{10}$CFU/mL (p < 0.0D0076), respectively. The in vivo data revealed that the mice treated with CIP-AuNPs significantly reduced bacterial colonization in the kidneys (Figure 7B,D) compared with those of CIP-treated mice. The CFU for the 10-fold serial dilutions in the infected, CIP treated and CIP AuNPs treated liver are shown in Figure 7C(a–c) respectively. Same dilutions for infected, CIP treated and CIP AuNPs treated kidneys are shown in Figure 7D(a–c) respectively.

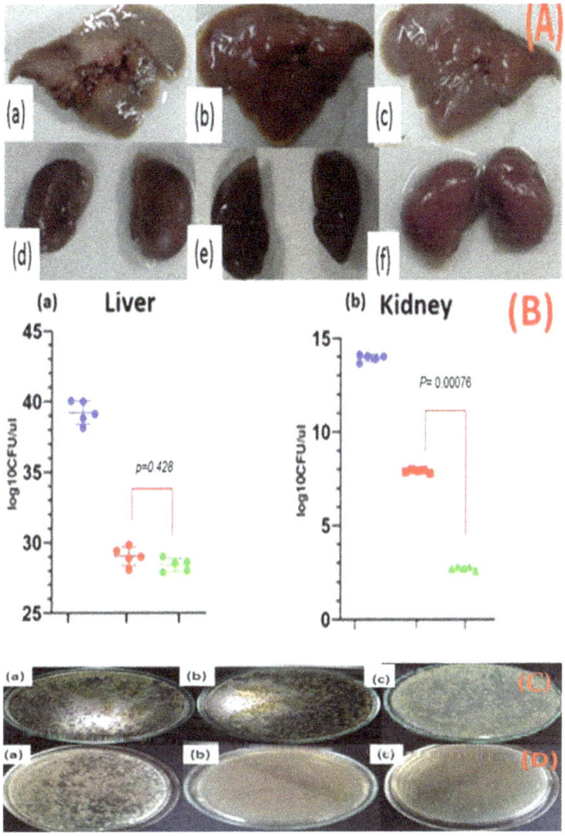

Figure 7. Effect of CIP (10 mg/Kg) and CIP-AuNPs (500 μg/Kg) on the colonization of *E. faecalis* in the (**A**) mouse organs—(**a**) infected liver, (**b**) CIP-treated liver, (**c**) CIP-AuNPs-treated liver, (**d**) infected kidney, (**e**) CIP-treated kidney, (**f**) CIP-AuNPs treated kidneys, infected kidneys, (**B**(**a**,**b**)) CFU *E. faecalis* in the infected liver and kidney; (**C**) CFU counting in the liver—(**a**) infected, (**b**) CIP-treated, and (**c**) CIP-AuNPs-treated; (**D**) CFU counting in the kidneys—(**a**) infected, (**b**) CIP-treated, and (**c**) CIP-AuNPs-treated.

3.10. Hemolytic Activity of CIP-AuNPs

Hemolytic activity results of the CIP, AuNPs, and CIP-AuNPs are presented in Figure 8. According to ISO/TR 7406, the percentage hemolysis considered to be safe is <5%. These results present the greater hemolytic activity of CIP (6.4, 7.2 and 10%) compared with AuNPs and CIP-AuNPs (2 mM). Since the intravenous route gave an escape to the drug from first-pass metabolism, results indicate a hundred percent availability of the drug or nanoparticles in plasma. Considering the average mouse weight was 28 g, the average injected doses of CIP and CIP-AuNPs in the mice were 280 μg and 14 μg, respectively. The results indicated that CIP had hemolytic activity at concentrations of 20 μg/mL and above.

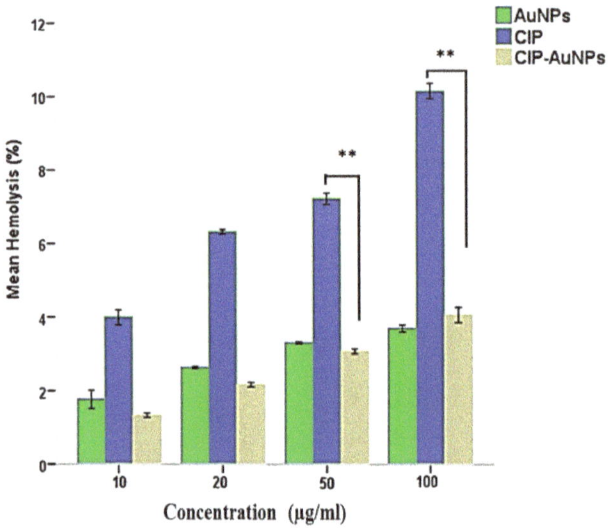

Figure 8. Hemolytic activity of CIP, AuNPs, and CIP-AuNPs, ** presents $p \leq 0.01$.

4. Discussion

Ciprofloxacin is a second-generation fluoroquinolone that is active against many Gram-negative and Gram-positive bacteria. It acts through inhibition of bacterial DNA gyrase and topoisomerase IV. There has been no cross-resistance reported for CIP and other fluoroquinolones; therefore, it is of high clinical value. However, there are certain side effects associated with CIP including low blood sugar, headache, nerve damage causing numbness, tendon rupture, and joint pains. Gold nanoparticles are efficient drug carriers that have the ability to improve the antibacterial effects of loaded antibiotics as well well as to reduce the amount of drug required to be effective because of their retention and penetration into bacterial biofilms and cell membranes at the infected sites.

This study reported the effectiveness of spherical CIP-AuNPs as an antibacterial platform against *E. faecalis* infection in the kidneys and liver of mice. The spherical CIP-AuNPs (Figures 1 and 3) were successfully prepared using a non-simple ionic interaction between CIP and negatively charged AuNPs, which maintained their therapeutic functions without chemical modification [28]. Drug adsorption or loading on the NP was determined by UV–Vis spectroscopy and FTIR. In CIP-AuNPs, the zeta potential values changing from −32.1 mV to −19.7 mV and −13.4 mV indicated the adsorption of CIP onto AuNPs [37]. The negative charge on the AuNPs and the positive charge of the CIP [26] (at 6.5 pH) resulted in electrostatic interactions; hence, enhanced CIP encapsulation efficiency and loading capacity was observed. The addition of different concentrations of the drug to the nanoparticles changed the color from red to purple to bluish purple, and finally, to blue. Increasing the concentration of CIP-AuNPs (>2 mM of CIP concentration) caused the zeta potential values to become less negative (Table 2). The zeta potential values ranging between −5 mV and 5 mV generally demonstrate fast aggregation [38]. The SPR spectra of CIP-AuNPs—2 mM (Figure 1) indicated an increase in the peak intensity of the AuNPs upon the addition of CIP. Moreover, there was a slight shift in the CIP peaks. These changes indicated loading of CIP onto the AuNPs. Considering the standard limits, the zeta potential values of CIP-AuNPs at a 2.5-mM CIP concentration showed fast aggregation. Optimization of the drug for loading is, thus, an important criterion for electrostatically loaded drugs. The encapsulation efficiency, which actually represents the amount of the drug that gets incorporated into a particle at a provided concentration, may help in this regard. The drug loading values present the percentage of drug in terms of weight of the nanoparticles. Considering this, we selected CIP-AuNPs (at a 2 mM CIP concentration) for the antibacterial and hemolysis studies.

The stability testing of the CIP-AuNPs illustrated a drastic shift in the stability of the CIP-AuNPs at 100 °C, which could be due to an increase in collision frequency between CIP-AuNPs. The activation energy produced from these collisions would be enough for CIP-AuNPs to react, which is fully depicted by the Arrhenius equation (Supplementary materials Figures S7–S12). From the observation of the effect of temperature on different CIP-AuNPs, the 2.5 mM ratio showed a change to dark blue. Thus broadening of the CIP-AuNPs peak at the 500 nm range shows particle aggregation. This is due to the fact that the particles destabilize causing a decrease in extinction peak intensity, resulting in a lower number of stable nanoparticles. In the case of pH, the CIP-AuNPs with 2 mM of CIP showed a peak shift from 537 nm to 521 nm at pH 7 and 10 (Figure S15). This is because of the fact that the CIP acquired a negative charge. This negative charge occurred due to deprotonation of the amine group, which then further led to drug unloading. The effect of salt showed the broadening of CIP-AuNPs peaks, this could be attributed to the fact of a very high molar concentrations of the salt used in this experiment. The submolar salt (100 mM and above) (Supplementary materials S17–S21) led to AuNP aggregation and drug unloading.

In the current study, the in vitro drug release was assessed by various kinetic models consisting of the zero-order, first-order, Higuchi, and Korsmeyer–Peppas models. A first-order release was seen for CIP release from the CIP-AuNPs. This shows that CIP release from CIP-AuNPs was directly dependent on the drug concentration (Supplementary materials Figures S2–S6). Overall, the release studies demonstrated the dependence of CIP-AuNPs on particle size. Smaller particles release drugs in a very slow fashion. The most suitable drug release kinetics was obtained by CIP-AuNPs (2 mM).

A higher level of antimicrobial activity is afforded by the CIP-AuNPs. This is due to the large surface-to-volume ratio of AuNPs, which facilitated the adsorption of CIP molecules onto the AuNP surfaces via electrostatic interaction between the amine groups of CIP and AuNPs. The MIC of the CIP-AuNPs was half to that of CIP. Considering that the overall particle seed was gold, the drug amount administered was much lower; thus, carrier driven therapies require less amount of the drug to achieve the same therapeutic outcome. It is well established that silver nanoparticles prepared via a chemical reduction procedure exhibited antibacterial activity. However, for the AuNPs in this study, no inherent antibacterial activity was obtained. This depicts that the surface-area-to-volume ratio of AuNPs served its purpose as a drug delivery system and supported the CIP approach to the target site, thus improving antibacterial activity. Studies revealed that both positively and negatively charged gold nanoparticles get localized in the bacterial membranes. For entry, the binding of negative charged particles to the bacteria's lipid bilayer results in local gelation. In contrast, the positively charged NPs get along the fluidity in the gelled regions of the bilayers [39–41]. Once CIP-AuNPs have reached the bacterial surface, they might have entered or fused with the cell membrane of *E. faecalis* and served as a reserve for the continuous release of CIP, which must have then diffused into the interior of the *E. faecalis*. It has also been reported that while acquiring resistance, *E. faecalis* has cationic membrane proteins on their surface, which alter the membrane fluidity and potential. At moderate levels of *E. faecalis* resistance, the cell surface becomes neutralized, while at further higher resistance levels, the cell surface acquires a positive charge. This positive charge on the membrane may further permit CIP-AuNPs to interact and enter the bacterial cell [39–41].

The in vivo antibacterial activity showed significant reduction in CFU by CIP-AuNPs compared with CIP. However, in the case of the liver, the difference was not significant but comparable to CIP; this could be attributed to the fact that the *E. faecalis* presented a very high hepatic load. Moreover, since the nanoparticles did not have a polymer coating such as polyethylene glycol, many of them can be uptaken by the reticuloendothelial system and could have been taken for detoxification. In the case of the kidneys, the nanoparticles may have stayed or been entrapped at the glomerular filtration barrier (GFB), which is present within the glomerulus and sandwiched between the glomerular capillary and Bowman's space. Different layers of the GFB have pore sizes between 8–100 nm and, thus, act as a sieve for the nanoparticles. The nanoparticles, especially the anionic nanoparticles, have higher retention time in the kidneys and, in general, the nanoparticles disintegrate before passing through the GFB. The accumulation of nanoparticles with this size range could be the reason behind the efficiency of the significant decrease in bacterial count in the mice treated with CIP-AuNPs over mice treated with CIP.

Ciprofloxacin has also been associated with hemolytic anemia [42]. In addition to improved antibacterial activity, this study (Figure 7) also reported limited hemolytic activity of the CIP-AuNPs (2 mM) on human red blood cells (RBCs). These results presented the greater hemolytic activity of CIP (6.4, 7.2, and 10.6%) as compared to AuNPs and CIP-AuNPs. It has been reported that a positive charge on the particle's surface possesses a membrane-destabilizing and concomitantly destructive effect that results from the interaction of a positive charge and a negative charge of the cell membrane. Therefore, the limited hemolytic activity of the CIP-AuNPs (with a 2-mM CIP concentration) is attributable to the negative charge on its surface and the lack of free CIP in the medium. Considering the average mouse weight as 28 g, the average injected doses of CIP and CIP-AuNPs in the mice were 280 μg and 14 μg, respectively. The results indicated CIP to be hemolytic at a concentration of 20 μg/mL and above (Figure 8). Due to this, the mice experienced hyperactivity and irritation at the site of injection soon after the administration of CIP, and decreased eating after the second day of treatment.

Antibiotics, such as CIP, are an attraction because of their broad spectrum and lack of resistance development. *E. faecalis* has resistance acquisition capacity and this challenge can be sealed with CIP. Despite of the enormous work conducted on the development of CIP nanocarriers, there is a lack of studies investigating the effects of these nanoparticles in animal models. These studies are extremely important as bacteria residing in the tissue and forming biofilms might not be easily accessed by circulating nanoparticles and therefore they can obtain a colonization advantage. This in vivo study concluded showing efficiency of the CIP-loaded gold nanoparticles against intra-abdominal and urinary tract infections caused by *E. faecalis*. Further investigations for the biocompatibility and biofilm formation in the host tissues are necessary and will be vital for the further assessment of these novel nanoparticles.

Supplementary Materials: The following are available online at https://www.mdpi.com/article/10.3390/nano11113152/s1. Figure S1: EDX analysis of the CIP-AuNPs; Figure S2: Kinetic analysis of the drug release for CIP-AuNPs (0.5 mM CIP)—(A) zero-order kinetics, (B) Korsmeyer–Peppas plot, (C) first-order kinetics, and (D) Higuchi kinetics; Figure S3: Kinetic analysis of the drug release for CIP-AuNPs (1 mM CIP)—(A) zero-order kinetics, (B) Korsmeyer–Peppas plot, (C) first-order kinetics, and (D) Higuchi kinetics; Figure S4: Kinetic analysis of the drug release for CIP-AuNPs (1.5 mM CIP)—(A) zero-order kinetics, (B) Korsmeyer–Peppas plot, (C) first-order kinetics, and (D) Higuchi kinetics; Figure S5: Kinetic analysis of the drug release for CIP-AuNPs (2 mM CIP)—(A) zero-order kinetics, (B) Korsmeyer–Peppas plot, (C) first-order kinetics, and (D) Higuchi kinetics; Figure S6: Kinetic analysis of the drug release for CIP-AuNPs (2.5 mM CIP)—(A) zero-order kinetics, (B) Korsmeyer–Peppas plot, (C) first-order kinetics, and (D) Higuchi kinetics; Figure S7: Temperature effect on CIP-AuNPs formulated with 0.5 mM CIP; Figure S8: Temperature effect on CIP-AuNPs formulated with 1 mM CIP; Figure S9: Temperature effect on CIP-AuNPs formulated with 1.5 mM CIP; Figure S10. Temperature effect on CIP-AuNPs formulated with a 2-mM CIP concentration; Figure S11: Temperature effect on CIP-AuNPs formulated with 2.5 mM CIP; Figure S12: pH effect on CIP-AuNPs formulated with 0.5 mM CIP; Figure S13: pH effect on CIP-AuNPs formulated with 1 mM CIP; Figure S14: pH effect on CIP-AuNPs formulated with 1.5 mM CIP; Figure S15: pH effect on CIP-AuNPs formulated with 2 mM CIP. Figure S16: pH effect on CIP-AuNPs formulated with 2.5 mM CIP; Figure S17: Effect of salt concentrations on CIP-AuNPs formulated with 0.5 mM CIP; Figure S18: Effect of salt concentrations on CIP-AuNPs formulated with 1 mM CIP; Figure S19: Effect of salt concentrations on CIP-AuNPs formulated with 1.5 mM CIP; Figure S20: Effect of salt concentrations on CIP-AuNPs formulated with 2 mM CIP; Figure S21: Effect of salt concentrations on CIP-AuNPs formulated with 2.5 mM CIP; Table S1. Spectral assignments for AuNPs, CIP-AuNPs, and free CIP.

Author Contributions: Conceptualization, A.N., S.M.A., S.R. and N.F.R.; methodology, A.N., S.M.A., T.T., N.F.R., A.B., T.J.W. and F.M.; software, S.M.A. and A.N.; validation, A.N., S.M.A., T.T., N.F.R., A.B., T.J.W., T.M. and F.M.; formal analysis, A.N., S.M.A., T.T., N.F.R., A.B., T.J.W. and F.M.; investigation, A.N., S.M.A., T.T. and A.B.; resources, N.F.R.; data curation, A.N., S.M.A., T.T., N.F.R., A.B., T.J.W., M.F., F.B. and F.M.; writing—original draft preparation, writing—review and editing, A.N., S.M.A., T.T., F.M., A.B., Z.R., I.S., A.S. and A.I.; visualization, A.N., S.M.A., T.T. and A.B.; supervision, N.F.R.; project administration, A.N., S.M.A., T.T. and A.B.; funding acquisition, N.F.R. All authors have read and agreed to the published version of the manuscript.

Funding: This research received funding from the "National Research Program for Universities (NRPU), Higher Education Commission (HEC)", Pakistan; grant number 6035.

Data Availability Statement: The data presented in this study are available on requests from corresponding author.

Acknowledgments: We are indebted to Higher Education Commission (HEC) of Pakistan, for the financial support to the project. We would like to express our gratitude to Irshad Hussain, (Lahore University of Management Sciences) Lahore, Pakistan for his words of wisdom and technical support.

Conflicts of Interest: The authors declare no conflict of interest.

References

1. Wu, D.; Fan, W.; Kishen, A.; Gutmann, J.L.; Fan, B. Evaluation of the Antibacterial Efficacy of Silver Nanoparticles against *Enterococcus faecalis* Biofilm. *J. Endod.* **2014**, *40*, 285–290. [CrossRef]
2. Sedgley, C.; Buck, G.; Appelbe, O. Prevalence of *Enterococcus faecalis* at Multiple Oral Sites in Endodontic Patients Using Culture and PCR. *J. Endod.* **2006**, *32*, 104–109. [CrossRef]
3. Saleh, I.M.; Ruyter, I.E.; Haapasalo, M.; Ørstavik, D. Survival of *Enterococcus faecalis* in infected dentinal tu-bules after root canal filling with different root canal sealers in vitro. *Int. Endod. J.* **2004**, *37*, 193–198. [CrossRef]
4. Liu, H.; Wei, X.; Ling, J.; Wang, W.; Huang, X. Biofilm Formation Capability of *Enterococcus faecalis* Cells in Starvation Phase and Its Susceptibility to Sodium Hypochlorite. *J. Endod.* **2010**, *36*, 630–635. [CrossRef] [PubMed]
5. Gilmore, M.S.; Clewell, D.B., Ike, Y.; Shankar, N. *Enterococci: From Commensals to Leading Causes of Drug Resistant Infection*; Eye and Ear Infirmary: Boston, MA, USA, 2014.
6. Livermore, D.M.; Blaser, M.; Carrs, O.; Cassell, G.; Fishman, N.; Guidos, R.; Levy, S.; Powers, J.; Norrby, R.; Tillotson, G.; et al. Discovery research: The scientific challenge of finding new antibiotics. *J. Antimicrob. Chemother.* **2011**, *66*, 1941–1944. [CrossRef] [PubMed]
7. Pinto, R.; Lopes, D.; Martins, M.C.L.; Van Dijck, P.; Nunes, C.; Reis, S. Impact of nanosystems in Staphylococcus aureus biofilms treatment. *FEMS Microbiol. Rev.* **2019**, *43*, 622–641. [CrossRef] [PubMed]
8. Vassallo, A.; Silletti, M.F.; Faraone, I.; Milella, L. Nanoparticulate Antibiotic Systems as Antibacterial Agents and Antibiotic Delivery Platforms to Fight Infections. *J. Nanomater.* **2020**, *2020*. [CrossRef]
9. Pugazhendhi, A.; Kumar, S.S.; Manikandan, M.; Saravanan, M. Photocatalytic properties and antimicrobial ef-ficacy of Fe doped CuO nanoparticles against the pathogenic bacteria and fungi. *Microb. Pathog.* **2018**, *122*, 84–89. [CrossRef]
10. Uzair, B.; Liaqat, A.; Iqbal, H.; Menaa, B.; Razzaq, A.; Thiripuranathar, G.; Rana, N.F.; Menaa, F. Green and Cost-Effective Synthesis of Metallic Nanoparticles by Algae: Safe Methods for Translational Medicine. *Bioengineering* **2020**, *7*, 129. [CrossRef]
11. Wang, L.; Hu, C.; Shao, L. The antimicrobial activity of nanoparticles: Present situation and prospects for the future. *Int. J. Nanomed.* **2017**, *12*, 1227–1249. [CrossRef]
12. Dykman, L.; Khlebtsov, N. Gold nanoparticles in biomedical applications: Recent advances and perspectives. *Chem. Soc. Rev.* **2012**, *41*, 2256–2282. [CrossRef] [PubMed]
13. Abdel-Raouf, N.; Al-Enazi, N.M.; Ibraheem, I.B. Green biosynthesis of gold nanoparticles using Galaxaura elongata and characterization of their antibacterial activity. *Arab. J. Chem.* **2017**, *10*, S3029–S3039. [CrossRef]
14. Li, W.; Geng, X.; Liu, D.; Li, Z. Near-infrared light-enhanced protease-conjugated gold nanorods as a photo-thermal antimicrobial agent for elimination of exotoxin and biofilms. *International journal of nanomedicine. Int. J. Nanomed.* **2019**, *14*, 8047. [CrossRef] [PubMed]
15. Rhazouani, A.; Gamrani, H.; El Achaby, M.; Aziz, K.; Gebrati, L.; Uddin, S.; Aziz, F. Synthesis and Toxicity of Graphene Oxide Nanoparticles: A Literature Review of In vitro and In vivo Studies. *BioMed Res. Int.* **2021**, *2021*. [CrossRef] [PubMed]
16. Kaushik, S.N.; Scoffield, J.; Andukuri, A.; Alexander, G.C.; Walker, T.; Kim, S.; Choi, S.C.; Brott, B.C.; Eleazer, P.D.; Lee, J.-Y.; et al. Evaluation of ciprofloxacin and metronidazole encapsulated biomimetic nanomatrix gel on *Enterococcus faecalis* and Treponema denticola. *Biomater. Res.* **2015**, *19*, 9. [CrossRef]

17. Azad, A.; Rostamifar, S.; Modaresi, F.; Bazrafkan, A.; Rezaie, Z. Assessment of the Antibacterial Effects of Bismuth Nanoparticles against *Enterococcus faecalis*. *BioMed Res. Int.* **2020**, *2020*. [CrossRef]
18. Parolia, A.; Kumar, H.; Ramamurthy, S.; Davamani, F.; Pau, A. Effectiveness of chitosan-propolis nanoparticle against *Enterococcus faecalis* biofilms in the root canal. *BMC Oral Heal.* **2020**, *20*, 339. [CrossRef]
19. Arias-Moliz, M.T.; Baca, P.; Solana, C.; Toledano, M.; Medina-Castillo, A.L.; Toledano-Osorio, M.; Osorio, R. Doxycycline-functionalized polymeric nanoparticles inhibit *Enterococcus faecalis* biofilm formation on dentine. *Int. Endod. J.* **2021**, *54*, 413–426. [CrossRef]
20. Maliszewska, I.; Wróbel, J.; Wanarska, E.; Podhorodecki, A.; Matczyszyn, K. Synergistic effect of methylene blue and biogenic gold nanoparticles against *Enterococcus faecalis*. *Photodiagnosis Photodyn. Ther.* **2019**, *27*, 218–226. [CrossRef]
21. Balto, H.; Bukhary, S.; Al-Omran, O.; BaHammam, A.; Al-Mutairi, B. Combined Effect of a Mixture of Silver Nanoparticles and Calcium Hydroxide against *Enterococcus faecalis* Biofilm. *J. Endod.* **2020**, *46*, 1689–1694. [CrossRef]
22. Halkai, K.; Mudda, J.A.; Shivanna, V.; Rathod, V.; Halkai, R. Antibacterial efficacy of biosynthesized silver nanoparticles against *Enterococcus faecalis* Biofilm: An in vitro study. *Contemp. Clin. Dent.* **2018**, *9*, 237–241. [CrossRef] [PubMed]
23. Wu, S.; Liu, Y.; Zhang, H.; Lei, L. Nano-graphene oxide with antisense walR RNA inhibits the pathogenicity of *Enterococcus faecalis* in periapical periodontitis. *J. Dent. Sci.* **2019**, *15*, 65–74. [CrossRef] [PubMed]
24. Martini, C.; Longo, F.; Castagnola, R.; Marigo, L.; Grande, N.M.; Cordaro, M.; Cacaci, M.; Papi, M.; Palmieri, V.; Bugli, F.; et al. Antimicrobial and Antibiofilm Properties of Graphene Oxide on *Enterococcus faecalis*. *Antibiotics* **2020**, *9*, 692. [CrossRef] [PubMed]
25. Herizchi, R.; Abbasi, E.; Milani, M.; Akbarzadeh, A. Current methods for synthesis of gold nanoparticles. *Artif. Cells Nanomed. Biotechnol.* **2016**, *44*, 596–602. [CrossRef]
26. Tom, R.T.; Suryanarayanan, V.; Reddy, P.G.; Baskaran, S.; Pradeep, T. Ciprofloxacin-protected gold nanopar-ticles. *Langmuir* **2004**, *20*, 1909–1914. [CrossRef]
27. Topal, G.R.; Devrim, B.; Eryilmaz, M.; Bozkir, A. Design of ciprofloxacin-loaded nano-and micro-composite particles for dry powder inhaler formulations: Preparation, in vitro characterisation, and antimicro-bial efficacy. *J. Microencapsul.* **2018**, *35*, 533–547. [CrossRef]
28. Arafa, M.G.; Mousa, H.A.; Afifi, N.N. Preparation of PLGA-chitosan based nanocarriers for enhancing anti-bacterial effect of ciprofloxacin in root canal infection. *Drug Deliv.* **2020**, *27*, 26–39. [CrossRef]
29. Albuquerque, M.T.; Valera, M.C.; Moreira, C.S.; Bresciani, E.; de Melo, R.M.; Bottino, M.C. Effects of ciprof-loxacin-containing scaffolds on *Enterococcus faecalis* biofilms. *J. Endod.* **2015**, *41*, 710–714. [CrossRef]
30. Assali, M.; Zaid, A.N.; Abdallah, F.; Almasri, M.; Khayyat, R. Single-walled carbon nanotubes-ciprofloxacin nanoantibiotic: Strategy to improve ciprofloxacin antibacterial activity. *Int. J. Nanomed.* **2017**, *12*, 6647–6659. [CrossRef]
31. Rana, N.F.; Sauvageot, N.; Laplace, J.-M.; Bao, Y.; Nes, I.; Rincé, A.; Posteraro, B.; Sanguinetti, M.; Hartke, A. Redox Balance via Lactate Dehydrogenase Is Important for Multiple Stress Resistance and Virulence in *Enterococcus faecalis*. *Infect. Immun.* **2013**, *81*, 2662–2668. [CrossRef]
32. Dash, S.; Murthy, P.N.; Nath, L.; Chowdhury, P. Kinetic modeling on drug release from controlled drug delivery systems. *Acta Pol. Pharm. Drug Res.* **2010**, *67*, 217–223.
33. Bettini, R.; Catellani, P.L.; Santi, P.; Massimo, G.; Peppas, N.A.; Colombo, P. Translocation of drug particles in HPMC matrix gel layer: Effect of drug solubility and influence on release rate. *J. Control. Release* **2001**, *70*, 383–391. [CrossRef]
34. Naz, S.S.; Islam, N.U.; Shah, M.R.; Alam, S.S.; Iqbal, Z.; Bertino, M.; Franzel, L.; Ahmed, A. Enhanced biocidal activity of Au nanoparticles synthesized in one pot using 2, 4-dihydroxybenzene carbodithioic acid as a reducing and stabilizing agent. *J. Nanobiotechnol.* **2013**, *11*, 13. [CrossRef] [PubMed]
35. Muhammad, Z.; Raza, A.; Ghafoor, S.; Naeem, A.; Naz, S.S.; Riaz, S.; Ahmed, W.; Rana, N.F. PEG capped methotrexate silver nanoparticles for efficient anticancer activity and biocompatibility. *Eur. J. Pharm. Sci.* **2016**, *91*, 251–255. [CrossRef]
36. Riaz, S.; Rana, N.F.; Hussain, I.; Tanweer, T.; Nawaz, A.; Menaa, F.; Janjua, H.A.; Alam, T.; Batool, A.; Naeem, A.; et al. Effect of Flavonoid-Coated Gold Nanoparticles on Bacterial Colonization in Mice Organs. *Nanomaterials* **2020**, *10*, 1769. [CrossRef]
37. Bhalerao, S.R.; Rote, A.R. Application of UV spectrophotometric methods for estimation of Ciprofloxacin and tinidazole in combined tablet dosage form. *Int. J. Pharmacol. Pharm. Sci.* **2012**, *4*, 4–7.
38. Bhattacharjee, S. DLS and zeta potential—What they are and what they are not? *J Control. Release* **2016**, *235*, 337–351. [CrossRef]
39. Canepa, E.; Salassi, S.; Simonelli, F.; Ferrando, R.; Rolandi, R.; Lambruschini, C.; Canepa, F.; Dante, S.; Relini, A.; Rossi, G. Non-disruptive uptake of anionic and cationic gold nanoparticles in neutral zwitterionic membranes. *Sci. Rep.* **2021**, *11*, 1–11. [CrossRef]
40. Okoampah, E.; Mao, Y.; Yang, S.; Sun, S.; Zhou, C. Gold nanoparticles-biomembrane interactions: From fundamental to simulation. *Colloids Surf. B Biointerfaces* **2020**, *196*, 111312. [CrossRef]
41. Dreaden, E.C.; Austin, L.; Mackey, M.A.; El-Sayed, M.A. Size matters: Gold nanoparticles in targeted cancer drug delivery. *Ther. Deliv.* **2012**, *3*, 457–478. [CrossRef]
42. Lim, S.; Alam, M.G. Ciprofloxacin-Induced Acute Interstitial Nephritis and Autoimmune Hemolytic Anemia. *Ren. Fail.* **2003**, *25*, 647–651. [CrossRef] [PubMed]

nanomaterials MDPI

Article

Local and Systemic In Vivo Responses to Osseointegrative Titanium Nanotube Surfaces

Erin A. Baker [1,2,3,*], Mackenzie M. Fleischer [1], Alexander D. Vara [1], Meagan R. Salisbury [1], Kevin C. Baker [1,3], Paul T. Fortin [1,3] and Craig R. Friedrich [2]

1 Departments of Orthopaedic Research and Surgery, Beaumont Health, Royal Oak, MI 48073, USA; mackenzie.fleischer@beaumont.org (M.M.F.); alexdvara@gmail.com (A.D.V.); meagan.salisbury@beaumont.org (M.R.S.); kevin.baker@beaumont.org (K.C.B.); paul.fortin@beaumont.org (P.T.F.)

2 Department of Mechanical Engineering-Engineering Mechanics, Michigan Technological University, Houghton, MI 49931, USA; craig@mtu.edu

3 Department of Orthopaedic Surgery, Oakland University William Beaumont School of Medicine, Rochester, MI 48309, USA

* Correspondence: erin.baker@beaumont.org

Abstract: Orthopedic implants requiring osseointegration are often surface modified; however, implants may shed these coatings and generate wear debris leading to complications. Titanium nanotubes (TiNT), a new surface treatment, may promote osseointegration. In this study, in vitro (rat marrow-derived bone marrow cell attachment and morphology) and in vivo (rat model of intramedullary fixation) experiments characterized local and systemic responses of two TiNT surface morphologies, aligned and trabecular, via animal and remote organ weight, metal ion, hematologic, and nondecalcified histologic analyses. In vitro experiments showed total adherent cells on trabecular and aligned TiNT surfaces were greater than control at 30 min and 4 h, and cells were smaller in diameter and more eccentric. Control animals gained more weight, on average; however, no animals met the institutional trigger for weight loss. No hematologic parameters (complete blood count with differential) were significantly different for TiNT groups vs. control. Inductively coupled plasma mass spectrometry (ICP-MS) showed greater aluminum levels in the lungs of the trabecular TiNT group than in those of the controls. Histologic analysis demonstrated no inflammatory infiltrate, cytotoxic, or necrotic conditions in proximity of K-wires. There were significantly fewer eosinophils/basophils and neutrophils in the distal region of trabecular TiNT-implanted femora; and, in the midshaft of aligned TiNT-implanted femora, there were significantly fewer foreign body giant/multinucleated cells and neutrophils, indicating a decreased immune response in aligned TiNT-implanted femora compared to controls.

Keywords: orthopedic; nanomedicine; nanomodified surfaces; animal model; immune response

Citation: Baker, E.A.; Fleischer, M.M.; Vara, A.D.; Salisbury, M.R.; Baker, K.C.; Fortin, P.T.; Friedrich, C.R. Local and Systemic In Vivo Responses to Osseointegrative Titanium Nanotube Surfaces. *Nanomaterials* **2021**, *11*, 583. https://doi.org/10.3390/nano11030583

Academic Editor: Angelo Ferraro

Received: 9 January 2021
Accepted: 15 February 2021
Published: 26 February 2021

1. Introduction

Titanium, both commercially pure and alloyed, has been used for decades in various biologic environments, including orthopedic implant designs [1–4]. With a combination of corrosion resistance, biocompatibility, mechanical properties approximating bone, and low cost, titanium continues to be a common material for fracture plates and screws as well as components requiring solid bone–implant fixation (e.g., knee arthroplasty tibial tray, hip arthroplasty femoral stem) [1–4]. To promote osseointegration, macroscale coatings have been applied to titanium implant surfaces via titanium plasma spray (TPS) as well as hydroxyapatite (HA) coating or sintering for powder- or bead-based coatings [5–8]. These coatings, however, are subjected to shear loads during surgical implantation, contact with surgical tools, and eventually micromotion at the bone–implant interface in vivo [5–8]. These coatings may then separate from the substrate, generating third-body wear debris that increases mechanical wear of bearing surfaces. Additionally, local phagocytic cells encountering this debris may initiate a biologic cascade leading to periprosthetic osteolysis [9]. The body's immune response to wear debris is dependent on particulate composition, concentration, and morphology and may result in periprosthetic joint infection, bone fracture, catastrophic implant fracture, as well as osteoclastic bone resorption around the implant, component loosening, and, ultimately, revision surgery [5,10,11].

Titanium nanotube (TiNT) surfaces, which are electrochemically etched from the titanium implant substrate instead of applied as a coating, are an emerging technology that may enhance osseointegration of orthopedic implants [12–17], although literature is scarce regarding the in vivo immune response and toxicity to these nanostructured materials as implantable devices [18,19]. Due to improved interfacial strength, decreased thickness, and potential for more rapid osseointegration, TiNT surfaces may generate less third-body wear particulate compared to current coating technologies; however, thorough characterization of the immune and inflammatory response to new implantable materials is needed.

Two TiNT surface morphologies, termed aligned and trabecular, have been developed for orthopedic applications necessitating osseointegrative properties [17,20–22]. The aligned TiNT morphology comprises arrays of vertically oriented nanotubes, while the trabecular TiNT morphology exhibits a random surface resembling the three-dimensional porous network of trabecular bone. In this study, local and systemic responses to aligned and trabecular TiNT surfaces were assessed through in vitro cellular response on each material as well as in vivo performance in a clinically relevant, rodent model of long-term femoral intramedullary implant to simulate joint arthroplasty, including assessment of longitudinal animal weights, remote organ weights, metal ion levels in remote organs and whole blood, hematology, and nondecalcified histology [23,24]. Following on survivability testing and finite element modeling indicating that compression and shear loading to the TiNT surfaces were below yield failure strength, this study hypothesizes that TiNT surfaces will demonstrate equivalent local and systemic responses, equivalent to unmodified titanium alloy substrate surfaces [22]. Further, we hypothesize that aligned and trabecular TiNT surfaces yield comparable in vitro and in vivo results, as both morphologies provide additional surface area for cell attachment as well as resultant bony ongrowth and ingrowth.

2. Materials and Methods

2.1. Implant Fabrication

Samples for in vitro and in vivo studies were fabricated with titanium alloy (Ti-6Al-4V ELI) sheet and wire (wire: Custom Wire Technologies, Inc., Port Washington, WI, USA), respectively, using a well-established electrochemical anodization process (Appendix A—Supplemental Methods, Figure A1) [17,21]. Following the initial etching process, aligned TiNT surfaces were then ultrasonicated in deionized (DI) water for 2 minutes; however, trabecular TiNT surfaces received no additional treatment (Figure 1). Control samples were rinsed with DI water and air-dried. On average, inside diameter (ID) of individual nanotubes was 60 nm and 1 μm length on aligned TiNT surfaces. Similarly, trabecular TiNT surfaces contained a 1 μm layer of 60 nm ID aligned nanotubes, on average, covered with an additional 1 μm of over-etched nanotubes. The over-etched layer mimics the porous structure and topography of trabecular bone.

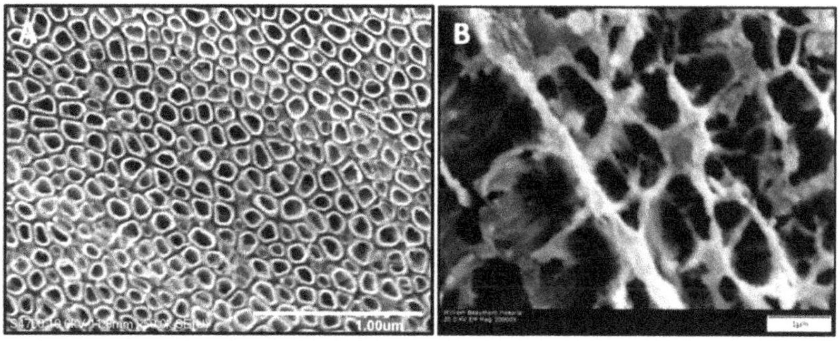

Figure 1. Aligned titanium nanotubes (TiNT) (**A**) and trabecular TiNT (**B**) surfaces. (**A**,**B**) Scale Bar: 1 μm.

2.2. *In Vitro Experimentation to Assess Cell Attachment and Morphology*

Cells for all in vitro experiments were isolated from the intramedullary cavities of the femora of 14-week-old female Sprague Dawley rats (SD; Charles River Laboratories, Wilmington, MA, USA). Following isolation and culture using previously described methods, the plastic-adherent fraction of bone marrow cells (BMC) was collected and used for all subsequent experiments (Appendix A—Supplemental Methods, Figure A2) [17]. This cell population was selected for experimentation due to the potential of mesenchymal stem cells (MSC) to differentiate toward numerous cell types, including osteoblasts.

Early cell attachment and morphology were compared between aligned TiNT, trabecular TiNT, and control surfaces, using three samples per group per timepoint (coupons sectioned from sheet: 10 mm × 10 mm; timepoints: 30 min, 2 h, 4 h) [15,25]. To increase attachment potential, all sample coupons were soaked in fetal bovine serum (FBS) for 30 min prior to drop-seeding (density: 40,000 BMCs/coupon), followed by a 6 h incubation before adding the remaining volume of media and incubating overnight. At each timepoint, cells were fixed and stained with Actin Green (Actin Green 488 ReadyProbes Reagent, Life Technologies, Carlsbad, CA, USA) for cytoskeleton visualization and morphology (1 drop stain; 40 min incubation) as well as 4′, 6-Diamidino-2-Phenylindole, dihydrochloride (DAPI; Life Technologies, Carlsbad, CA, USA) for nucleus visualization (0.5 mL stain; 20 min incubation). Following cell fixation, fluorescence imaging (IX71, Olympus America, Center Valley, PA, USA) was performed at thirteen standardized regions of interest to quantify the total number of adherent cells, cell equivalent diameter, and cell eccentricity. Because all non-adherent cells were removed prior to imaging, Actin Green and DAPI also demonstrated cell adhesion to the surfaces. Environmental scanning electron microscopy (SEM; Vega3XMU, Tescan USA, Warrendale, PA, USA) was used to further document cell morphology; specifically, cells were fixed in 4% glutaraldehyde on coupons, and then coupon was sputter-coated with a thin layer of gold–palladium alloy. Samples were imaged at 20.0 kV and a magnification of 300×. Total cell number, cell equivalent diameter, and cell eccentricity (measured using onboard microscope measurement tool) were statistically compared using a one-way analysis of variance (ANOVA) model, with a Tukey post-hoc test and $\alpha = 0.05$.

2.3. *In Vivo Experimentation to Assess Biologic Response to Nanotube Surfaces*

After preparing titanium alloy Kirschner wires (K-wire; Custom Wire Technologies, Inc., Port Washington, WI; USA, material: Ti6Al4V ELI Hard, diameter: 1.25 mm, single trocar tip for insertion), aligned TiNT-etched, trabecular TiNT-etched, or unetched titanium (control) K-wire implants (n = 6 per group) were inserted retrograde into the femoral intramedullary canals of SD rats for a single, long-term endpoint of 12 weeks [24,26,27]. Three naïve/nonoperative animals were housed for the same duration, in order to establish baseline characteristics. Treatment groups were randomized just prior to surgery by nonoperative staff. Rats received daily veterinary care to identify complications and were weighed weekly; at endpoint, hematologic, metal ion, and histologic analyses were performed.

2.4. Surgical Procedure for Unilateral Intramedullary Implantation

Under an Institutional Animal Care and Use Committee (IACUC)-approved protocol, 14-week-old, female SD rats underwent unilateral femoral implants via retrograde insertion. After anesthetization, rats were placed supine on a sterile, heated operating table with the knee in maximum flexion. Following final sterile preparation, a lateral–patellar skin incision allowed access to the distal femur. After drilling a shallow, pilot hole in the intercondylar groove, a K-wire was inserted into the intramedullary canal to the greater trochanter (proximal femur). Placement of the implant was confirmed via fluoroscopy. The K-wire was then clipped and recessed at the insertion site, followed by closure of arthrotomy and skin incisions with appropriate suture materials. Animals were then recovered and allowed ad libitum activity under endpoint.

2.5. Postoperative Care and Health Assessments

All rats were evaluated daily for clinical signs of complications and overall health, including pain level, activity level, and food/water consumption. Rats were weighed preoperatively and weekly throughout the experiment. At endpoint, remote organs were collected and weighed, as another assessment of animal health. A one-way ANOVA model was used to compare longitudinal animal weights and endpoint organ weights between treatment groups, with $\alpha = 0.05$.

2.6. Hematologic Analysis

At endpoint, each anesthetized rat underwent antemortem cardiac puncture to collect blood for hematologic analyses (CBC; complete blood count with differential) to quantify the following thirteen parameters associated with either systemic inflammation/infection or anemia status: white blood cell count (WBC), lymphocyte concentration (Lymph), monocyte concentration (Mono), granulocyte concentration (Gran), hematocrit (HCT), mean cell volume of red blood cells (MCV), red blood cell distribution width (RDWa), hemoglobin concentration (Hgb), mean cell hemoglobin concentration (MCHC), mean cell hemoglobin (MCH), red blood cell count (RBC), total platelet count (PLT), and mean and platelet volume (MPV) (HemaTrue Hematology Analyzer, Heska, Loveland, CO, USA). Blood was collected in vacutainer tubes with Ethylenediaminetetraacetic acid (EDTA) and immediately processed. A one-way ANOVA model was used to compare hematologic parameters between treatment groups, with $\alpha = 0.05$.

2.7. Metal Ion Analysis of Whole Blood and Remote Organs

Inductively coupled plasma mass spectrometry (ICP-MS; HP 4500, Agilent Technologies, Santa Clara, CA, USA) was used to measure the titanium, aluminum, and vanadium concentrations in each remote organ and a whole blood sample from each animal (Appendix A—Supplemental Methods). ICP-MS calibration was performed using both ASTM standards of each target element as well as non-implanted samples of the rods from each testing condition. At endpoint, remote organs (i.e., spleen, liver, lungs, kidneys, brain) were harvested, weighed, and stored (4 °C in ultra-low leachable sample tubes) until analysis. Samples were analyzed in tandem with a laboratory reagent blank and duplicate laboratory fortified blanks to quantify background metal levels as well as process accuracy and precision, respectively. A one-way analysis of variance model was used to compare metal ion levels between treatment groups, with $\alpha = 0.05$.

2.8. Nondecalcified Histologic Analysis

At study endpoint, implanted femora were also harvested for nondecalcified histologic analysis. Femora from each animal were stripped of soft tissues and 10% zinc-buffered formalin for 96 h, followed by three rinses in phosphate buffered saline, and 70% ethanol storage to prepare for histologic processing. Femora were subsequently embedded in methyl methacrylate, then ground and polished longitudinally to the approximate center of the implant, followed by cutting and applying one 5 μm thick section to a slide. Each slide was stained with Stevenel's Blue and van Gieson picrofuchsin (SBVG) for visualization of presence of immune-related cellular activity. SBVG is a widely-used stain to assess bone formation, due to visualization of both mature bone and osteoid as well as fibrous tissue, which may signify inferior osseointegration and/or increased inflammatory response, especially at the bone–implant interface [28–30]. Stained sections were manually scanned at a magnification of 20× to facilitate both high-magnification and whole-mount analyses (90i, Nikon Instruments, Inc., Melville, NY, USA). Three 20× regions of interest (ROI) per location (i.e., distal, midshaft, proximal) per section were captured. Within each ROI, five cell types were identified: foreign body giant (FBG)/multinucleated cell, granulocyte (non-neutrophilic, including eosinophils and basophils), neutrophil, monocyte, and lymphocyte. Cell constituents within each ROI were then graded on a scale of 0 to 2, with 0 = cells comprising 0–25% of the field, 1 = cells comprising 25–50% of field, and 2 = cells comprising greater than 50% of field. For FBG/multinucleated or granulocyte, a Grade 2 was defined as three or more cells per field. Grades were statistically compared between TiNT groups and control, using a Mann–Whitney rank sum test, with $\alpha = 0.05$.

3. Results

3.1. In Vitro Experimentation to Assess Cell Attachment and Morphology

DAPI staining indicated that cells were viable and adhered on both the TiNT and control surfaces at the three early timepoints of 0.5, 2, and 4 h (Figure 2). The total number of adherent cells was significantly greater on the TiNT surfaces than on control surfaces, demonstrating more rapid cell attachment on both TiNT surfaces compared to control (Figure 3); specifically, total adherent cells on trabecular TiNT and aligned TiNT surfaces were significantly greater than control at 0.5 h ($p = 0.014$ and $p = 0.018$, respectively) and 4 h ($p = 0.008$ and $p = 0.044$). Over the 3.5-hour period, the number of cells increased on all surfaces, and the total number of adherent cells was equivalent or slightly greater on trabecular TiNT surfaces compared to aligned TiNT surfaces. Analysis of total cell count was unfeasible at later timepoints (3, 7, 14, 21 days) due to cell coalescence or superimposition. Actin Green staining showed active cell spreading on TiNT and control surfaces at all timepoints (Figure 2).

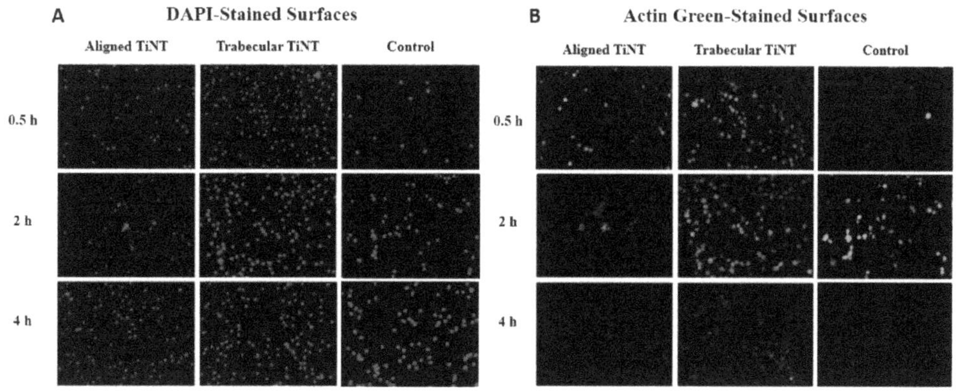

Figure 2. Representative fluorescent images of dihydrochloride (DAPI)-stained (**A**) and Actin Green-stained (**B**) aligned TiNT, trabecular TiNT, and control surfaces at three early timepoints, 0.5 h, 2 h, and 4 h.

At the 2- and 4-h timepoints, increased spreading was observed on TiNT surfaces compared to control surfaces. Additional imaging via SEM showed differential cell morphology patterns as a function of sample topography (Figure 4). On the TiNT surfaces, cells exhibited a globular shape, compared to the elongated, fibrillar cell morphology on control surfaces.

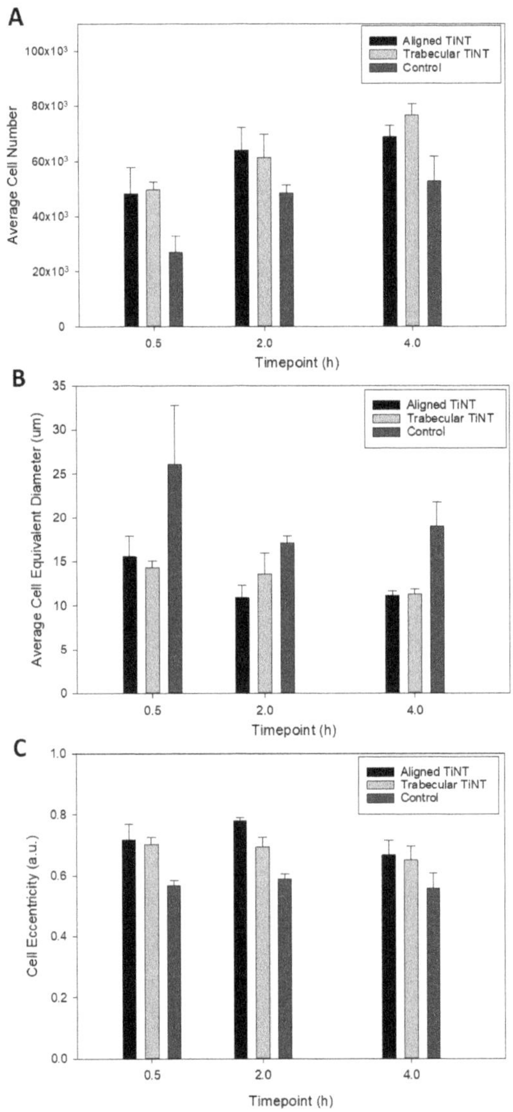

Figure 3. Average total cell number (**A**), cell equivalent diameter (**B**), and cell eccentricity (**C**) on TiNT and control surfaces at 0.5, 2, and 4 h. For Cell Number, significant comparisons were: aligned TiNT vs. control, $p = 0.018$; trabecular TiNT vs. control, $p = 0.008$; aligned TiNT vs. control, $p = 0.044$. For Cell Diameter, significant comparisons were: aligned TiNT vs. control, $p = 0.048$; trabecular TiNT vs. control, $p = 0.031$; aligned TiNT vs. control, $p = 0.008$; aligned TiNT vs. control, $p = 0.003$; trabecular TiNT vs. control, $p = 0.003$. For Cell Eccentricity, significant comparisons were: aligned TiNT vs. control, $p = 0.004$; trabecular TiNT vs. control, $p = 0.007$; aligned TiNT vs. control, $p < 0.001$; trabecular TiNT vs. control, $p = 0.003$.

Subsequent analysis of cell morphology images yielded cell equivalent diameter and eccentricity (Figure 3). Quantification of the cell equivalent diameter indicated that BMC on the TiNT surfaces were smaller (in diameter) than BMC on control surfaces. The discrepancy in cell diameter was significant between trabecular TiNT and aligned TiNT surfaces versus control at 0.5 h ($p = 0.031$ and $p = 0.048$, respectively). At the 2-hour timepoint, there was a significant difference in diameter between only the aligned TiNT and control surfaces ($p = 0.008$). At 4 h, the difference diameter was significant for between both trabecular TiNT and aligned TiNT, compared to control (both $p = 0.003$). Cells on the TiNT surfaces had significantly greater eccentricity than on the control surfaces, with cells on aligned TiNT demonstrating the greatest eccentricity (Figure 3). At 0.5 h, there was a significant difference in eccentricity between the trabecular TiNT and aligned TiNT, compared to control ($p = 0.007$ and $p = 0.004$, respectively). There was a significant difference between both the trabecular TiNT and aligned TiNT groups versus control again at 2 h ($p = 0.003$ and $p < 0.001$, respectively). There were no significant differences in eccentricity between groups at the 4-hour timepoint.

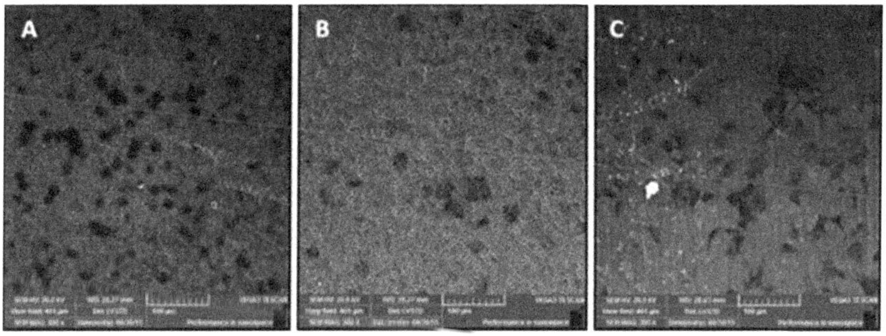

Figure 4. SEM micrographs demonstrating differential cell spreading patterns, via extracellular matrix aggregates, between aligned TiNT (**A**) and trabecular TiNT (**B**) versus control (**C**) surfaces at 3-day attachment timepoint.

3.2. *In Vivo Experimentation to Assess Biologic Response to Nanotube Surfaces*

Following the in vitro study, which demonstrated cell viability on the TiNT surfaces, the planned in vivo experiment was performed. Over the 12-week study, animals gained an average of 148 g, with control animals gaining the most weight on average (179 g) and trabecular TiNT animals gaining the least on average (118 g) (Figure 5). No animals lost more the 10% body weight, our institutional trigger for intervention. In week 9, the aligned TiNT group weighed significantly more than the trabecular TiNT group ($p = 0.013$), and in week 11, the control group weighed significant more than the trabecular TiNT group ($p = 0.023$). There were no significant differences at any other weekly time points. Mass of the spleen, brain, kidneys, lungs, and liver were obtained, and there were no significant differences between any groups for the specimens (Figure 5).

Hematologic analysis was performed, and no significant associations were found between the three groups for the thirteen parameters, and there were no significant differences between any groups for any parameters (Tables 1 and 2). Additionally, titanium, aluminum, and vanadium levels in each implant were analyzed to quantify the effect of the etching process on chemistry. Compared to control, aligned TiNT had 2.0 times the concentration of aluminum, approximately 1.8 times the titanium, and 2.9 times the vanadium; similarly, trabecular TiNT had 2.5 times the aluminum, 2.2 times the titanium, and 3.4 times the vanadium.

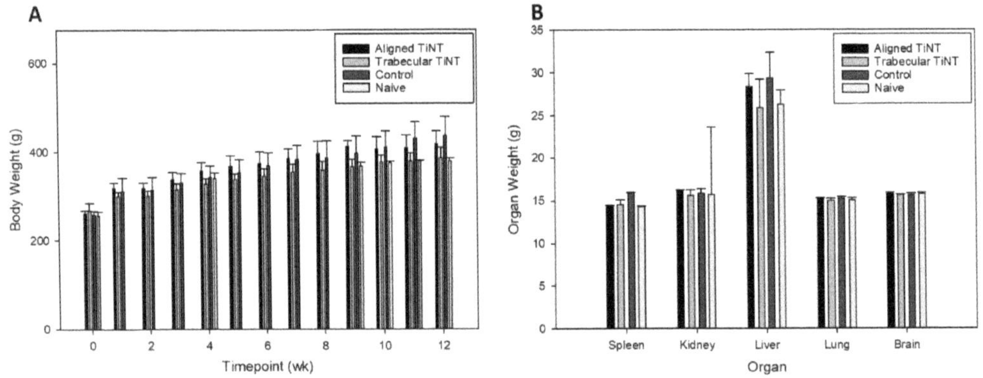

Figure 5. Average animal body weight over 12-week study (**A**) and organ weight at endpoint (**B**) per group. For Week 9, significant comparison: aligned TiNT vs. trabecular TiNT, $p = 0.013$; For Week 11, significant comparison: trabecular TiNT vs. control, $p = 0.023$.

Table 1. Hematologic analysis of white blood cell function of each treatment group at endpoint *.

Group	WBC	Lymph	Mono	Gran	Lymph%	Mono%	Gran%
Aligned TiNT	9.5 (2.7)	7.3 (1.8)	0.4 (0.2)	1.7 (0.7)	77.9 (2.7)	3.9 (1.3)	18.2 (2.6)
Trabecular TiNT	8.02 (3.5)	6.2 (2.7)	0.3 (0.1)	1.5 (0.8)	77.9 (3.9)	3.3 (1.6)	18.8 (2.9)
Control	8.9 (2.2)	7.0 (1.6)	0.3 (0.1)	1.5 (0.6)	79.7 (2.5)	3.1 (0.8)	17.3 (2.2)
Naïve	11.0 (6.1)	8.4 (3.9)	0.5 (0.4)	2.1 (1.8)	79.1 (6.1)	3.1 (0.8)	17.8 (5.3)

* Standard deviation listed in parentheses.

Table 2. Hematologic analysis of red blood cell function of each treatment group at endpoint *.

Group	HCT	MCV	RDWa	RDW%	Hgb	MCHC	MCH	RBC	PLT	MPV
Aligned TiNT	40.1 (3.2)	49.7 (1.1)	32.8 (1.0)	16.4 (0.5)	15.3 (1.2)	38.3 (0.5)	19.0 (0.3)	8.1 (0.6)	318.7 (149.4)	6.4 (0.8)
Trabecular TiNT	38.3 (1.5)	50.0 (0.9)	32.9 (0.7)	16.4 (0.6)	14.6 (0.4)	38.3 (0.6)	19.1 (0.5)	7.7 (0.4)	275.8 (151.6)	6.7 (0.6)
Control	40.1 (3.5)	50.9 (1.2)	34.0 (1.6)	16.4 (0.4)	15.1 (1.2)	37.7 (0.5)	19.1 (0.4)	7.9 (0.6)	335.0 (89.7)	6.4 (0.5)
Naïve	41.4 (2.8)	50.6 (0.8)	33.7 (0.2)	16.5 (0.4)	15.6 (1.1)	37.7 (0.2)	19.1 (0.3)	8.2 (0.5)	305.7 (93.5)	6.1 (0.8)

* Standard deviation listed in parentheses.

After weighing each organ, metal concentrations were assessed (Figure 6). Aluminum levels in the lungs were significantly greater in the trabecular TiNT group compared to control ($p = 0.022$). No other organs exhibited significantly increased titanium, aluminum, or vanadium levels compared to control. Histologic analysis demonstrated a lack of inflammatory infiltrate in proximity to the nails within the intramedullary space, for all groups (Table 3; Appendix A—Supplemental Results, Table A1). There were significantly fewer granulocytes and neutrophils in the distal ROI of the femora implanted with trabecular TiNT-etched implants ($p = 0.040$ and $p = 0.019$, respectively). In the midshaft ROI, there were significantly fewer foreign body giant/multinucleated cells and neutrophils in the aligned TiNT group ($p = 0.039$ and $p = 0.019$, respectively). There were no observed necrotic or cytotoxic events observed within the marrow cavity of any of the animals (Figure 7).

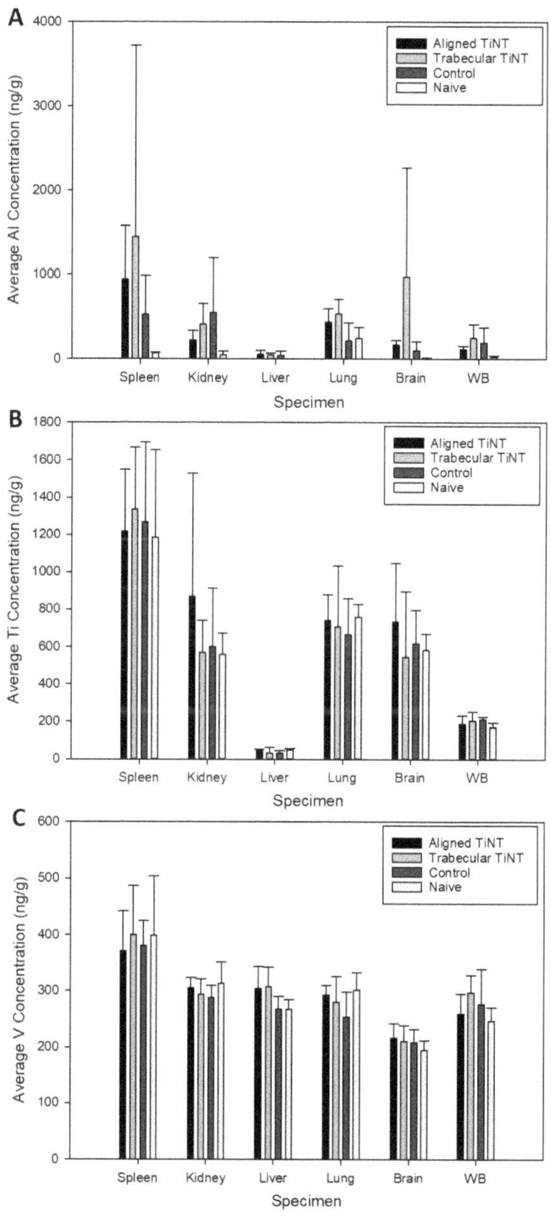

Figure 6. Concentration of aluminum (**A**), titanium (**B**), and vanadium (**C**) in remote organs and whole blood for each group. (WB = whole blood). For aluminum concentration, significant comparison: trabecular TiNT vs. control, $p = 0.022$.

Table 3. Statistical comparisons (p-values) between implant groups with control of average histologic grade for three regions of interest *.

ROI	Group (vs. Control)	Foreign Body Giant/Multinucleated	Granulocyte	Neutrophil	Monocyte	Lymphocyte
Distal	Trabecular TiNT	0.126	0.040	0.019	0.720	0.391
	Aligned TiNT	0.287	0.476	0.345	0.720	0.260
Midshaft	Trabecular TiNT	0.142	0.603	1.000	0.314	0.982
	Aligned TiNT	0.039	0.682	0.019	0.736	0.260
Proximal	Trabecular TiNT	0.527	0.646	0.345	0.652	0.871
	Aligned TiNT	0.939	0.219	0.345	0.747	0.314

* Bolded p-values indicate significant results; ROI = region of interest.

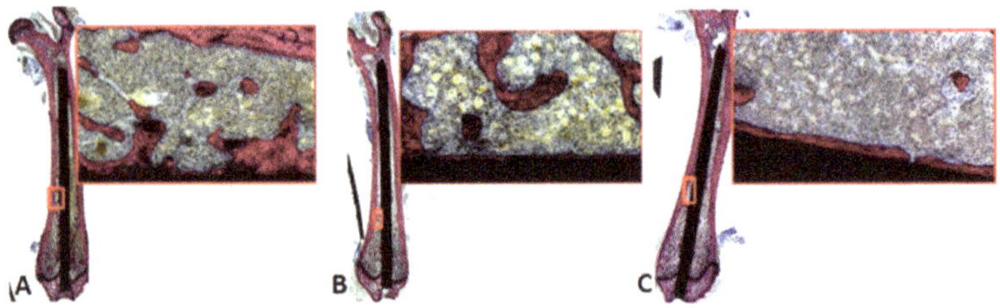

Figure 7. Longitudinal histologic sections and regions of interest of representative femora implanted with aligned TiNT (**A**), trabecular TiNT (**B**), and (**C**) control K-wires.

4. Discussion

Local and systemic responses to aligned and trabecular TiNT surfaces were evaluated via an in vitro study focused on cell morphological and attachment behavior, followed by a clinically relevant in vivo study of biologic response to implant materials. The in vivo study included multiple characterization methods, including a general health assessment (e.g., body weight) as well as hematologic, metal ion, and histologic analyses.

DAPI and Actin Green staining demonstrated cell attachment and spreading from 0.5 h through 21 days on all surfaces, an indication of the suitable environment of both TiNT surfaces and control. Total cell counts were greater on TiNT surfaces compared to unetched controls at the three early timepoints, which corresponds with current reports of more rapid cell attachment on nanotube surfaces [31–33]. Additionally, the diameter of BMC on TiNT surfaces was smaller compared to cells seeded onto controls, and cells on TiNT surfaces were more eccentric than cells on controls. These data correspond with SEM imaging of attached BMC, which showed a rounded cell morphology on TiNT surfaces and fibrillar-shaped cells on control surfaces. The diameter, eccentricity, and imaging findings may be indicative of cell development in different planes on the TiNT surfaces compared to controls. Shokuhfar et al. sectioned TiNT arrays seeded with osteoblasts (MC3T3-E1; mouse osteoblasts) and observed cell attachment on nanotube arrays as well as cell filopodia stretching downward into the hollow portion of individual nanotubes [20]; therefore, differences in cell morphologic behavior between groups may be explained by the BMC interaction with TiNT surfaces, as the BMC extend into nanotubes and voids within the array, instead of across the top of the surface. Rangamani et al. performed modeling of cell shapes and determined that greater cell eccentricity was associated with enhanced signal modulation due to temporary collections of activated receptors in sections of greater curvature of cells, specifically in growth factor receptor pathways; therefore, greater cell eccentricity on nanotube surfaces may demonstrate amplified signals resulting in downstream effects, such as bone formation [34].

In the in vivo study, there were several significant differences in body weight between groups at two weeks during the experiment; however, the body weights recovered were not an established trend, and no interventions were required at any time for weight loss. At endpoint, remote organ masses were compared between groups, and these also showed no significant difference. Despite increased concentrations of aluminum, titanium, and vanadium in the aligned TiNT and trabecular TiNT wires versus control wires, only aluminum levels were significantly increased in the lungs of the rats implanted with trabecular TiNT-etched wires compared to controls. No other organs or whole blood samples showed significantly elevated metal levels, when the experimental groups were compared to control. A study of intraarticular injection of TiO_2 nanoparticles (45 nm average diameter; 0.2, 2, 20 mg/kg TiO_2 in suspension) into rat knee joints showed nanoparticle migration to remote organs as well as pathologic changes in the heart, lung, liver, and knee at 7 days post-injection [35,36]. Although our study showed metal ion concentration in remote organs via ICP-MS and did not histologically assess remote organs, we theorize that the amount of TiO_2 particulate debris shed in the rat femora was likely less than 0.2 mg/kg. At study endpoint, no hematologic markers of white and red blood cell function, which would signal systemic inflammation-, infection-, or anemia-related complications, were significantly elevated in the experimental groups compared to control. Histologic analysis showed several significant differences in cell populations between the TiNT groups and control, with less foreign body giant/multinucleated, eosinophil/basophil, and neutrophil cell activity in TiNT-implanted femora than control. There were no significant differences between monocyte and lymphocyte activity.

In in vitro and in vivo studies of vascular toxicity, Bayat et al. established that titanium nanotubes with 30 nm diameter as well as ultra-small TiO_2 nanoparticles (1−3 nm) were not cytotoxic, and nanoparticles did not possess oxidative potential [37]. In vivo studies in rat models have also shown that TiO_2 nanoparticles are not genotoxic [38,39]. Neacsu et al. showed that after seeding murine macrophages onto TiNT and unetched titanium surfaces under pro-inflammatory and standard conditions, inflammatory activity related to cytokine and chemokine gene expression, foreign body giant cell products, and nitric oxide release all decreased on TiNT surfaces but not on unetched controls; therefore, the authors suggested that the TiNT surfaces may regulate macrophage response, thereby diminishing the overall inflammatory cascade [40]. Studies by Ostberg et al. and Latha et al., using models of leukocyte-seeded TiNT surface and mixed lymphocyte reaction (MLR), respectively, concurred with these findings [41,42]; specifically, Latha et al. showed a 30–35% suppression of splenocyte proliferation of titanium nanotubes ($H_2Ti_3O_7$) [42]. Assessing hemocompatbility, Smith et al. showed increased adsorption of blood serum protein, platelet adhesion and activation, and clotting of whole blood as well as no evidence of monocyte activation and cytokine secretion on TiNT surfaces versus control [43]. Radizun et al. showed that aluminum nanoparticles, in concentrations up to 400 μg/mL, have no significant toxic effect on mammalian cell viability, and Alshatwi et al. found human MSC cytotoxicity via Al_2O_3 nanoparticles at 40 μg/mL; in this study, aluminum content in all specimens was substantially below this threshold (maximum aluminum content within range of standard deviation; Figure 6: 4.24 μg/mL) [44,45]. In a clinical, in vivo study, Swiatkowska et al. described titanium levels ranging between 2.20 and 2.56 μg/L in blood and plasma, measured via ICP-MS, in well-functioning unilateral hip implants, indicating some level of titanium may be tolerable by patients, without eliciting a whole-body response [46].

Study limitations included difficulties with staining TiNT surfaces, especially at timepoints greater than 4 h. Because of the nanotopography and voids, which were ideal for cell attachment, stain penetration of the cells was complicated and incomplete. While three naïve animals were included to assess background metal levels, inclusion of additional animals may have been beneficial to further study and quantify these levels. Additionally, a full assessment of metal levels of animal food, bedding, enrichment materials, etc. may have allowed further reduction and elimination of background aluminum, titanium, and vanadium levels in all animals [47,48]. Based on the results of this study, a full systematic characterization of this material's biocompatibility and toxicity properties should be conducted in accordance with applicable standards, such as ISO 10993-11 (Biological Evaluation of Medical Devices—Part 11: Tests for Systemic Toxicity).

5. Conclusions

Previous studies have demonstrated the biocompatibility of TiNT surfaces and correspond with the findings of this study showing TiNT surfaces support cell attachment and proliferation and do not initiate systemic effects in an in vivo model of intramedullary implantation. The presented in vitro data demonstrated that cells cultured on bare TI6Al4V surfaces were spindle shaped, while those cultured on TiNT surfaces demonstrated a more rounded appearance, which may confer benefit with respect to in vivo bone formation. In vivo data demonstrated that the accumulation of metal ions in filtering organs was largely similar between the two morphologies of TiNT surfaces and control bare Ti6Al4V surfaces, aside from an increased aluminum concentration in the lungs of rats with aligned TiNT implants. This increased aluminum concentration in lung specimens was not associated with alterations in overall health of the animal or observed pathology. As such, it can be concluded that at the time points studied, metal accumulation associated with TiNT surfaces was similar to control surfaces, and no systemic or local complications were observed in TiNT- or control-implanted animals.

Author Contributions: Conceptualization, E.A.B., K.C.B., P.T.F. and C.R.F.; methodology, E.A.B., M.M.F., A.D.V., K.C.B., M.R.S., P.T.F. and C.R.F.; formal analysis, E.A.B., K.C.B., A.D.V. and M.R.S.; investigation, E.A.B., M.M.F., A.D.V., M.R.S., K.C.B., P.T.F. and C.R.F.; resources, E.A.B., K.C.B., P.T.F. and C.R.F.; writing—original draft preparation, E.A.B. and M.R.S.; writing—review and editing, E.A.B., M.M.F., A.D.V., K.C.B., P.T.F. and C.R.F.; supervision, P.T.F. and C.R.F.; project administration, E.A.B. and K.C.B. All authors have read and agreed to the published version of the manuscript.

Funding: This research received no external funding.

Data Availability Statement: The data presented in this study are available on request from the corresponding author. The data are not publicly available due to institutional policies.

Conflicts of Interest: The authors declare no conflict of interest specific to this work.

Appendix A. Supplemental Methods and Results

Appendix A.1. Titanium Nanotube-Etched Specimen Fabrication

Titanium nanotube-etched specimens were prepared for in vitro and in vivo experiments, based on previous methods developed and tested by our group [17,20–22]. Titanium alloy (Ti-6Al-4V ELI) sheet (in vitro experiments) or wire (in vivo study) were polished with 600 grit abrasive sheet and deionized (DI) water, then rinsed in DI water and air-dried. All sample materials were cleaned with acetone just prior to etching, followed by air drying. On wire samples, the trocar tip was coated to prevent nanotube formation on the site of insertion. To etch, samples (sheet or wire) were suspended vertically toward one side of a glass beaker and then connected as the anode to a variable DC power supply. Additionally, in the beaker, a small diameter graphite rod was suspended, diametrically opposed to the sample, and connected to ground (cathode). An electrolyte solution (98 vol.% ethylene glycol, 2 vol.% deionized water, and 0.6 wt.% NH4F) was added to the beaker, with a final volume just below the electrical connections to the sample and graphite. Before adding NH4F to the electrolyte solution, it was first completely dissolved in DI water; following this addition, the electrolyte solution was mixed to ensure all constituents were fully combined. Then, the power supply was engaged (+60 VDC) and etching was allowed to proceed for 40 minutes. After etching, the power supply was disengaged and sample(s) were immediately removed from the beaker, followed by a 1-minute rinse under DI water and air drying. After drying, the coating on each trocar tip was removed after etching and sonication (aligned TiNT samples were sonicated for 2 min; trabecular TiNT samples were not sonicated). Sheet samples were then sectioned into coupons (10 mm × 10 mm); wires were net-shape and did not require sectioning. To convert the amorphous titania oxide layer to the crystalline anatase phase, which increases the hydrophilicity of the TiNT surfaces, samples were placed in a programmable annealing oven. The temperature was increased by 7.5 °C per minute to a steady state of 450°C for a total heating time of three hours. Samples were removed after the oven was completely cooled, which was approximately 5 h. Specimens were stored in cushioned, sealed storage cases until use (Figure A1).

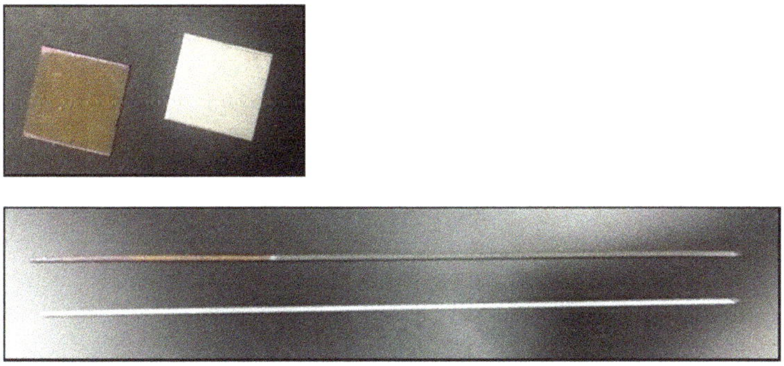

Figure A1. Digital photographs showing the final appearance of 10 mm × 10 mm coupon samples (top) and Kirschner wires (bottom). TiNT-etched surfaces appear golden and smooth, compared to unmodified control surfaces.

Appendix A.2. In Vitro Experimentation to Assess Adherent Cell Count, Cell Equivalent Diameter, and Cell Eccentricity

Rat bone marrow cells (BMCs) were used for all in vitro experimentation, with cell harvest and preparation conducted based on previously developed laboratory protocols by our group [49]. To obtain BMCs, rats were euthanized via CO2 asphyxiation, and then each femur and tibia were aseptically harvested. After removing both the proximal and distal ends of each bone, marrow cavities were flushed with warm, sterile phosphate-buffered saline. Whole bone marrow was plated (T-25 culture flasks) and cultured (37 °C and 5% CO2) in a sterile, copper-lined CO2 incubator. Non-adherent cells were removed after 24 h of incubation via thorough rinsing with warm, sterile saline. Then, the plastic-adherent fraction of bone marrow cells (BMCs), an enriched source of mesenchymal stem cells (MSC) capable of differentiating toward numerous cell types, was obtained.

For in vitro experimentation, sample coupons were placed into individual wells of 12-well polystyrene culture plates and soaked in fetal bovine serum (FBS) for 30 min prior to seeding to increase attachment. Rat BMCs (P2-3; density = 40,000 cells per coupon; suspended in 50 µL of media) were drop-seeded on each sample and incubated (37 °C and 5% CO2) for 6 h before adding the remaining volume of media (DMEM; Dulbecco's Modified Eagle Medium; supplementation = 10% fetal bovine serum, and 1% penicillin-streptomycin). Incubation (same conditions) continued for an additional 20 h to ensure attachment. Following incubation, fluorescence imaging at 13 standardized regions of interest per coupon was performed, approximately 70% of the specimen surface, followed by subsequent quantification of (Figure A2).

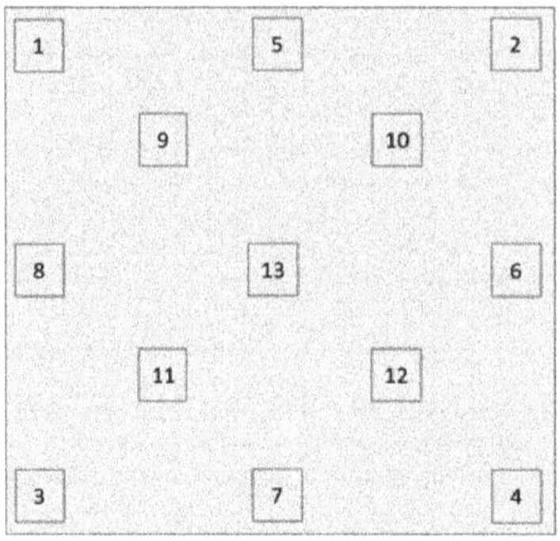

Figure A2. Imaging convention for each coupon in the experiment. Regions were chosen to reflect the corners (1–4), edges (5–8), central area (9–12), and absolute center (13) of each sample.

Appendix A.3. Metal Ion Analysis of Remote Organs and Whole Blood

Inductively coupled plasma mass spectrometry (ICP-MS; HP 4500, Agilent Technologies, Santa Clara, CA) was performed to quantify aluminum, titanium, and vanadium content in whole-blood and remote organ samples. Sample tubes, made of ultra-low leachable material to avoid metal contamination, were labeled and weighed to three significant figures before sample collection. At endpoint, whole blood ($\geq$2 mL) was collected, before proceeding with dissection of remote organs (i.e., spleen, liver, lungs, kidneys, brain). Organs were weighed before being placed into tubes, and each filled sample tube was weighed again to determine the actual/net weight of the collected specimen.

All samples were transported and held on ice (~4 °C) until analysis. Samples were digested to dissolve the metals into an aqueous solution by placing a combination of highly pure solutions of nitric acid, hydrochloric acid, hydrogen peroxide, and water into the tubes. The tubes were heated at 95 °C for 2 h per cycle. Some of the samples went through multiple digestion cycles for thorough digestion. For ICP-MS analysis, samples were included in analytical batches of 20 samples, which includes a laboratory reagent blank (LRB) to establish a reference to the laboratory background value, a laboratory fortified blank (LFB) to measure accuracy of the analytical procedure, and a duplicate LFB to measure the precision of the analytical procedure. The final volume of digestion was set to 40 mL per specimen digested, and the digested solutions were further diluted by 10 times and analyzed for aluminum, titanium, and vanadium contents using a standard operating procedure (UL SOP# 114) based on the EPA guidelines (Method 200.8).

The total weight of each metal per specimen and concentration of each metal in a specimen were recorded; weights and concentrations were reported to three significant figures. Accuracy and precision data as part of the quality control carried through the analytical batches were also documented. Chain of custody forms containing sample descriptions, sampling dates, etc. were completed and retained.

Appendix A.4. Histologic Analysis of Bone-Implant Interface

Average cell count data of five cell types (foreign body giant/multinucleated, granulocyte, neutrophil, monocyte, lymphocyte) for the three implant groups (trabecular TiNT, aligned TiNT, control) in the three regions of interest (ROI; distal, midshaft, and proximal femora) have been provided (Table A1).

Table A1. Average histologic grade for three regions of interest in each treatment group *.

ROI	Group	Foreign Body Giant/Multinucleated	Granulocyte	Neutrophil	Monocyte	Lymphocyte
Distal	Trabecular TiNT	0.8 (1.0)	0.1 (0.0)	0.7 (1.0)	0.8 (1.0)	1.1 (1.0)
	Aligned TiNT	0.7 (1.0)	0.2 (0.0)	1.1 (1.0)	0.8 (1.0)	1.1 (1.0)
	Control	0.6 (0.5)	0.3 (0.0)	1.0 (1.0)	0.7 (1.0)	1.2 (1.0)
Midshaft	Trabecular TiNT	0.5 (0.5)	0.6 (1.0)	1.0 (1.0)	0.6 (1.0)	1.2 (1.0)
	Aligned TiNT	0.4 (0.0)	0.8 (1.0)	0.7 (1.0)	0.7 (1.0)	1.1 (1.0)
	Control	0.8 (1.0)	0.7 (1.0)	1.0 (1.0)	0.7 (1.0)	1.2 (1.0)
Proximal	Trabecular TiNT	0.8 (1.0)	0.6 (1.0)	0.9 (1.0)	0.8 (1.0)	1.4 (1.0)
	Aligned TiNT	0.7 (1.0)	0.4 (0.0)	1.1 (1.0)	0.9 (1.0)	1.3 (1.0)
	Control	0.7 (1.0)	0.7 (1.0)	1.0 (1.0)	0.8 (1.0)	1.4 (1.0)

* Standard deviation listed in parentheses.

References

1. Branemark, P.I.; Adell, R.; Albrektsson, T. Osseointegrated titanium fixtures in the treatment of edentulousness. *Biomaterials* **1983**, *4*, 25–28. [CrossRef]
2. Albrektsson, T.; Brånemark, P.-I.; Hansson, H.-A.; Lindström, J. Osseointegrated Titanium Implants:Requirements for Ensuring a Long-Lasting, Direct Bone-to-Implant Anchorage in Man. *Acta Orthop. Scand.* **1981**, *52*, 155–170. [CrossRef] [PubMed]
3. Ionescu, F.; Reclaru, L.; Ardelean, L.C.; Blatter, A. Comparative Analysis of the Corrosion Resistance of Titanium Alloys Intended to Come into Direct or Prolonged Contact with Live Tissues. *Materials* **2019**, *12*, 2841. [CrossRef]
4. Ratner, B.D.; Schoen, F.J.; Lemons, J.E. *Biomaterials Science: An Introduction to Materials in Medicine*, 2nd ed.; Elsevier Academic Press: San Diego, CA, USA, 2004.
5. Goodman, S.B.; Gallo, J. Periprosthetic Osteolysis: Mechanisms, Prevention and Treatment. *J. Clin. Med.* **2019**, *8*, 2091. [CrossRef] [PubMed]
6. Goodman, S.B.; Gallo, J.; Gibon, E.F.; Takagi, M. Diagnosis and management of implant debris-associated inflammation. *Expert Rev. Med. Devices* **2019**, *17*, 41–56. [CrossRef]
7. Hergel, C.A.; Acar, Y.B.; Ates, M.; Kucukkeles, N. In-vitro evaluation of the effects of insertion and sterilization procedures on the mechanical and surface characteristics of mini screws. *Eur. Oral Res.* **2019**, *53*, 25–31. [CrossRef] [PubMed]
8. Wang, J.; Wang, L.; Fan, Y. Adverse Biological Effect of TiO_2 and Hydroxyapatite Nanoparticles Used in Bone Repair and Replacement. *Int. J. Mol. Sci.* **2016**, *17*, 798. [CrossRef]
9. Hallab, N.J.; Jacobs, J.J. Biologic effects of implant debris. *Bull. NYU Hosp. Jt. Dis.* **2009**, *67*, 182–188. [PubMed]
10. Hallab, N.J. Biologic Responses to Orthopedic Implants: Innate and Adaptive Immune Responses to Implant Debris. *Spine* **2016**, *41*, S30–S31. [CrossRef] [PubMed]
11. Reddy, A.; Caicedo, M.S.; Samelko, L.; Jacobs, J.J.; Hallab, N.J. Implant debris particle size affects serum protein adsorption which may contribute to particle size-based bioreactivity differences. *J. Long Term Eff. Med.* **2014**, *24*, 77–88. [CrossRef] [PubMed]
12. Bjursten, L.M.; Rasmusson, L.; Oh, S.; Smith, G.C.; Brammer, K.S.; Jin, S. Titanium dioxide nanotubes enhance bone bondingin vivo. *J. Biomed. Mater. Res. Part A* **2009**, *92*, 1218–1224. [CrossRef]
13. Von Wilmowsky, C.; Bauer, S.; Roedl, S.; Neukam, F.W.; Schmuki, P.; Schlegel, K.A. The diameter of anodic TiO_2 nanotubes affects bone formation and correlates with the bone morphogenetic protein-2 expression in vivo. *Clin. Oral Implant. Res.* **2012**, *23*, 359–366. [CrossRef] [PubMed]
14. Brammer, K.S.; Frandsen, C.J.; Jin, S. TiO2 nanotubes for bone regeneration. *Trends Biotechnol.* **2012**, *30*, 315–322. [CrossRef] [PubMed]
15. Xing, H.; Komasa, S.; Taguchi, Y.; Sekino, T.; Okazaki, J. Osteogenic activity of titanium surfaces with nanonetwork structures. *Int. J. Nanomed.* **2014**, *9*, 1741–1755. [CrossRef] [PubMed]

16. Rani, V.D.; Vinoth-Kumar, L.; Anitha, V.; Manzoor, K.; Deepthy, M.; Shantikumar, V.N. Osteointegration of titanium implant is sensitive to specific nanostructure morphology. *Acta Biomater.* **2012**, *8*, 1976–1989. [CrossRef] [PubMed]
17. Baker, E.A.; Vara, A.D.; Salisbury, M.R.; Fleischer, M.M.; Baker, K.C.; Fortin, P.T.; Roberts, R.V.; Friedrich, C.R. Titania nanotube morphologies for osseointegration via models of in vitro osseointegrative potential and in vivo intramedullary fixation. *J. Biomed. Mater. Res. Part B Appl. Biomater.* **2020**, *108*, 1483–1493. [CrossRef]
18. Wang, J.; Meng, F.; Song, W.; Jin, J.; Ma, Q.; Fei, D.; Fang, L.; Chen, L.; Wang, Q.; Zhang, Y. Nanostructured titanium regulates osseointegration via influencing macrophage polarization in the osteogenic environment. *Int. J. Nanomed.* **2018**, *13*, 4029–4043. [CrossRef] [PubMed]
19. Bai, L.; Liu, Y.; Du, Z.; Weng, Z.; Yao, W.; Zhang, X.; Huang, X.; Yao, X.; Crawford, R.; Hang, R.; et al. Differential effect of hydroxyapatite nano-particle versus nano-rod decorated titanium micro-surface on osseointegration. *Acta Biomater.* **2018**, *76*, 344–358. [CrossRef] [PubMed]
20. Shokuhfar, T.; Hamlekhan, A.; Chang, J.-Y.; Choi, C.K.; Sukotjo, C.; Friedrich, C. Biophysical evaluation of cells on nanotubular surfaces: The effects of atomic ordering and chemistry. *Int. J. Nanomed.* **2014**, *9*, 3737–3748. [CrossRef]
21. Shin, D.H.; Shokuhfar, T.; Choi, C.K.; Lee, S.-H.; Friedrich, C. Wettability changes of TiO2nanotube surfaces. *Nanotechnology* **2011**, *22*, 315704. [CrossRef]
22. Friedrich, C.R.; Kolati, M.; Moser, T.; Sukotjo, C.; Shokuhfar, T. Survivability of TiO2nanotubes on the surface of bone screws. *Surf. Innov.* **2014**, *2*, 60–68. [CrossRef]
23. Gupta, A.; Liberati, T.A.; Verhulst, S.J.; Main, B.J.; Roberts, M.H.; Potty, A.G.R.; Pylawka, T.K.; Iii, S.F.E.-A. Biocompatibility of single-walled carbon nanotube composites for bone regeneration. *Bone Jt. Res.* **2015**, *4*, 70–77. [CrossRef] [PubMed]
24. Back, D.A.; Pauly, S.; Rommel, L.; Haas, N.P.; Schmidmaier, G.; Wildemann, B.; Greiner, S.H. Effect of local zoledronate on implant osseointegration in a rat model. *BMC Musculoskelet. Disord.* **2012**, *13*, 42. [CrossRef] [PubMed]
25. Su, Y.; Komasa, S.; Sekino, T.; Nishizaki, H.; Okazaki, J. Nanostructured Ti6Al4V alloy fabricated using modified alkali-heat treatment: Characterization and cell adhesion. *Mater. Sci. Eng. C* **2016**, *59*, 617–623. [CrossRef] [PubMed]
26. Pauly, S.; Back, D.A.; Kaeppler, K.; Haas, N.P.; Schmidmaier, G.; Wildemann, B. Influence of Statins locally applied from orthopedic implants on osseous integration. *BMC Musculoskelet. Disord.* **2012**, *13*, 208. [CrossRef]
27. Virdi, A.S.; Irish, J.; Sena, K.; Liu, M.; Ke, H.Z.; McNulty, M.A.; Sumner, D.R. Sclerostin Antibody Treatment Improves Implant Fixation in a Model of Severe Osteoporosis. *J. Bone Jt. Surg. Am. Vol.* **2015**, *97*, 133–140. [CrossRef] [PubMed]
28. Betoni, W.; Queiroz, T.P.; Luvizuto, E.R.; Valentini-Neto, R.; Garcia-Júnior, I.R.; Bernabé, P.F.E. Evaluation of Centrifuged Bone Marrow on Bone Regeneration Around Implants in Rabbit Tibia. *Implant. Dent.* **2012**, *21*, 481–485. [CrossRef] [PubMed]
29. Freire, A.R.; Rossi, A.C.; Queiroz, T.P.; Gulinelli, J.L.; Souza, F.A.; Margonar, R.; Garcia-Júnior, I.R.; Hochuli-Vieira, E.; Okamoto, R. Histometric Analysis of Bone Repair in Bone-Implant Interface Using a Polylactic/Polyglycolic Acid Copolymer Associated With Implants in Rabbit Tibia. *J. Oral Implant.* **2012**, *38*, 449–457. [CrossRef]
30. Nich, C.; Marchadier, A.; Sedel, L.; Petite, H.; Vidal, C.; Hamadouche, M. Decrease in particle-induced osteolysis in ovariectomized mice. *J. Orthop. Res.* **2009**, *28*, 178–183. [CrossRef]
31. Popat, K.C.; Leoni, L.; Grimes, C.A.; Desai, T.A. Influence of engineered titania nanotubular surfaces on bone cells. *Biomaterials* **2007**, *28*, 3188–3197. [CrossRef] [PubMed]
32. Liu, X.; Zhou, X.; Li, S.; Lai, R.; Zhou, Z.; Zhang, Y.; Zhou, L. Effects of titania nanotubes with or without bovine serum albumin loaded on human gingival fibroblasts. *Int. J. Nanomed.* **2014**, *9*, 1185–1198. [CrossRef]
33. Minagar, S.; Wang, J.; Berndt, C.C.; Ivanova, E.P.; Wen, C. Cell response of anodized nanotubes on titanium and titanium alloys. *J. Biomed. Mater. Res. Part A* **2013**, *101*, 2726–2739. [CrossRef]
34. Rangamani, P.; Lipshtat, A.; Azeloglu, E.U.; Calizo, R.C.; Hu, M.; Ghassemi, S.; Hone, J.; Scarlata, S.; Neves, S.R.; Iyengar, R. Decoding Information in Cell Shape. *Cell* **2013**, *154*, 1356–1369. [CrossRef] [PubMed]
35. Wang, J.-X.; Fan, Y.-B.; Gao, Y.; Hu, Q.-H.; Wang, T.-C. TiO_2 nanoparticles translocation and potential toxicological effect in rats after intraarticular injection. *Biomaterials* **2009**, *30*, 4590–4600. [CrossRef] [PubMed]
36. Fan, X.; Feng, B.; Liu, Z.; Tan, J.; Zhi, W.; Lu, X.; Wang, J.; Weng, J. Fabrication of TiO_2 nanotubes on porous titanium scaffold and biocompatibility evaluation in vitro and in vivo. *J. Biomed. Mater. Res. Part A* **2012**, *100*, 3422–3427. [CrossRef] [PubMed]
37. Bayat, N.; Lopes, V.R.; Schölermann, J.; Jensen, L.D.; Cristobal, S. Vascular toxicity of ultra-small TiO_2 nanoparticles and single walled carbon nanotubes in vitro and in vivo. *Biomaterials* **2015**, *63*, 1–13. [CrossRef] [PubMed]
38. Ames, B.N.; Lee, F.D.; Durston, W.E. An Improved Bacterial Test System for the Detection and Classification of Mutagens and Carcinogens. *Proc. Natl. Acad. Sci. USA* **1973**, *70*, 782–786. [CrossRef] [PubMed]
39. Naya, M.; Kobayashi, N.; Ema, M.; Kasamoto, S.; Fukumuro, M.; Takami, S.; Nakajima, M.; Hayashi, M.; Nakanishi, J. In vivo genotoxicity study of titanium dioxide nanoparticles using comet assay following intratracheal instillation in rats. *Regul. Toxicol. Pharmacol.* **2012**, *62*, 1–6. [CrossRef]
40. Neacsu, P.; Mazare, A.; Cimpean, A.; Park, J.; Costache, M.; Schmuki, P.; Demetrescu, I. Reduced inflammatory activity of RAW 264.7 macrophages on titania nanotube modified Ti surface. *Int. J. Biochem. Cell Biol.* **2014**, *55*, 187–195. [CrossRef]
41. Östberg, A.-K.; Dahlgren, U.; Sul, Y.-T.; Johansson, C.B. Inflammatory cytokine release is affected by surface morphology and chemistry of titanium implants. *J. Mater. Sci. Mater. Electron.* **2015**, *26*, 1–9. [CrossRef]
42. Latha, T.S.; Reddy, M.C.; Durbaka, P.V.R.; Muthukonda, S.V.; Lomada, D. Immunomodulatory properties of titanium dioxide nanostructural materials. *Indian J. Pharmacol.* **2017**, *49*, 458–464. [CrossRef]

43. Smith, B.S.; Yoriya, S.; Grissom, L.; Grimes, C.A.; Popat, K.C. Hemocompatibility of titania nanotube arrays. *J. Biomed. Mater. Res. Part A* **2010**, *95*, 350–360. [CrossRef] [PubMed]
44. Radziun, E.; Wilczyńska, J.D.; Książek, I.; Nowak, K.; Anuszewska, E.; Kunicki, A.; Olszyna, A.; Ząbkowski, T. Assessment of the cytotoxicity of aluminium oxide nanoparticles on selected mammalian cells. *Toxicol. Vitr.* **2011**, *25*, 1694–1700. [CrossRef] [PubMed]
45. Alshatwi, A.A.; Subbarayan, P.V.; Ramesh, E.; Al-Hazzani, A.A.; AlSaif, M.A.; Alwarthan, A.A. Al2O3Nanoparticles Induce Mitochondria-Mediated Cell Death and Upregulate the Expression of Signaling Genes in Human Mesenchymal Stem Cells. *J. Biochem. Mol. Toxicol.* **2012**, *26*, 469–476. [CrossRef] [PubMed]
46. Swiatkowska, I.; Martin, N.G.; Henckel, J.; Apthorp, H.; Hamshere, J.; Hart, A.J. Blood and plasma titanium levels associated with well-functioning hip implants. *J. Trace Elements Med. Biol.* **2020**, *57*, 9–17. [CrossRef]
47. Agency for Toxic Substances and Disease Registry (ATSDR). *Toxicological Profile for Aluminum*; Department of Health and Human Services, Public Health Service: Atlanta, GA, USA, 2008.
48. Agency for Toxic Substances and Disease Registry (ATSDR). *Toxicological Profile for Vanadium*; Department of Health and Human Services, Public Health Service: Atlanta, GA, USA, 2012.
49. Maerz, T.; Fleischer, M.; Newton, M.D.; Davidson, A.; Salisbury, M.; Altman, P.; Kurdziel, M.D.; Anderson, K.; Bedi, A.; Baker, K.C. Acute mobilization and migration of bone marrow-derived stem cells following anterior cruciate ligament rupture. *Osteoarthritis Cartilage* **2017**, *25*, 1335–1344. [CrossRef]

nanomaterials

Article

Potential Production of Theranostic Boron Nitride Nanotubes (^{64}Cu-BNNTs) Radiolabeled by Neutron Capture

Wellington Marcos Silva [1], Helio Ribeiro [2] and Jose Jaime Taha-Tijerina [3,4,*]

1 Departamento de Química, Universidade Federal de Minas Gerais, Avenida Presidente Antônio Carlos, 6.627, Belo Horizonte 31270-901, MG, Brazil; wellingtonmarcos@yahoo.com.br
2 Escola de Engenharia, Mackenzie Presbyterian University, Rua da Consolação 896, São Paulo 01302-907, SP, Brazil; helio.ribeiro1@mackenzie.br
3 Departamento de Ingeniería, Universidad de Monterrey, Av. Ignacio Morones Prieto 4500 Pte., San Pedro Garza García 66238, NL, Mexico
4 Engineering Technology Department, University of Texas Rio Grande Valley, Brownsville, TX 78520, USA
* Correspondence: jose.taha@udem.edu

Abstract: In this work, the radioisotope ^{64}Cu was obtained from copper (II) chloride dihydrate in a nuclear research reactor by neutron capture, (^{63}Cu(n,γ)^{64}Cu), and incorporated into boron nitride nanotubes (BNNTs) using a solvothermal process. The produced ^{64}Cu-BNNTs were analyzed by TEM, MEV, FTIR, XDR, XPS and gamma spectrometry, with which it was possible to observe the formation of^{64}Cu nanoparticles, with sizes of up to 16 nm, distributed through nanotubes. The synthesized of ^{64}Cu nanostructures showed a pure photoemission peak of 511 keV, which is characteristic of gamma radiation. This type of emission is desirable for Photon Emission Tomography (PET scan) image acquisition, as well as its use in several cancer treatments. Thus, ^{64}Cu-BNNTs present an excellent alternative as theranostic nanomaterials that can be used in diagnosis and therapy by different techniques used in nuclear medicine.

Keywords: theranostic nanomaterials; boron nitride; neutron capture reaction; nuclear medicine

Citation: Silva, W.M.; Ribeiro, H.; Taha-Tijerina, J.J. Potential Production of Theranostic Boron Nitride Nanotubes (^{64}Cu-BNNTs) Radiolabeled by Neutron Capture. *Nanomaterials* **2021**, *11*, 2907. https://doi.org/10.3390/nano11112907

Academic Editors: Félix Zamora and Angelo Ferraro

Received: 1 September 2021
Accepted: 17 October 2021
Published: 30 October 2021

1. Introduction

The discovery and development of new materials are often the catalysts for technological advances, particularly when they can be applied to various areas of research. A historic milestone in the search for new materials occurred in 1995 when boron nitride nanotubes (BNNTs) emerged as a key material in nanotechnology science [1]. BNNTs have cylindrical structures (one-dimensional—1D), formed only by atoms of boron (B) and nitrogen (N), with diameters in the order of nanometers and lengths in the order of microns [2]. Similar to single-wall carbon nanotubes (SWCNT), depending on the angle θ in which the sheet is rolled, nanotube structures are formed with armchair (θ = 30°), zig-zag (θ = 0°) and chiral ($0 < /\theta/ < 30°$) forms [3]. Otherwise, the multi-walled BNNTs are formed by several sheets of hexagonal boron nitride (h-BN) concentrically wrapped. Both BNNTs have excellent chemical, physical and mechanical properties [4–8] and a band gap of approximately 5.5 eV [3]. Several methods have been used for the synthesis of BNNTs [5]; however, chemical vapor deposition (CVD) is the most used method and requires a simple apparatus which produces BNNTs with excellent structural quality [9].

The interest in BNNT-based technologies in recent years can be measured by the large number of scientific works that have been published, in addition to the increase in large-scale production [10]. Within the class of nanostructured materials, the BNNTs have great potential for several biomedical applications. Some early studies demonstrated that nanotubes did not show toxicity at concentrations below 50 μg/mL [11]. Furthermore, they favor the reuptake of molecules into the cell interior and can be functionalized with different biological protein epitopes [12]. Recently, BNNTs have been used as nanovectors for DNA, drugs and radioisotopes, and as boosters for biomaterials. In 2012, Soares et al. [13] used BNNTs radiolabeled with ^{99m}Tc to investigate the cell-distribution behavior *in vivo* through a process of passive accumulation in solid tumors. Diverse studies applying BNNTs to cancer treatment have been reported. For example, when linked to target molecules, BNNTs could be used as therapeutic agents capable of killing cancer cells by boron neutron capture therapy. This medical approach is generally applied in brain cancer treatments, and it is based on the capture of the neutron reaction ^{10}B (n,α) ^{7}Li, where a ^{10}B atom captures a low-energy thermal neutron and then decays to produce ^{4}He (alpha particles) and ^{7}Li, resulting in a dense ionizing radiation which is capable of destroying the cells where the reaction takes place [12]. Another potential application of BNNTs is in diagnostic medicine. In this sense, BNNTs doped with rare earth beta-emitters with short half-lives, such as ^{153}Sm and ^{159}Gd, can also be used as radioisotopes for imaging [14].

In this context, nanotechnology has revolutionized so-called traditional medicine by introducing novel concepts and methods that had never been imagined. Thus, nanomedicine has improved the diagnosis of various diseases through techniques based on magnetism or nuclear reactions with different electronic devices, using biosensors or radioisotope-doped nanomaterials. In this way, the study of a more accurate diagnostic method using novel technologies is as relevant a goal as the prevention and treatment of oncological diseases. Therefore, a class of new nanomaterials, in which boron nitride nanotubes (BNNTs) stand out, has been the target of studies that have led to an understanding of the correlation between their structure and properties, which enables their use in diagnostic medicine. Due to their empty internal spaces, BNNTs can be filled by different chemical species, such as enzymes, noble metals, rare earths, and radioisotopes, especially copper-64 (64-Cu), which allows this type of material to be applied as a biological marker and in diagnostic medicine. For instance, copper-64 ($T_{1/2}$ = 12.7 h; $\beta+$, 0.653 MeV (17.8%); β^-, 0.579 MeV (38.4%)) has decay characteristics that allow it to be applied to obtain images of positron emission tomography (PET-scan) and in cancer-directed radiotherapy. Copper, for instance, has already well-established coordination chemistry that allows its reaction with an extensive variety of chelating systems that could potentially be linked to peptides and other interesting biological molecules such as antibodies, proteins, and nanoparticles. Its specific half-life expands the ability to image molecules of various dimensions, mainly including the slower compensating proteins and nanoparticles. Due to the versatility of applications of 64-Cu, a significant increase in scientific and technical publications has been seen over the last 2 decades, mainly in PET-scan imaging, but also in targeted cancer radiotherapy.

Thus, this work aimed to synthesize and characterize ^{64}Cu-BNNTs with appreciable properties that suggest numerous multifunctional applications, with advantages for cancer diagnosis and therapy, such as: (i) increased bioavailability; (ii) reduction in systemic adverse effects, thereby increasing patient comfort and adherence to treatment; (iii) improved osteogenic differentiation response promoted by the ^{64}Cu-BNNTs system and targeting of tumor cells, among others. It is also important to mention that the combination of ^{64}Cu-BNNTs has not yet been reported in the literature.

2. Experiment

2.1. Raw Materials

Copper (II) chloride dihydrate (99.999%), iron (III) oxide nano powder (<50 nm particle size) and amorphous boron powder ($\geq$95%) were obtained from Sigma Aldrich Brazil-Ltda, Sao Paulo, Brazil (CAS Number 10125-13-0) and used as received.

2.2. Synthesis and Purification of Boron Nitride Nanotubes

BNNTs were processed from mixing amorphous boron and iron (III) oxide powder (ratio 0.02) in a horizontal tubular reactor. This reactor consisted of an alumina with an inlet and outlet for the flow of ammonia and nitrogen gases. The synthesis was carried out under a NH_3/N_2 atmosphere at a 150/20 sccm (standard cubic centimeters per minute) flow rate with a heating rate of 10 °C min^{-1} from room temperature up to 1200 °C. An isotherm was maintained for 2 h. After this step was completed, the reactor was cooled down to room temperature under a N_2 atmosphere.

The synthesized BNNTs were purified using sulfuric and nitric acids in the ratio of 3:1, respectively. The reaction mixture was kept under stirring and reflux conditions at 80 °C for 2 h, followed by the filtration process. The resulting solid was washed with deionized water and oven-dried for 4 h at 110 °C. In this process, hydroxyl groups (-OH) were introduced into the structure of the tubes.

2.2.1. Activation Process of ^{64}Cu Radioisotope

The radioisotope ^{64}Cu was obtained by neutron activation of the copper (II) chloride dihydrate sample in a nuclear research reactor (TRIGA Mark-1) at CDTN (Belo Horizonte, Brazil) by the neutron capture reaction ^{63}Cu(n,γ)^{64}Cu. The irradiation was performed on 20 mg samples over 8 h under a thermal neutron flux of 6.6 × 1011 cm^{-2}s^{-1}. The theoretical induced activities were estimated according to the research of Zangirolami et al. [15]. The calculations were carried out while considering the amount of Cu in the sample and using the thermal neutron capture cross-sections as a reference, in accordance with an IAEA (International Atomic Energy Agency) publication [16].

2.2.2. Incorporation of Cu and ^{64}Cu to the BNNT Samples

The BNNT (100 mg) sample was dispersed in anhydrous ethanol. With the aid of an autoclave with a polytetrafluoroethylene (PTFE) vessel, the Cu and ^{64}Cu radioisotope were incorporated into the BNNTs. The incorporation reaction was carried out in an oven at a temperature of 180 °C for two hours. After this period, the material was cooled to room temperature and filtered. The radiochemical purity of the sample was assessed by gamma spectroscopy, using an HP-Ge detector (Ortec Ametek, Oak Ridge, TN, USA) with 25% efficiency, and analyzed using the Canberra Genie 2000 software, Meriden, CT, USA [17]. The evaluation of the specific activity was carried out using a CRC 15R activimeter that had been previously calibrated for copper-64 emission. Figure 1a,d schematically show all stages of ^{64}Cu-BNNT production.

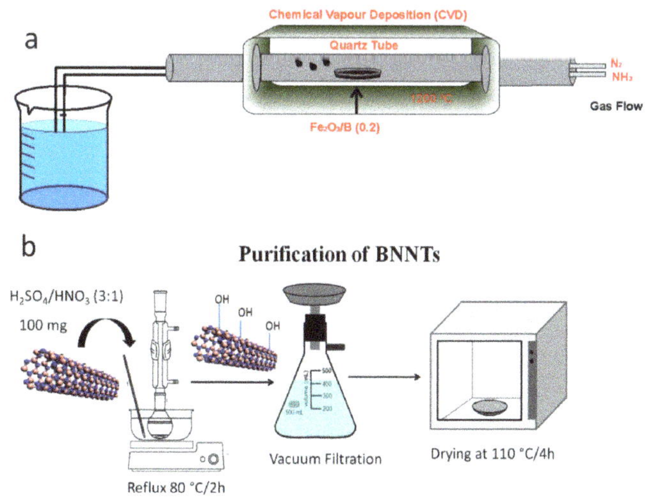

Figure 1. *Cont.*

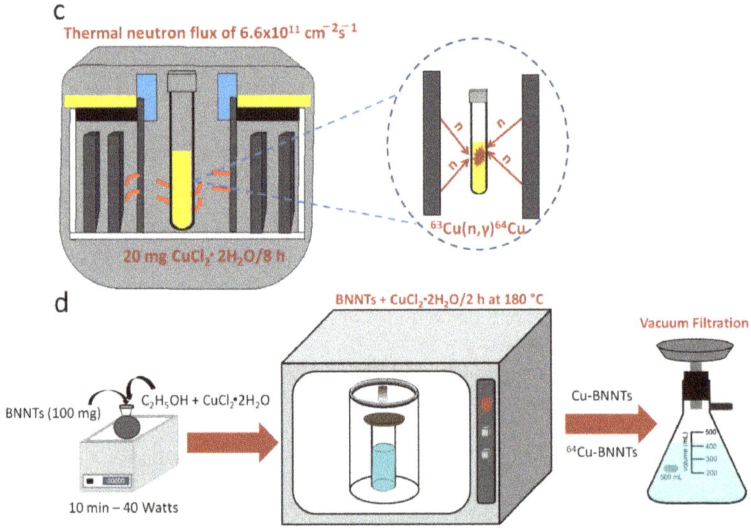

Figure 1. Schematic representation of synthesis of the BNNTS (**a**), purification process (**b**), activation process of $CuCl_2{\cdot}2H_2O$ to obtainment of ^{64}Cu radioisotope (**c**) and incorporation of Cu and ^{64}Cu into BNNTs samples (**d**).

3. Characterization

FTIR measurements of the BNNT and Cu-BNNT samples were performed with a Bruker model Vertex 70v instrument (Belo Horizonte, Brazil). The spectra were collected in ATR mode with 64 accumulations, a resolution of 4 cm^{-1}, and in the 4500–300 cm^{-1} region in transmission mode and then were systematically adjusted; baseline corrections were considered for this analysis. An ultima IV Rigaku Diffractometer with Cu-Kα radiation was employed to study the main crystalline phases in the synthesized BNNT and Cu-BNNT samples (Belo Horizonte, Brazil). The Bragg's angle values were measured in the 10–80° range, with a scanning rate of 0.02° min^{-1}. XPS spectra were obtained using monochromatic Al Kα radiation (1486.6 eV) with an electron energy analyzer (Specs, Phoibos-150) that enabled high-energy resolution and an excellent signal-to-noise ratio (Belo Horizonte, Brazil). The signal of adventitious carbon (C 1 s at 284.6 eV) was used to correct the binding-energy scale of the survey and the high-resolution spectra. High-resolution spectra in the regions of interest were fitted assuming its shape as a convolution of Lorentzian and Gaussian functions of different components, and the background contribution was removed by the Shirley method [18,19]. SEM analysis was performed with a Carl Zeiss Field Emission Scanning Electron Microscope, model sigma VP (Belo Horizonte, Brazil), operating in vacuum with an electron-beam-acceleration voltage between 5 and 30 kV. The BNNT and Cu-BNNT powders were deposited directly onto the carbon tape. The Transmission Electron Microscopy (TEM) images were obtained on a FEI TEM-LaB6 TECNAI G2 microscope (Belo Horizonte, Brazil), with a tungsten-filament electron gun operating at 200 kV. Samples were dispersed in acetone for 30 min using a water bath sonicator and one drop was deposited onto a 200-mesh holey carbon–copper grid. The activity of the ^{64}Cu-BNNTs after irradiation was obtained from the gamma spectrum, using an HP-Ge detector (Belo Horizonte, Brazil), with a nominal efficiency of 25%, and the Canberra Genie 2000 software.

4. Results and Discussion

The FTIR spectrum was obtained in order to identify the vibrational modes in the BNNT samples (Figure 2). The absorption peaks between 3400 and 3200 cm^{-1}could be attributed to the vibrational modes of the hydroxyl groups (-OH) from water molecules adsorbed on the sample surface [20,21]. However, it could also be attributed to the presence of copper hydroxide. The region between 2000 and 60 cm^{-1} has several peaks (Figure 2b,c). The well-known longitudinal (LO) vibrations along the axis resonate sharply around 1369 cm^{-1}, and a second signal (1545 cm^{-1}) appears for tangential (T) circumferential in-plane modes (νB-N). These T modes should be dependent on the diameter (curvature) but seem to only be visible in highly pure, crystalline BNNTs [22]. Another typical absorption peak for BNNTs is located around 790 cm^{-1} and is related to out-of-plane B-N-B bending (δB-N-B) vibrations [20,21,23]. In both spectra, the peaks between the 1100 and 880 cm^{-1} regions give an account of the anti-symmetrical and symmetrical stretching vibrations of B-O bonds in BO_3 and BO_4 groups formed from B-OH, and peaks at 701.3, 685.8 and 453.3 cm^{-1} are assigned to the bend vibrations of B-O bonds in BO_3 and BO_4 groups [24]. The peak at 426.0 cm^{-1} is assigned to the stretching vibrations νCu(II)-O of copper oxide CuO [25].

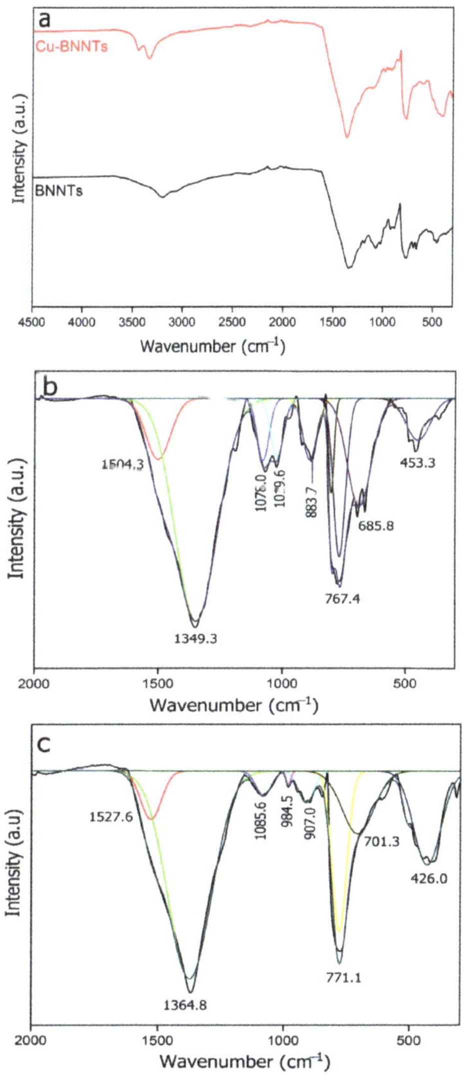

Figure 2. Infrared spectra of (**a**) BNNTs and Cu-BNNTs in the region between 4500 and 60 cm^{-1}. (**a**) Highlighted regions between 2000 and 60 cm^{-1} for (**b**) BNNTs and (**c**) Cu-BNNTs.

The XRD of BNNTs and Cu-BNNTs is shown in Figure 3. An intense peak close to $2\theta = 26.65°$ (Figure 3a) corresponds to the plane (002) and is attributed to the main peak of the h-BN structure. Peaks assigned to h-BN are also observed at $2\theta = 41.78°$, 42.81°, 50.16°, 55.09°, 59.40° and 76.05°, which correspond to the (100), (101), (102), (004), (103) and (110) planes, respectively [14,23,26]. After the introduction of Cu nanoparticles to the BNNTs, new diffraction peaks were observed (Figure 3b),so the region between 30° and 80° was highlighted. The presence of CuO and Cu_2O nanoparticles were identified at 36.89°, 39.71° and 65.3°, which may have occurred due to the exposure of the nanoparticles to the surrounding environment during characterization [27]. The characteristic diffraction peaks of copper nanoparticles located at 32.42° and 44.81° were observed. They correspond to the (110) and (200) crystallographic planes of face-center cubic (fcc), respectively [27,28]. Debye–Scherrer's equation, i.e., $D = 0.9 \times \lambda/(\beta \times \cos\theta)$, was used to calculate the size of copper nanoparticles, where D represents crystalline size, 0.9 is Scherrer's constant, λ is the wavelength of the X-ray, β is the full width at the half-maximum of the diffraction peak (FWHM) and θ represents Bragg's angle [27,29]. The calculations were performed using the mean values of the FWHM of the peaks, with 2θ of 32.4° and 44.8°. The average size of the Cu nanoparticles was 16 nm. This is a dimension in which nanoparticles can penetrate tumor cells through the Increased Permeability and Retention Effect (EPR), thus they can be used as a theranostic nanomaterials.

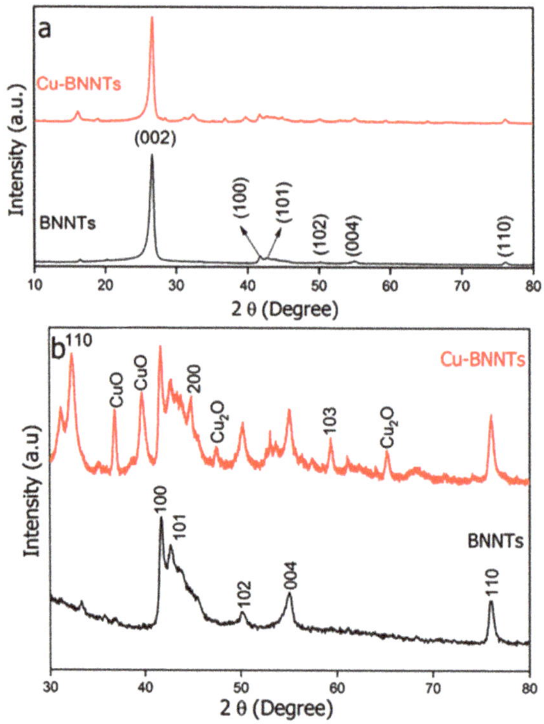

Figure 3. XRD of BNNTs and Cu-BNNTs in the region between 10° and 80° (**a**). Highlighted regions between 30° and 80° (**b**).

Figure 4 shows the XPS survey of the samples. In all the survey spectra, only the presence of B (190.10 eV), N (398.12 eV), C (284.49 eV) and O (532.51 eV) was identified. The C and O signals were also identified on the BNNT surface. The presence of C is related to the surface contamination that usually occurs during the preparation process and to the exposure of the specimens to air, and the presence of O is due to the purification process and is commonly observed in XPS measurements. The presence of F (689.51 eV) and Si (102.53 eV) is due to residues from the vessel that was used for synthesis and the Cl (198.53 eV) comes from the reagent $CuCl_2 \cdot 2H_2O$ that was used for the synthesis of copper nanoparticles.

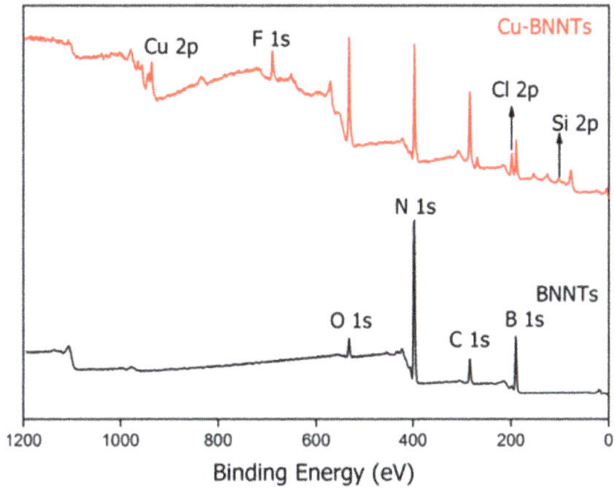

Figure 4. Survey XPS spectra for BNNTs and Cu-BNNTs.

Table 1 shows the data obtained from the XPS survey. The stoichiometry rate of boron and nitrogen atoms (B:N) is confirmed by the peak areas of the XPS survey spectra. The obtained value for the B:N ratio is 1.09. A similar result was obtained by Silva et al. and Juan Li et al. [21,30]. According to these studies, the non-stoichiometric BN nanotubes present an excess of B atoms, and oxygen-doping in the h-BN network leads to the formation of ternary BN_xO_y species [31]. Hydroxyl groups increase in modulus to negative charges on the tubes' surfaces. As copper nanoparticles have positively charged surfaces, we believe that an electrostatic interaction occurs between the two nanostructures.

Table 1. Surface composition (at.%) and B:N ratio, as determined by XPS for Cu BNNT sample.

Elements	Bending Energy (eV)	Atomic Percentage (at.%)
O 1s	532.51	16.047
C 1s	284.49	26.010
N 1s	398.12	21.914
F 1s	689.51	4.179
Cu 2p	936.50	2.771
Cl 2p	198.53	3.004
B 1s	190.10	23.936
Si 2p	102.53	2.138
B:N	-	1.090

The B 1s, N 1s and O 1s core-level photoemission spectra for all samples are shown in Figure 5. The B 1s peak (Figure 5a,b) at 190.1 eV and N 1s peak (Figure 5c,d) at 398.0 eV correspond to the B–N bonding, matching the BE values reported for bulk h-BN [21]. The component at 188.3 eV (Figure 5a,b) is attributed to B bonded to Fe from the catalyst (Fe_2B) [32]. A minor contribution of boron oxide (B_2O_3) was also identified at 192.4 eV (Figure 5a). After the synthesis of Cu nanoparticles, two new contributions were observed at 200.4 and 197.0 eV (Figure 5b); the first is attributed to Cl II and the second to the presence of metallic copper (Cu 0). In addition, the binding energies at 396.6 eV and 400.0 eV are attributed to B–N–B bonding [33] and O-B-N bonding [14], respectively. The O 1s spectra are also shown in Figure 5e,f. Both samples show peaks at 535.5, 532.9 and 530.2 eV, respectively. The first is associated with oxygen atoms bonded to O-H from the purification process, the second is characteristic of B–O bonds (B_2O_3) and the third is associated with oxygen atoms bonded to Fe-O (Fe_2O_3), which is related to the synthesis process [26].

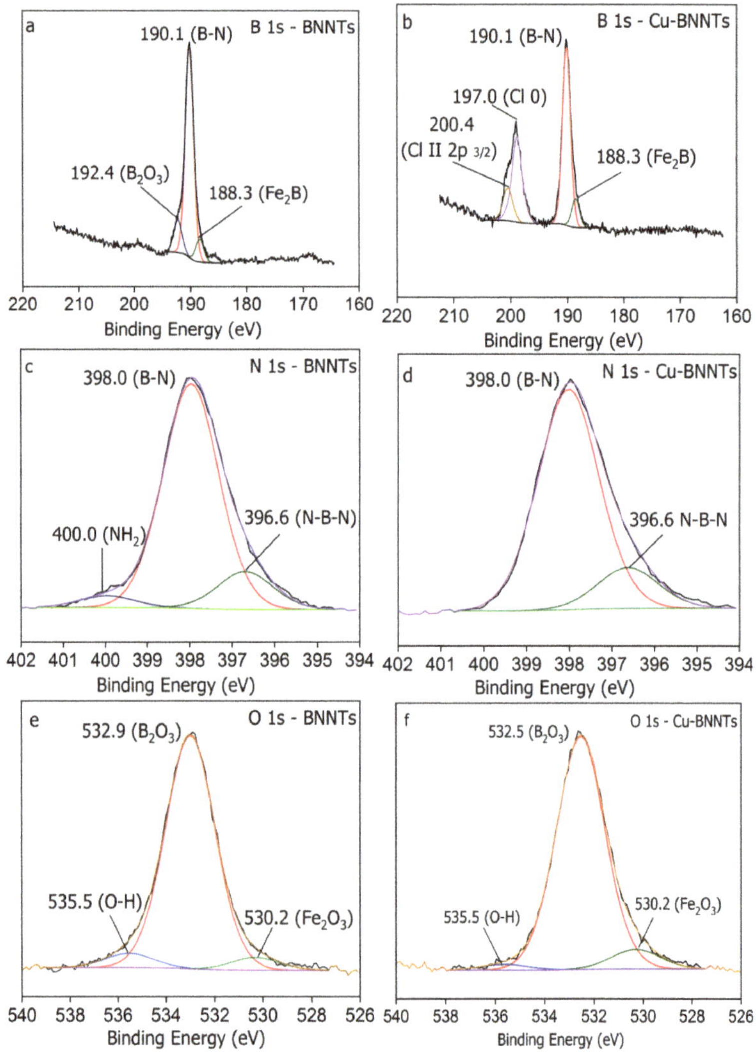

Figure 5. B 1s, N 1s and O 1s high-resolution XPS spectra for (**a**,**c**,**e**) BNNTs, (**b**,**d**,**f**) Cu-BNNTs.

Figure 6 depicts the high-resolution spectrum XPS for Cu-BNNTs. The dominant Cu $2p_{3/2}$ peaks at 933.8 and Cu $2p_{1/2}$ at 953.6 eV, which demonstrate the successful reduction of $CuCl_2$, can be assigned to metallic Cu(0), whereas the small peak at 936.2 eV can be assigned to Cu(II). Meanwhile, the peaks of Cu $2p_{1/2}$ at 955.9 eV, in combination with the satellite peak at 944.3 eV, are typical characteristics of CuO, implying the uniform surface oxidation of Cu nanoclusters that were exposed to air under ambient conditions [34,35]. Cu(0) and Cu(I) are hard to differentiate since they have a ~0.3 eV separation in binding energy, whereas Cu(0) and Cu(II) have more than a 2eV separation. The Cu $2p_{5/2}$ peak around 958.1 eV indicates a Cu(II) oxidation state [36,37].

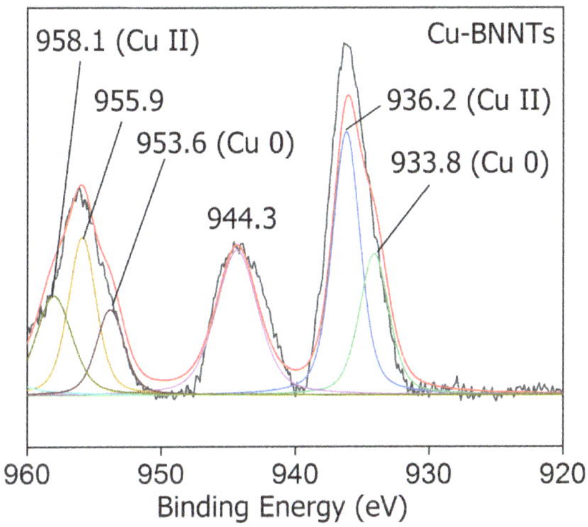

Figure 6. High-resolution spectrum XPS for Cu-BNNTs.

The morphological characteristics of the produced samples were studied using SEM and TEM microscopy, shown in Figures 7 and 8, respectively. It was observed in both samples that the produced BNNTs have non-uniform lengths and diameters. This feature is due to the nanotubes entanglement during their growth process, which is very common during the synthesis process. This nanotube characteristic was also observed in our previous work [23].

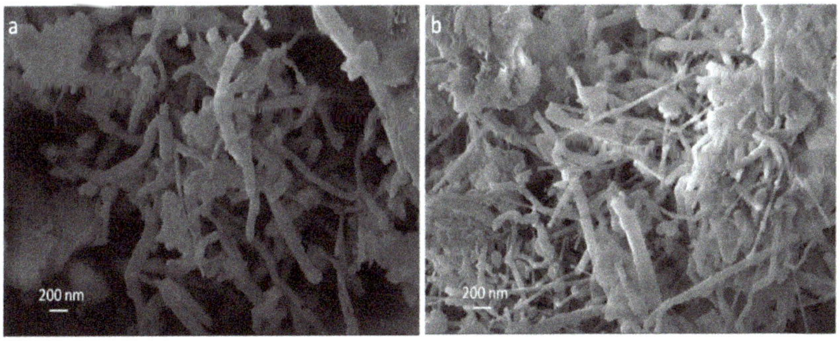

Figure 7. SEM images of BNNTs (**a**) and Cu-BNNTs (**b**).

TEM images are shown in Figure 8a–f. The highlighted regions in Figure 8a,c,e are shown in high-resolution images in Figure 8b,d,f. All images represent typical nanotubes with defined internal channels and external walls that are structurally well-organized [14].

When comparing Figure 8a,b with Figure 8c–f, particles with higher electron absorptions, appearing as darker sites isolated from each other, can be associated with the presence of an electron-conducting nanostructure, indicating the presence of metallic Cu [33]. It is expected that combining these metal nanostructures with BNNTs can improve the stability of the nanoparticles within the solution, allowing their application as theranostic nanomaterials. After increasing image resolution (Figure 8d–f) it was observed that Cu nanoparticles are present in great numbers on the tube surfaces and inside the channels. Referring to the scale bar, it is important to note that the size of nanoparticles is close to 16 nm, as determined by the Debye–Scherrer´s equation.

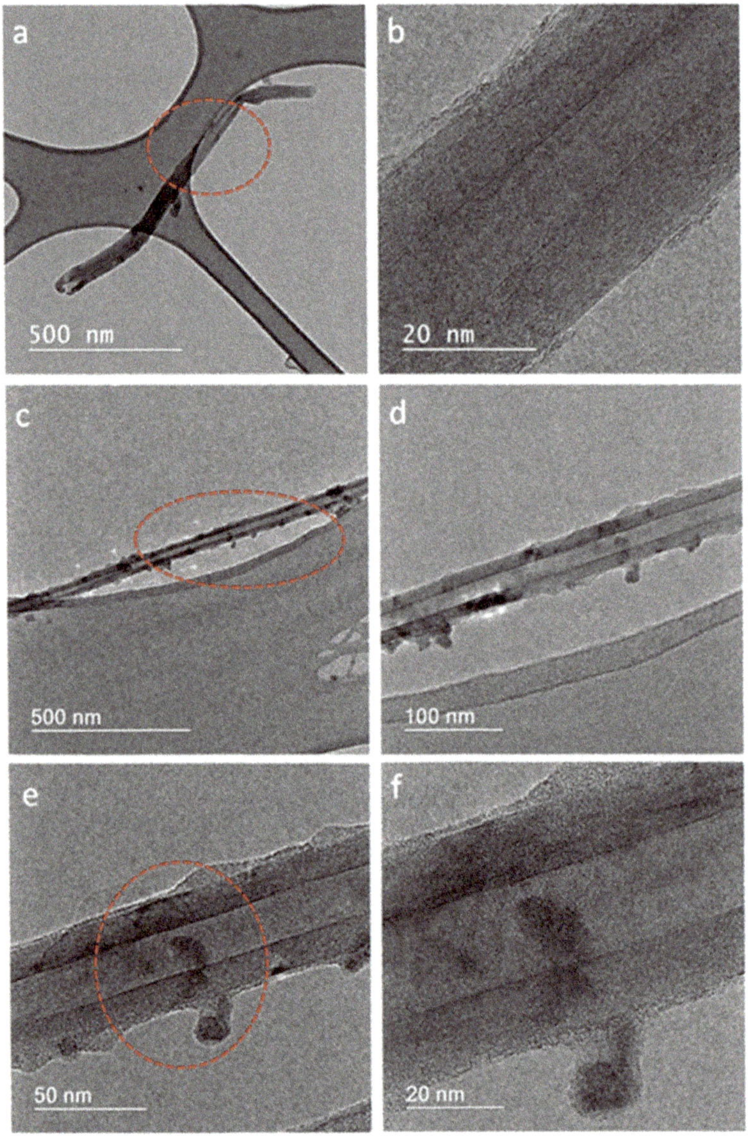

Figure 8. TEM images of BNNTs (**a**,**b**) and Cu-BNNTs (**c**–**f**).

The ^{64}Cu nanoparticles, synthesized from the high-purity $CuCl_2{\cdot}H_20$ by using a solvothermal method, produced stable and contaminant-free radioisotopes, as shown in the gamma spectrum in Figure 9.

The pure photopeak of the ^{64}Cu-BNNTs is 511 keV [38,39]; this energy is compatible with gamma rays for the obtention of images by Photon Emission Tomography (PET-scan), and for cancer treatment due to its β-emissions, with an energy of 579 keV. This result illustrates that the ^{64}Cu-BNNTs can be used as a potential nanomaterial that is able to produce images as well as promote several cancer treatments.

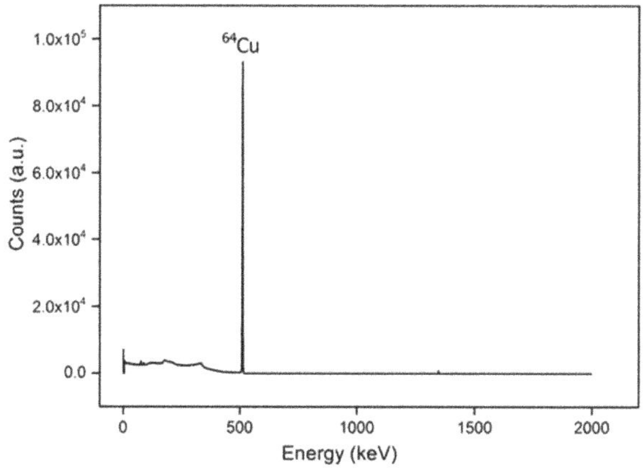

Figure 9. Gamma spectrum of ^{64}Cu-BNNTs.

5. Conclusions

The ^{64}Cu nanostructures were incorporated within the BNNTs structure by a solvothermal method which produces stable and contaminant-free radioisotopes. Using XDR data and Debye–Scherrer's equation, it was possible to determine that metallic Cu nanoparticles have sizes of about 16 nm. The TEM images showed that the BNNTs are structurally well-organized, presenting Cu nanoparticles in their internal channels with an even distribution on their surfaces. The ^{64}Cu nanoparticles in the BNNTs also showed a pure photoemission peak of 511 keV, which is characteristic of gamma radiation. These results corroborate the fact that the studied system has high potential to be used in nuclear medicine as a theranostic material. However, this subject needs to be further explored.

Author Contributions: W.M.S. and H.R. contributed to the conceptualization, methodology, measuring campaign, literature research, project administration, data interpretation, data analysis, validation, formal analysis, resources, investigation, figures, study design, supervision and writing. J.J.T.-T., contributed to the methodology, resources, data interpretation, validation, formal analysis, investigation, figures, and writing. All authors have read and agreed to the published version of the manuscript.

Funding: This research was funded by Mackenzie Research Fund (MackPesquisa, ProjectNo. 181009). Supported by the National Council for Scientific and Technological Development (CNPq), the Coordination for the Improvement of Higher Education Personnel—Brazil (CAPES), and the Universidad de Monterrey.

Institutional Review Board Statement: Not applicable.

Informed Consent Statement: Not applicable.

Data Availability Statement: Not applicable.

Acknowledgments: The authors would like to thank Universidad de Monterrey, Mexico, the Centro de Microscopia da Universidade Federal de Minas Gerais, Belo Horizonte, Brazil, and the Centro de Desenvolvimento da Tecnologia Nuclear—CDTN, Belo Horizonte, Brazil National Council for Scientific and Technological Development (CNPq), and Coordination for the Improvement of Higher Education Personnel—Brazil (CAPES).

Conflicts of Interest: The authors declare no conflict of interest or financial intention.

References

1. Chopra, N.G.; Luyken, R.J.; Cherrey, K.; Crespi, V.H.; Cohen, M.L.; Louie, S.G.; Zettl, A. Boron Nitride Nanotubes. *Science* **1995**, *269*, 966–967. [CrossRef]
2. Terrones, M.; Romo-Herrera, J.M.; Cruz-Silva, E.; López-Urías, F.; Muñoz-Sandoval, E.; Velázquez-Salazar, J.J.; Terrones, H.; Bando, Y.; Golberg, D. Pure and dopedboronnitridenanotubes. *Mater. Today* **2007**, *10*, 30–38. [CrossRef]

3. Golberg, D.; Bando, Y.; Huang, Y.; Terao, T.; Mitome, M.; Tang, C.; Zhi, C. Boron nitride nanotubes and nanosheets. *ACS Nano* **2010**, *4*, 2979–2993. [CrossRef] [PubMed]
4. Chen, Y.; Zou, J.; Campbell, S.J.; le Caer, G. Boron nitride nanotubes: Pronounced resistance to oxidation. *Appl. Phys. Lett.* **2004**, *84*, 2430–2432. [CrossRef]
5. Zhi, C.; Bando, Y.; Tang, C.; Golberg, D. Boron nitride nanotubes. *Mater. Sci. Eng. R Rep.* **2010**, *70*, 92–111. [CrossRef]
6. Golberg, D.; Bando, Y.; Tang, C.; Zhi, C. Boron Nitride Nanotubes. *Adv. Mater.* **2007**, *19*, 2413–2432. [CrossRef]
7. Cohen, M.L.; Zettl, A. The physics of boron nitride nanotubes. *Phys. Today* **2010**, *63*, 34–38. [CrossRef]
8. Roy, S.; Zhang, X.; Puthirath, A.B.; Meiyazhagan, A.; Bhattacharyya, S.; Rahman, M.M.; Babu, G.; Susarla, S.; Saju, S.K.; Tran, M.K.; et al. Structure, Properties and Applications of Two-Dimensional Hexagonal Boron Nitride. *Adv. Mater.* **2021**, *2021*, 2101589. [CrossRef]
9. Ahmad, P.; Khandaker, M.U.; Amin, Y.M.; Khan, Z.R. Influence of growth duration on size and morphology of boron nitride nanotubes grown via chemical vapor deposition technique. *J. Phys. Chem. Solids* **2015**, *85*, 226–232. [CrossRef]
10. Guo, L.; Singh, R.N. Catalytic growth of boron nitride nanotubes using gas precursors. *Phys. E Low-Dimens. Syst. Nanostruct.* **2009**, *41*, 448–453. [CrossRef]
11. Ciofani, G.; Raffa, V.; Menciassi, A.; Cuschieri, A. Boron nitride nanotubes: An innovative tool for nanomedicine. *In Vitro* **2009**, *4*, 8–10. [CrossRef]
12. Ciofani, G.; Danti, S.; Genchi, G.G.; Mazzolai, B.; Mattoli, V. Boron nitride nanotubes: Biocompatibility and potential spill-over in nanomedicine. *Small* **2013**, *9*, 1672–1685. [CrossRef] [PubMed]
13. Soares, D.C.F.; Ferreira, T.H.; Ferreira, C.D.A.; Cardoso, V.N.; de Sousa, E.M.B. Boron nitride nanotubes radiolabeled with 99mTc: Preparation, physicochemical characterization, biodistribution study, and scintigraphic imaging in Swiss mice. *Int. J. Pharm.* **2012**, *423*, 489–495. [CrossRef] [PubMed]
14. da Silva, W.M.; Silva, R.H.d.A.e.; Cipreste, M.F.; Andrade, G.F.; Gastelois, P.L.; Macedo, W.A.d.; de Sousa, E.M.B. Boron nitride nanotubes radiolabeled with 153Sm and 159Gd: Potential application in nanomedicine. *Appl. Radiat. Isot* **2020**, *2020*, 157. [CrossRef] [PubMed]
15. Zangirolami, D.M.; Oliveira, A.H.; Ferreira, A.V. Thermal and Epithermal Neutron Fluence Rates in the Irradiation Facilities of the TRIGA IPR-R1 Nuclear Reactor. *Brazilian J. Phys.* **2010**, *2021*, 40. [CrossRef]
16. Mughabghab, S.F. Resonance Integrals and G-Factors. *Int. Nucl. Data Commun.* **2003**, *11376*, 7.
17. Hossain, I.; Sharip, N.; Viswanathan, K.K. Efficiency and resolution of HPGe and NaI(Tl) detectors using gamma-ray spectroscopy. *Sci. Res. Essays* **2012**, *7*, 86–89. [CrossRef]
18. Végh, J. The Shirley background revised. *J. Electron Spectros. Relat. Phenomena* **2006**, *151*, 159–164. [CrossRef]
19. Végh, J. The analytical form of the Shirley-type background. *J. Electron Spectros. Relat. Phenomena* **1988**, *46*, 411–417. [CrossRef]
20. Nazar, A.M.D.; Silva, W.M.; de Oliveira, E.F.; de Sousa, E.M.B. Boron nitride nanotubes decorated with magnetite nanoparticles for application as a multifunctional system in cancer treatment. *Nano-Struct. Nano-Objects* **2020**, *24*, 100616. [CrossRef]
21. da Silva, W.M.; Monteiro, G.A.A.; Gastelois, P.L.; de Sousa, R.G.; Macedo, W.A.d.; Sousa, E.M.B. Efficient sensitive polymer-grafted boron nitride nanotubes by microwave-assisted process. *Nano-Struct. Nano-Objects* **2018**, *15*, 186–196. [CrossRef]
22. Harrison, H.; Lamb, J.T.; Nowlin, K.S.; Guenthner, A.J.; Ghiassi, K.B.; Kelkar, A.D.; Alston, J.R. Quantification of hexagonal boron nitride impurities in boron nitride nanotubes: Via FTIR spectroscopy. *Nanoscale Adv.* **2019**, *1*, 1693–1701. [CrossRef]
23. da Silva, W.M.; Ribeiro, H.; Ferreira, T.H.; Ladeira, L.O.; Sousa, E.M.B. Synthesis of boron nitride nanostructures from catalyst of iron compounds via thermal chemical vapor deposition technique. *Phys. E Low-Dimens. Syst. Nanostruct.* **2017**, *89*, 177–182. [CrossRef]
24. Petrov, T.; Markova-Deneva, I.; Chauvet, O.; Nikolov, R.; Denev, I. Sem and ft-ir spectroscopy study of Cu, Sn and Cu-Sn nanoparticles. *J. Univ. Chem. Technol. Metall.* **2012**, *47*, 197–206.
25. Amelkovich, Y.A.; Nazarenko, O.B.; Sechin, A.I.; Visakh, P.M. Characterization of copper nanopowders after natural aging. *IOP Conf. Ser. Mater. Sci. Eng.* **2015**, *81*, 012072. [CrossRef]
26. da Silva, W.M.; Ferreira, T.H.; de Morais, C.A.; Leal, A.S.; Sousa, E.M.B. Samarium doped boron nitride nanotubes. *Appl. Radiat. Isot* **2018**, *131*, 30–35. [CrossRef]
27. Chauhan, P.S.; Shrivastava, V.; Tomar, R.S. Biofabrication of copper nanoparticles: A next-generation antibacterial agent against wound-associated pathogens. *Turkish J. Pharm. Sci.* **2018**, *15*, 238–247.
28. Betancourt-Galindo, R.; Reyes-Rodriguez, P.Y.; Puente-Urbina, B.A.; Avila-Orta, C.A.; Rodríguez-Fernández, O.S.; Cadenas-Pliego, G.; Lira-Saldivar, R.H.; García-Cerda, L.A. Synthesis of copper nanoparticles by thermal decomposition and their antimicrobial properties. *J. Nanomater.* **2014**, *2014*, 10. [CrossRef]
29. Shirui, Z.; Yabin, D.; Mingyan, Y.; Xiaolong, G.; Xinwei, X.; Yupeng, J.; Yang, X. Microwave annealing effects on ZnO films deposited by atomic layer deposition. *J. Semicond.* **2014**, *35*, 1–7.
30. Li, J.; Li, J.; Yin, Y.; Chen, Y.; Bi, X. Water-assisted chemical vapor deposition synthesis of boron nitride nanotubes and their photoluminescence property. *Nanotechnology* **2013**, *24*, 365605. [CrossRef]
31. Hubácek, M.; Sato, T.; Ishii, T. A Coexistence of Boron Nitride and Boric Oxide. *J. Sol.* **1994**, *109*, 384–390. [CrossRef]
32. Ferreira, T.H.; Marino, A.; Rocca, A.; Liakos, I.; Nitti, S.; Athanassiou, A.; Mattoli, V.; Mazzolai, B.; de Sousa, E.M.B.; Ciofani, G. Folate-grafted boron nitride nanotubes: Possible exploitation in cancer therapy. *Int. J. Pharm.* **2015**, *481*, 56–63. [CrossRef]

33. Monteiro, G.A.A.; de Sousa, R.G.; da Silva, W.M.; Gastelois, P.L.; Macedo, W.A.d.; de Sousa, E.M.B. Microwave radiation-assisted covalent functionalization of boron nitride nanotubes and their grafting with cationic thermo and pH-sensitive hydrogel. *Appl. Nanosci.* **2021**, *11*, 505–520. [CrossRef]
34. Jin, Z.; Liu, C.; Qi, K.; Cui, X. Photo-reduced Cu / CuO nanoclusters on TiO_2 nanotube arrays as highly efficient and reusable catalyst. *Sci. Rep.* **2017**, *7*, 39695. [CrossRef] [PubMed]
35. Ghosh, S.; Das, R.; Chowdhury, I.H.; Bhanja, P.; Naskar, M.K. RSC Advances. *RSC Adv.* **2015**, *5*, 101519–101524. [CrossRef]
36. Mondal, P.; Sinha, A.; Salam, N.; Roy, A.S.; Jana, N.R.; Islam, S.M. Enhanced catalytic performance by copper nanoparticle-graphene based composite. *RSC Adv.* **2013**, *3*, 5615–5623. [CrossRef]
37. Fernandez, V.; Kiani, D.; Fairley, N.; Felpin, F.X.; Baltrusaitis, J. Curve fitting complex X-ray photoelectron spectra of graphite-supported copper nanoparticles using informed line shapes. *Appl. Surf. Sci.* **2020**, *505*, 143841. [CrossRef]
38. Kim, J.Y.; Park, H.; Lee, J.C.; Kim, K.M.; Lee, K.C.; Ha, H.J.; Choi, T.H.; An, G.I.; Cheon, G.J. A simple Cu-64 production and its application of Cu-64 ATSM. *Appl. Radiat. Isot* **2009**, *67*, 1190–1194. [CrossRef]
39. Pérot, B.; Jallu, F.; Passard, C.; Gueton, O.; Allinei, P.-G.; Loubet, L.; Estre, N.; Simon, E.; Carasco, C.; Roure, C.; et al. The characterization of radioactive waste: A critical review of techniques implemented or under development at CEA. *EPJ Nucl. Sci. Technol.* **2018**, *4*, 3. [CrossRef]

nanomaterials MDPI

Article

Development of a Novel Methotrexate-Loaded Nanoemulsion for Rheumatoid Arthritis Treatment with Site-Specific Targeting Subcutaneous Delivery

Parvathy Suresh [1], Mounir M. Salem-Bekhit [2,3], Hafsa Palathum Veedu [1], Sultan Alshehri [4,*], Sreeja Chandrasekhar Nair [1], Sarah I. Bukhari [4], Vidya Viswanad [1,*], Ehab I. Taha [4], Ram Kumar Sahu [5,*], Mohammed M. Ghoneim [6] and Ibrahim Elbagory [7]

1 Department of Pharmaceutics, Amrita School of Pharmacy, Amrita Vishwa Vidyapeetham, AIMS Health Science Campus, Kochi 682041, India; sparvathy09@yahoo.in (P.S.); hafsapv1@gmail.com (H.P.V.); sreeju2u@gmail.com (S.C.N.)
2 Kayyali Chair for Pharmaceutical Industry, Department of Pharmaceutics, College of Pharmacy, King Saud University, P.O. Box 2457, Riyadh 11451, Saudi Arabia; mbekhet@ksu.edu.sa
3 Department of Microbiology and Immunology, Faculty of Pharmacy, Al-Azhar University, Cairo 11884, Egypt
4 Department of Pharmaceutics, College of Pharmacy, King Saud University, Riyadh 11451, Saudi Arabia; sbukhari@ksu.edu.sa (S.I.B.); eelbadawi@ksu.edu.sa (E.I.T.)
5 Department of Pharmaceutical Science, Assam University (A Central University), Silchar 788011, India
6 Department of Pharmacy Practice, College of Pharmacy, AlMaarefa University, Ad Diriyah 13713, Saudi Arabia; mghoneim@mcst.edu.sa
7 College of Pharmacy, Northern Border University, Arar 1321, Saudi Arabia; ibrahim.elbagory@nbu.edu.sa
* Correspondence: salshehri1@ksu.edu.sa (S.A.); vidyaviswanad@aims.amrita.edu (V.V.); ramkumar.sahu@aus.ac.in (R.K.S.)

Citation: Suresh, P.; Salem-Bekhit, M.M.; Veedu, H.P.; Alshehri, S.; Nair, S.C.; Bukhari, S.I.; Viswanad, V.; Taha, E.I.; Sahu, R.K.; Ghoneim, M.M.; et al. Development of a Novel Methotrexate-Loaded Nanoemulsion for Rheumatoid Arthritis Treatment with Site-Specific Targeting Subcutaneous Delivery. *Nanomaterials* **2022**, *12*, 1299. https://doi.org/10.3390/nano12081299

Academic Editor: Angelo Ferraro

Received: 8 March 2022
Accepted: 5 April 2022
Published: 11 April 2022

Abstract: Rheumatoid arthritis (RA) is a systemic, chronic autoimmune disease that causes disability due to progressive inflammation and destruction of the tissues around the joints. Methotrexate is mainly used to prevent the progression of joint destruction and reduce the deformity. The major challenge in treating RA with methotrexate is the systemic side effects that limit dose escalation. Hence, a novel formulation of a methotrexate-loaded nanoemulsion for subcutaneous administration was developed that aims to deliver methotrexate into the system via the lymph. The methotrexate-loaded nanoemulsion was prepared by using the aqueous-titration method. The prepared nanoemulsion was investigated for particle size, surface charge, surface morphology, entrapment efficiency, DSC (differential scanning colorimetry), drug release, hemocompatibility assay, and cytotoxicity, as well as anti-arthritic and stability studies. The vesicle size, zeta potential, PDI (polydispersity index), and entrapment efficiency of the optimized nanoemulsion were 87.89 ± 2.86 nm, 35.9 ± 0.73 mV, 0.27, and 87 ± 0.25%, respectively. The DSC study showed that the crystalline methotrexate was converted to an amorphous form and the drug was fully incorporated into the vesicles. After 72 h, the optimized nanoemulsion showed a drug release of 96.77 ± 0.63%, indicating a sustained-release dosage form. Cytocompatibility testing by MTT (3-(4,5-dimethylthiazol-2-yl)-2,5-diphenyl-2H-tetrazolium bromide) assay on macrophage cell lines showed that the nanoemulsion was non-toxic. The formulation showed significant anti-arthritic activity compared to the marketed drug solution. In addition, the nanoemulsion containing methotrexate remained stable for three months when stored at a low temperature. Since the nanoemulsion containing methotrexate has excellent physicochemical properties and lowers systemic side effects by targeted delivery, it is a desirable technology for subcutaneous drug delivery.

Keywords: rheumatoid arthritis; methotrexate; anti-arthritic activity; nanoemulsion; hemocompatibility; MTT assay; stability studies

1. Introduction

Rheumatoid arthritis (RA) is a systemic autoimmune disease characterized by chronic inflammation of the joints and associated tissues. Although the inflammation is primarily associated with the joints, other organs in the body are also affected [1,2]. RA has a prevalence of approximately 0.24%, with a higher prevalence in females than males [3]. There is no clear cause for the disease, but genetic factors, environmental factors, and lifestyle changes trigger the inflammatory response in the body. Hormonal imbalance alters cortisol and androgen levels triggered by stress and leads to inflammatory response [4,5]. It is believed that the influx of inflammatory cells (T cells and B cells) into the pannus and synovial fluid is the cause of the pathology of RA leading to complete tissue destruction [6]. Inflammation causes an increase in the production of cytokines, which results in joint damage, synovitis, and edema. Synovial fluid contains a large number of immune cells, most of which are monocytes and mast cells. In addition, there are a small number of adaptive immune cells, such as plasma cells, Th1 (T-helper type 1), Th17, and B cells, which are involved in the inflammatory process. Further, the joint fluid contains high concentrations of antibodies against citrullinated proteins, which stimulates the release of complement proteins, which in turn trigger the onset of the inflammatory response [7].

Treatment aims to reduce joint inflammation and pain as well as tissue abnormalities and deformities in order to restore life's quality. Corticosteroids, NSAIDs (non-steroidal anti-inflammatory drugs), and opioids are considered to be first-line drug therapies for inflammatory pain and swelling. These medications are given for a limited period and are intended to relieve the pain and swelling associated with inflammation. The prevention of further joint damage and reduction of deformity are the primary goals of therapy [8,9]. The nonbiologic drugs methotrexate, sulfasalazine, hydroxychloroquine, and leflunomide are the most commonly used disease-modifying antirheumatic drugs (DMARDs) [10]. Fewer people take gold salts, azathioprine, and cyclosporine, among other rare medications. In RA, biologic DMARDs consist of monoclonal antibodies and recombinant receptors that inhibit the production of cytokines that contribute to the prevention of inflammation [11,12].

Methotrexate is the most important component of DMARDs, and its structure is shown in Figure 1. It is a folate derivative that suppresses pyrimidine and purine production [12]. Methotrexate reduces inflammation in RA by reducing the proliferation of inflammatory cells and lymphocytes, resulting in a decrease in the number of active inflammatory cells. Methotrexate increases the level of intracellular adenosine, which interacts with certain cell-surface receptors to inhibit the synthesis of the chemokines and leukotriene B4 as well as the regulation of synovial collagenase genes [13].

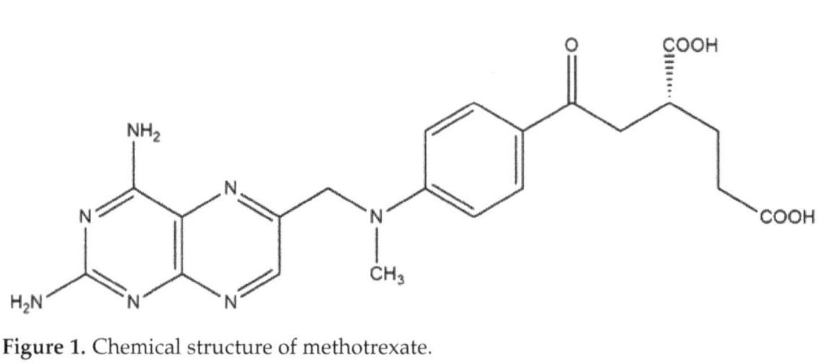

Figure 1. Chemical structure of methotrexate.

The systemic side effects of methotrexate, which limit dose escalation, are a significant obstacle in treating RA with this drug. The use of encapsulated carriers for targeted delivery may reduce the need for excessive and frequent dosing. When it comes to the pathophysiology of RA, the lymphatic system plays an important role [14,15]. Anti-inflammatory drugs can be delivered through the lymphatic system to improve bioavailability and therapeutic efficacy while reducing adverse effects [16]. Drugs can be more effectively targeted to the lymphatic system when encapsulated in carriers ranging in size from micro- to nanoscale.

The use of encapsulated carriers, particularly nanoparticles, can contribute to delayed release while reducing the amount of drugs used [17,18].

When nanoparticle carriers are administered intravenously, they are consumed by reticular endothelial cells, preventing the drug from reaching the lymphatic system. Subcutaneous administration can be used for lymphatic targeting because of the permeability of capillaries, which allows the transport of smaller molecules [19]. Subcutaneous delivery of drug-loaded nano-molecules and their uptake into the lymph make it a suitable route that has a positive impact on the ongoing treatment profile of RA. Because of the above statement, it was planned to prepare and characterize a novel methotrexate-loaded nanoemulsion for the treatment of RA by targeted subcutaneous administration with improved bioavailability and therapeutic efficacy. The use of the nanoemulsion as a carrier can improve the bioavailability of methotrexate. The toxicity of the nanoemulsion as a carrier was studied using the cytotoxic and hemolysis assays to ensure that it was safe to use. In addition, CFA (complete Freund adjuvant)-induced arthritis in rats was used as a disease paradigm to evaluate the antirheumatic effect of methotrexate-loaded nanoemulsion.

2. Results and Discussions

2.1. Particle Size Study

The size of the particles plays a significant influence in their uptake and retention in lymphoid tissue. When the particle size of different formulations was determined using a zetasizer by differential light scattering, a decrease in particle size was observed with the addition of a mixture of surfactant and co-surfactant. The optimized nanoemulsion loaded with methotrexate had a mean particle size of 87.89 ± 2.86 nm, whereas the particle size of the nanoemulsion without the drug was 76.39 ± 1.53 nm (Figure 2). The larger size of the nanoemulsion loaded with methotrexate compared to the nanoemulsion without the drug indicates that the drug was successfully loaded into the nanoemulsion. It has been documented that a nanoemulsion with a particle size smaller than 100 nm increases the transport of the drug through the lymphatic capillaries, which absorb the drug from the interstitial space [20]. Moreover, this particular size of the nanoemulsion can be easily absorbed and retained in the lymphoid tissue, resulting in improved therapeutic efficacy compared to conventional drugs The PDI of the optimized nanoemulsion loaded with methotrexate was 0.27 (Figure 2). The nanoemulsion has smaller vesicles with a low PDI. The PDI reflects the potential of the size distribution of the vesicles in the nanoemulsion. In addition, the PDI is one of the most effective techniques with which to evaluate the homogeneity and stability of the nanoemulsion vesicles [21,22]. A higher PDI value of about 0.7 shows that the formulation has a very wide particle-size distribution, while values of 0.2 and below are mostly agreeable formulations. A PDI value of 0.3 and below is considered acceptable and indicates a uniform distribution of lipiodol vesicles [22]. The size distribution of the optimized nanoemulsion was in agreement with the particle size.

Zeta potential is an indicator of the stability of nanoemulsion. Determination of the zeta potential gave a value of −35.9 ± 0.73 mV for the optimized nanoemulsion loaded with methotrexate, while the zeta potential for nanoemulsion without drug was −33.9 ± 1.03 mV (Figure 3). The values of the zeta potential confirmed that the prepared nanoemulsion was negatively charged droplets with the capacity of higher lymphatic uptake and longer retention time. It measures the electrical repulsion force between the particles. It has been documented a higher zeta potential (>−30 mV) is advantageous to vesicle's physical stability as it avoids accumulation between vesicles owing to electrostatic repulsion resulting in a stable emulsion. The low zeta potential of emulsions leads to coagulation or flocculation of vesicles, resulting in poor physical stability [23–25]. According to these findings, we concluded that the selected composition of drug, oil, and surfactant makes the nanoemulsion electrically stable.

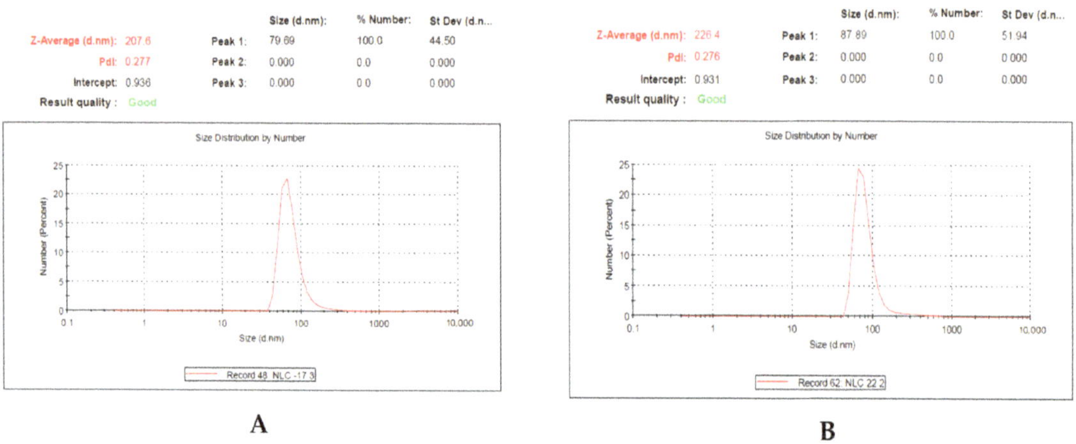

Figure 2. Data of particle-size distribution and zeta potential of the prepared nanoemulsion; (**A**): Nanoemulsion without drug; (**B**): Methotrexate-loaded nanoemulsion.

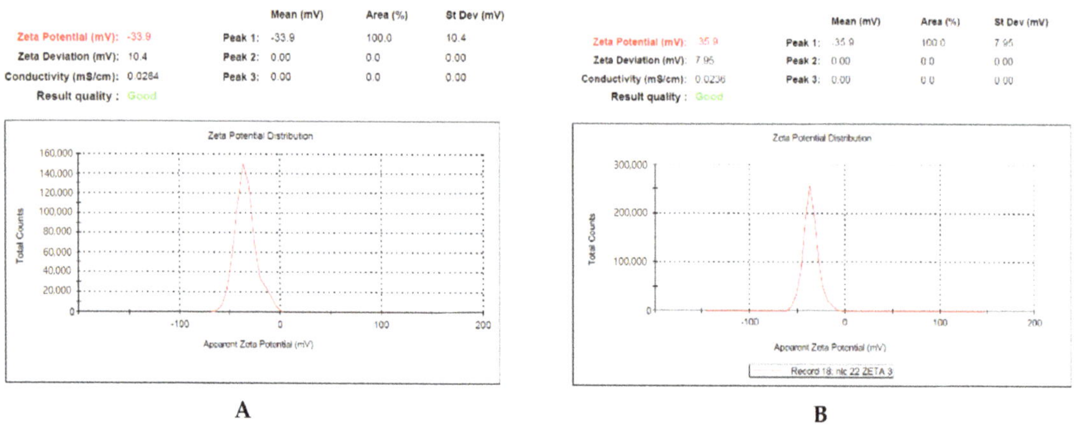

Figure 3. Data of zeta potential. (**A**): Nanoemulsion without drug; (**B**): Methotrexate-loaded nanoemulsion.

2.2. Morphology of Particle

The SEM (scanning electron microscope) photograph of the methotrexate-loaded nanoemulsion shows that the particles have almost spherical morphology, are uniform and monodispersed (Figure 4). A nanoemulsion is a kinetically stable, dispersed system with particles of small droplet size. The spherical shapes indicate the stability of the system without aggregation and gravitational separation. This stability can be attributed to the presence of surface-active agents. The study conducted by Zhou et al. shows that the spherical nanoemulsion can rapidly penetrate the cell [26]. The results from SEM are consistent with the result of particle-size characterization.

The TEM (transmission electron microscopy) image of methotrexate-loaded nanoemulsion shows a particle size of less than 100 nm with a spherical shape (Figure 5). The results of the particle size and SEM study support the findings of TEM. El-Refai et al. and Antil et al. documented a spherical and uniform distribution of vesicles in the nanoemulsion prepared with sesame oil and Tween 80 in their TEM study [27,28]. The results of TEM are consistent with the study in which the nanoemulsion was prepared with sesame oil and Tween 80.

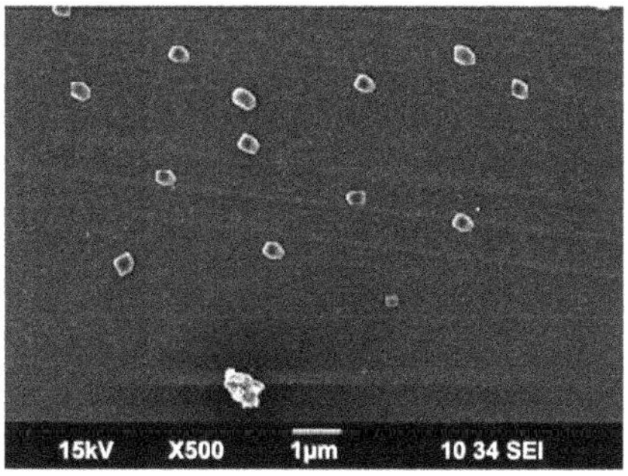

Figure 4. Representation of SEM findings of methotrexate-loaded nanoemulsion.

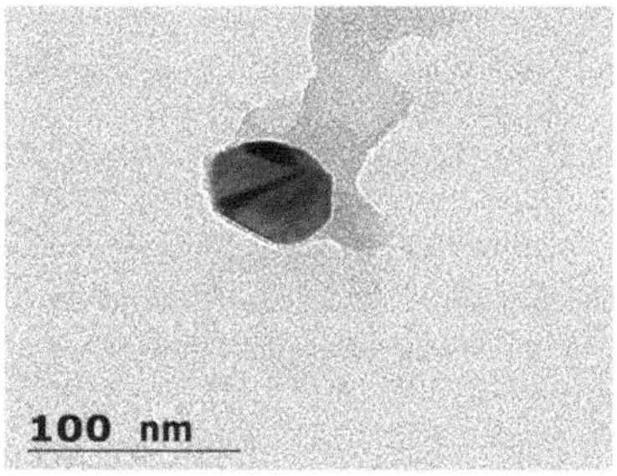

Figure 5. Representation of TEM findings of methotrexate-loaded nanoemulsion.

2.3. DSC Study

The DSC thermogram of methotrexate showed an endothermic peak at 104.74 °C, which is related to the melting point of the methotrexate, and a peak endothermic peak at 82.57 °C, which resembles the melting point of the methotrexate loaded nanoemulsion (Figure 6). The DSC thermogram shows a sharp endothermic peak of methotrexate, proving its crystalline state. The endothermic peak of methotrexate did not appear in the DSC thermogram of the nanoemulsion, and the broadened thermal peak of the nanoemulsion confirms that methotrexate was fully incorporated into the vesicles. The different thermal peak of methotrexate and the nanoemulsion justifies that the drug is in an amorphous state. Moreover, as the temperature increases, there is an interaction between methotrexate and oil, which forms a complex that has a lower melting point (82.57 °C) than pure methotrexate [29,30].

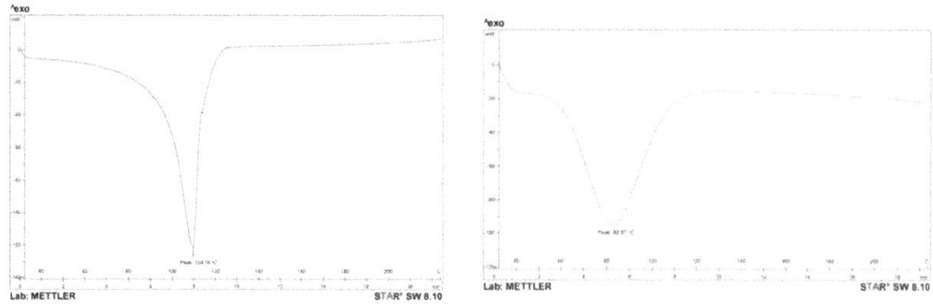

(A) (B)

Figure 6. Differential scanning colorimetry of pure-drug and methotrexate-loaded formulations: (**A**) Pure drug; (**B**) Methotrexate-loaded formulation.

2.4. Entrapment Efficiency of Nanoemulsion

To determine the quantity of drugs successfully incorporated into the nanoparticles, the entrapment efficiency must be measured. The entrapment efficiency of the nanoemulsion was calculated using an indirect method by measuring the free drug available in the aqueous phase. The entrapment efficiency of the methotrexate-loaded nanoemulsion was found to be 87.25%, indicating high efficacy. The higher inclusion of methotrexate may be due to its stronger incorporation into the hydrophobic core. Antil et al. reported a 78% encapsulation efficiency of the metaxalone-loaded nanoemulsion containing sesame oil and Tween 80 [28]. The encapsulation-efficiency results are in agreement with the results of Antil et al.

2.5. In Vitro-Release Study

In vitro release from the methotrexate-loaded nanoemulsion was performed using a dialysis membrane that showed a biphasic release pattern. In this formulation, the drug was released in a burst at the beginning of the test, followed by a sustained release. The methotrexate-loaded nanoemulsion exhibited a 96.77% release after 72 h (Figure 7). The rapid release of the drug of 13.39 ± 1.14 from the nanoemulsion was observed during the first 2 h of the experiment. The suppressed drug release was observed after 2 h, indicating prolonged drug release from the nanoemulsion. The results are in agreement with other studies that addressed the biphasic release of methotrexate from the nanoemulsion. In this regard, Rashid et al. and Rathee et al. showed drug release of 72% after 20 h and 55% after 24 h, respectively [31–33]. Initially, the rapid release of the drug from the nanoemulsion could be caused by the methotrexate that is adsorbed on the droplet surface or dispersed in the surfactant. The sustained release indicates that the nano-sized drug is retained within the nanoemulsion droplet. In addition, the drug release in the nanoemulsion is also controlled by the interactions of the drug with the surfactants and its distribution between the aqueous and oil phases. The delayed drug release is of great interest for systemic delivery, especially for arthritis. The results suggest that the sustained release of the methotrexate-loaded nanoemulsion is a better formulation for the treatment of arthritis.

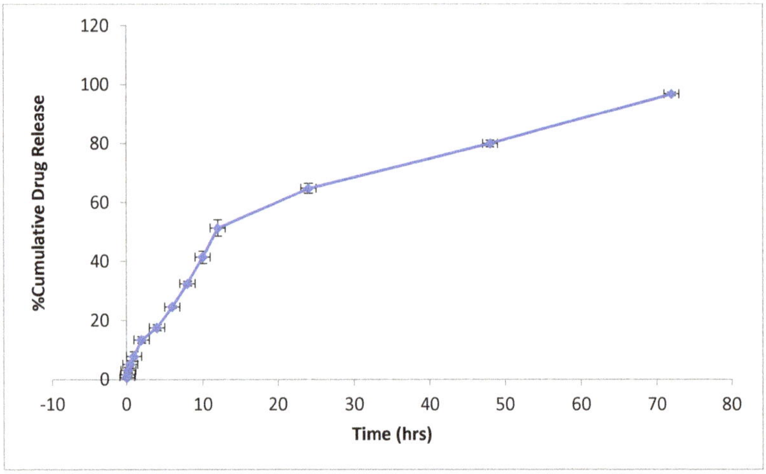

Figure 7. Graphical representation of drug release from the nanoemulsion loaded with methotrexate.

The order and mechanism of drug release from the nanoemulsion were investigated by applying the drug-release data to the zero-order, first-order, Higuchi and Korsmeyer–Peppas models. It was found that the drug release from the nanoemulsion was most closely associated with the Higuchi model, based on the R^2 values. The diffusion of methotrexate from the oily core and interface is retarded by the aqueous medium, which acts as a physical barrier to the release of the drug due to its poor solubility in water, a likely cause of prolonged release. It would appear that drug release is controlled by diffusion if the n value from the Korsmeyer–Peppas equation is used (Figure 8).

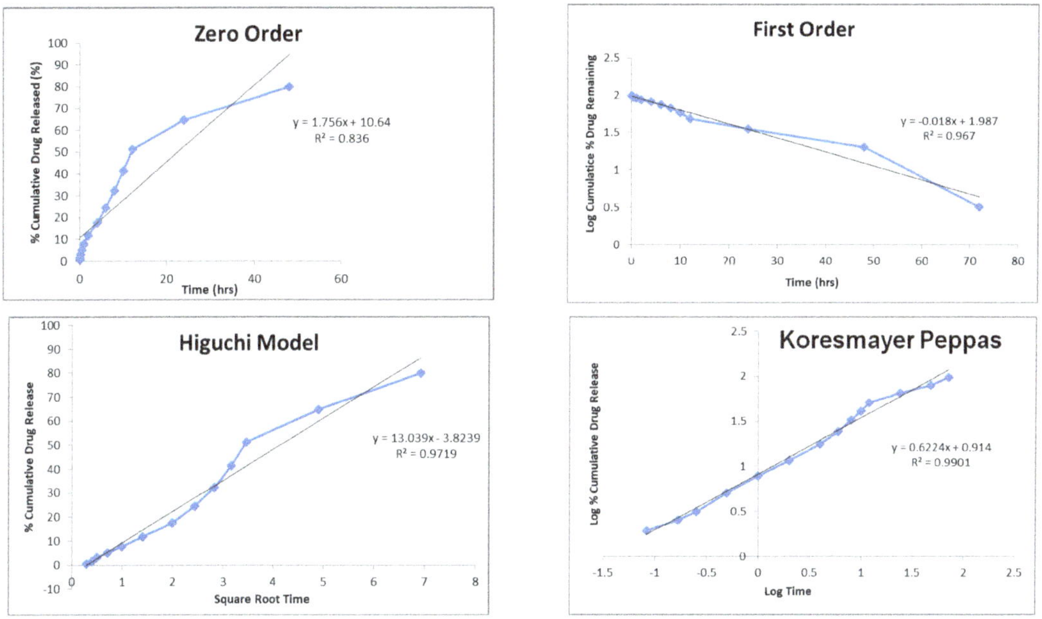

Figure 8. Graphical representation of kinetic-data analysis of methotrexate-loaded nanoemulsion.

2.6. Hemocompatibility Analysis of Nanoemulsion

The nanoemulsions loaded with methotrexate were tested and showed negligible hemolysis after a prescribed incubation period. The different concentrations of methotrexate-loaded nanoemulsion showed hemolysis ranging from 0.14 to 1.27% (Figure 9). The percentage of hemolysis in the samples does not increase significantly as the concentration of nanoemulsion increases. The marketed methotrexate drug solution exhibited 1.98% hemolysis. It has been documented that materials with hemolysis greater than 5% are considered hemolytic, while materials with hemolysis between 5 and 2% are termed as slightly hemolytic, and materials with less than 2% hemolysis are considered very hemocompatible [34]. The results indicate that the nanoemulsion is less than 2% hemolytic. This suggests that the prepared nanoemulsion is very hemocompatible and has no toxic effect on blood vessels. The nanoemulsion loaded with methotrexate showed less hemolysis compared to the marketed drug solution, indicating that the nanoemulsion is much safer compared to the conventional formulation.

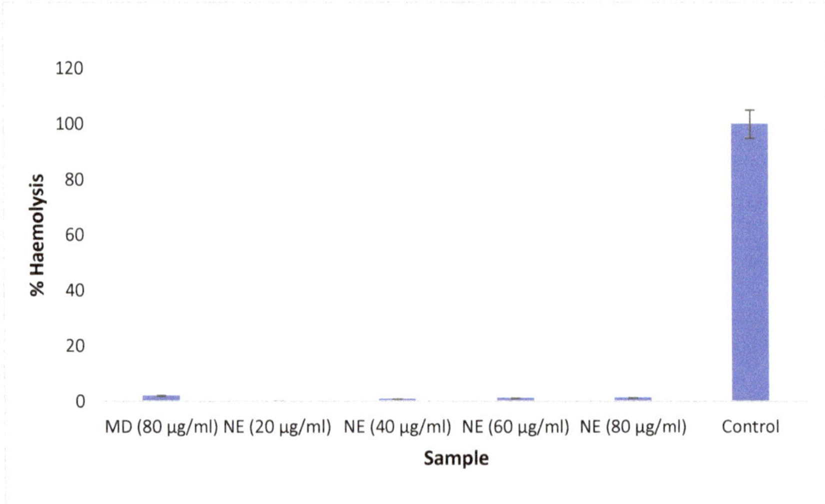

Figure 9. Percentage hemolysis of methotrexate-loaded nanoemulsion and pure methotrexate solution. NE: Methotrexate-loaded nanoemulsion; MD: Marketed drug.

2.7. Cytotoxic Assay

The MTT study was used to detect the living cell's metabolic activity in order to evaluate the cell viability of methotrexate-loaded nanoemulsion against the macrophage cell line RAW 264.7. The cell viability of RAW 264.7 cells at various concentrations of methotrexate-loaded nanoemulsion ranged from 95.79% to 60.21% (Figure 10). It was found that the cell viability of RAW 264.7 cells decreased in direct proportion to the concentration of methotrexate-loaded nanoemulsion. This indicates that the higher concentration of nanoemulsion induces mild cell cytotoxicity. The RAW 264.7 cell line exhibited a cell viability of less than 88% at concentrations of 6.25 and 12.5 µg/mL in the nanoemulsion. The results of hemolysis support the results of the cytotoxicity assay. In addition, the findings indicate that the higher dose of methotrexate-loaded nanoemulsion may cause mild toxicity to cells, while the lower dose is safe.

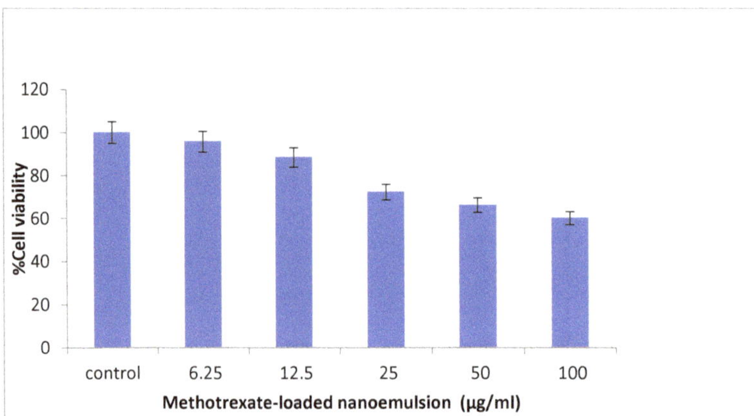

Figure 10. Graphical representation of in vitro cytotoxic effect of methotrexate-loaded nanoemulsion.

2.8. In Vivo Anti-Arthritic Activity of Nanoemulsion

The methotrexate-loaded nanoemulsion was intended to investigate the anti-arthritic efficacy by measuring the potential of the nanoemulsion to inhibit CFA-induced knee edema in rats. It showed a significant reduction in inflammation in the methotrexate-loaded nanoemulsion and pure methotrexate groups over 30 days compared to the control group. Figure 11 demonstrated a continuing increase in knee inflammation by the disease control animals. The animal treated with the methotrexate-loaded nanoemulsion (10 mg/mL per kg body weight) and marketed methotrexate significantly ($p < 0.05$) decreased the knee inflammation compared to control animals on days 6–30. The highest knee edema was noted during the 30 days of study in control animals (23.11 $\pm$ 0.14 mm), whereas the edema was controlled by the methotrexate-loaded-nanoemulsion-treated (8.12 $\pm$ 0.13 mm) and marketed-drug-treated (9.40 $\pm$ 0.39 mm) animals. The methotrexate-loaded nanoemulsion showed a greater reduction in swelling compared to the marketed drug, suggesting that the methotrexate-loaded nanoemulsion has improved anti-arthritic potential compared to the conventional dosage form.

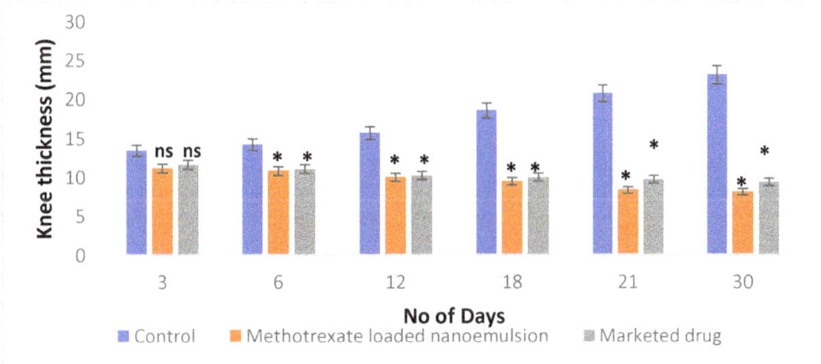

Figure 11. Knee-edema studies in complete-Freund-adjuvant-induced arthritis model. Results are shown as mean $\pm$ SEM ($n = 6$) and analyzed by one-way ANOVA followed by Tukey's test; ns: nonsignificant; * $p < 0.05$ significance difference to compared to control group.

The liver-protective properties of the methotrexate-loaded nanoemulsion were studied after the end of the experiment. When methotrexate is administered for a prolonged period or at higher doses, it causes liver toxicity, but in the present study, the formulations were administered to the rats only for a short time, so the possibility of liver toxicity was low. The function of hepatic marker enzymes SGOT (serum glutamic oxaloacetic transaminase), SGPT (serum glutamic pyruvic transaminase) and ALP (alkaline phosphatase) were used to evaluate the cellular architecture of the animals with CFA-induced arthritis. SGOT, SGPT, and ALP were elevated in the control group of animals. Hepatic marker enzymes were significantly ($p < 0.05$) decreased in methotrexate-loaded-nanoemulsion- and marketed-drug-treated animals (Figure 12). It was observed that the methotrexate-loaded nanoemulsion exhibited greater liver-protective properties than the marketed drug solution. The results indicated that the CFA-induced animals experienced a significant effect on hepatic enzyme activity, while the methotrexate-loaded nanoemulsion and marketed methotrexate significantly decreased hepatic enzyme activity compared with the control group. The results suggest that the nanoemulsion loaded with methotrexate has better liver-protective properties compared to the marketed drug, as the function of liver marker enzymes such as SGOT, SGPT and ALP is much closer to the healthy liver function. This confirms that the methotrexate-loaded nanoemulsion enables targeted drug delivery with minimal access of the drug to the liver and also prevents the metabolism of the drug by the liver cell, resulting in improved therapeutic efficacy. The in vivo anti-arthritic study confirms the capability of the methotrexate-loaded nanoemulsion to lessen systemic toxicity as well as improve anti-arthritic efficacy compared to the usual dosage forms.

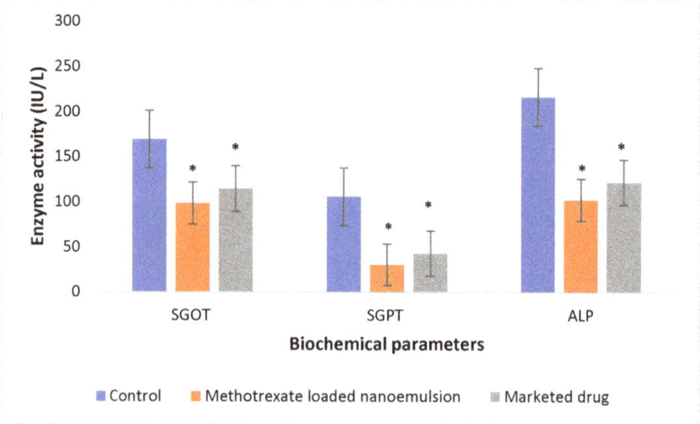

Figure 12. Liver-function tests for methotrexate-loaded nanoemulsion and pure methotrexate solution. Results are shown as mean ± SEM (n = 6) and analyzed by one-way ANOVA followed by Tukey's test; * $p < 0.05$ significance difference compared to control group.

A radiographic analysis of the animals on day 30 showed bone devastation and soft-tissue swelling along with joint-space narrowing in the CFA-induced control group, indicating subchondral erosion in the arthritic state. The methotrexate-loaded nanoemulsion and marketed drug solution showed no bone destruction and normal soft-tissue swelling in the animals (Figure 13). When RA occurs, radiography is an important diagnostic technique that can be used to determine the extent of arthritis. The early stages of arthritis are characterized by soft-tissue swelling and inflammation, whereas the later stages are characterized by subchondral erosions and narrowing of the joint space. The methotrexate-loaded nanoemulsion and marketed drug solution showed a significant reduction in joint destruction and soft-tissue damage in the animals. The potent anti-arthritis effect of the methotrexate-loaded nanoemulsion was subsequently validated and confirmed by the radiographic study of the knee joints of the animals.

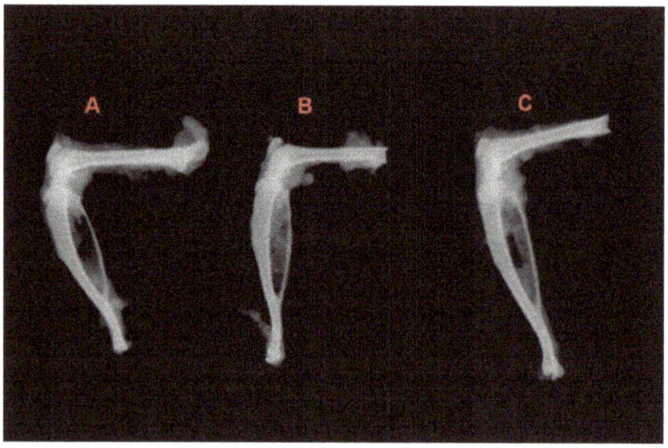

Figure 13. Radiographic analysis of the knees with the administration of the formulation. (**A**): FCA-induced arthritis, control group; (**B**): Methotrexate-loaded nanoemulsion; (**C**): marketed drug solution.

2.9. Stability Studies

To verify the stability properties of the methotrexate-loaded nanoemulsion, the stability studies of the nanoemulsion were carried out at refrigerated, room and elevated temperatures for three months. The particle size and entrapment efficiency changed significantly during storage at room temperature and elevated temperature. The possible mechanism for this was the polymorphic transformation of methotrexate. No changes in particle size and entrapment efficiency were observed at cooling temperatures (Figure 14). This indicates that the methotrexate-loaded nanoemulsion formulation was more stable under cooling temperature. Therefore, the methotrexate-loaded nanoemulsion is an excellent formulation for curing RA as it mitigates systemic side effects and improves treatment efficacy compared to the conventional dosage form.

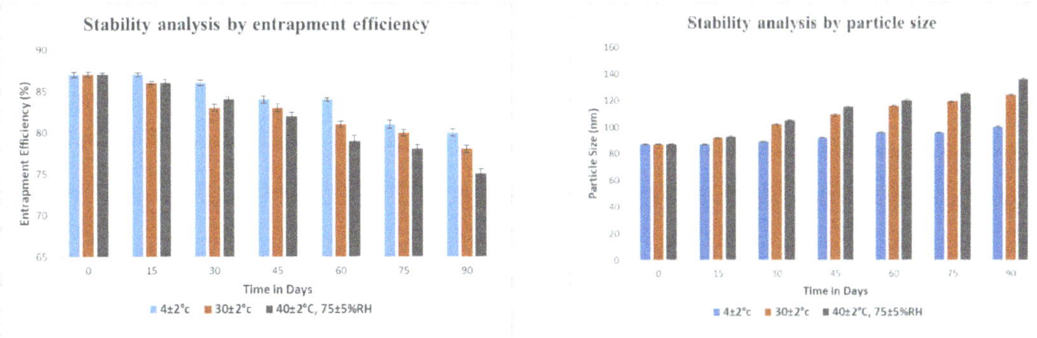

Figure 14. Graphical representation of stability study of methotrexate-loaded nanoemulsion according to particle size and entrapment efficiency.

3. Materials and Methods

3.1. Materials

The sample of the methotrexate injection was acquired from the leading IP Pharmacy center, AIMS, Kochi, Kerala, India. The surfactant, namely sesame oil, was purchased from the local grocery. Nice Chemicals, located in Ernakulam, Kerala, India, provided the Tween 80 for this study. The DMEM (Dulbecco's modified Eagle's medium) was procured from the well-known supplier thermo fisher scientific India Pvt Ltd., Ernakulum, Kerala, India.

3.2. Methodology

3.2.1. Formulation of Methotrexate-Loaded Nanoemulsion

Aqueous-phase titration was used to prepare the methotrexate-loaded nanoemulsion. In this procedure, the distilled water was gently added to the oil mixture and mixed dropwise with dynamic stirring until the resulting formulation appeared transparent and clear. Methotrexate solution at the concentration of 50 mg/mL was prepared and 0.2 mL solution was mixed into 3 mL of sesame oil, and then a mixture of 7 mL of Tween 80 and 2 mL of ethanol was slowly added while stirring. The resultant solution was vortexed for about 5 min using a vortex mixer. The aqueous phase, which consisted of distilled water, was added dropwise while constantly stirring at 500 rpm until a nanoemulsion was formed [35].

3.2.2. Particle Characteristics

The droplet size, polydispersity index (PDI), and zeta potential of the prepared nanoemulsion were determined using the Malvern Zetasizer (Nano ZS Malvern Instruments Ltd., Mavern, UK). After appropriate dilution, the nanoemulsions were sonicated and the droplet size was measured. The goal of this study was to assess the stability of the prepared nanoemulsion by measuring the droplet size, distribution, and zeta potential [35,36].

3.2.3. Particle Surface Morphology

The morphology and size were also determined by scanning electron microscopy (SEM) (VEGA 3, TESCAN, Brno-city, Kohoutovice, Czech Republic) and transmission electron microscopy (TEM) (TALOS, Thermo fisher scientific India Private Limited, Mumbai, India), respectively. The methotrexate-loaded nanoemulsion was diluted 10-fold with distilled water and dried at room temperature on 200 mesh film grids. For SEM analysis, samples were fixed in the sample holder with double-sided adhesive tape, and pictures were taken at a voltage of 5 kV. For TEM, samples were stained with a 2% phosphotungstic-acid solution and then dried for 2 min before being viewed under an electron microscope at 100 kV [37,38].

3.2.4. Differential Scanning Colorimetry (DSC)

The DSC (DSC 204F1 Phoenix, NETZSCH-Gerätebau GmbH, Selb, Germany) method was used to determine the change in physical properties and temperature of the optimized nanoemulsion. The sample was placed in an aluminum pan sealed with perforated lids and heated in the temperature range of 40–300 °C at a constant rate of 10 °C per minute. In addition, the inert atmosphere was created by a nitrogen purge of 50 mL/min. It assesses the stability of the formulation [39].

3.2.5. Entrapment Efficiency

In a slight modification of the method reported by Sarheed et al., a cellulose dialysis membrane technique was used to estimate the encapsulation of methotrexate in the nanoemulsion by assessing the free drug available in the aqueous phase. Before the experiment, the membrane was soaked overnight in phosphate-buffer solution (pH 7.4). By capping both ends, the nanoemulsion with 10 mg of drug incorporated was kept in the dialysis membrane. Phosphate-buffer solution and ethanol were mixed in a ratio of 80:20, and 100 mL was added into the receptor compartment. The dialysis membrane was placed in this 100 mL receptor compartment and shaken with a mechanical stirrer for 24 h. Aliquots were withdrawn from the receptor compartment and the amount of free drug crossing the dialysis membrane was measured by UV spectroscopy (UV-1700, Shimadzu, Kyoto, Japan) at 303.5 nm [40]. The following equation was used to measure the entrapment efficiency:

$$\text{Entrapment Efficiency} = \frac{\text{Quantification sample drug} - \text{Quantification of free drug in the solution}}{\text{Quantification sample drug}} \times 100$$

3.2.6. In Vitro Drug Release and Kinetic-Modeling Study

The cellulose dialysis membrane technique was used for in vitro drug release from the methotrexate-loaded nanoemulsion. The membrane was well cleaned and soaked overnight in phosphate buffer (pH 7.4) before use. Then, 10 mg of the drug was added to the membrane and both ends were tightly sealed. The dialysis membrane was kept in a beaker containing a phosphate-buffer solution with a pH of 7.4 and a temperature of 37 ± 1 °C throughout the study. The solution in the beaker was continuously stirred with a magnetic stirrer at 50 rpm. Samples were withdrawn from the beaker at specific intervals and the volume was maintained by adding the same amount of medium. The withdrawn aliquots were analyzed spectrophotometrically at 303.5 nm. Various kinetic models were developed to measure the release order and mechanism of release based on these results [41].

3.2.7. Hemocompatibility Analysis

The hemocompatibility study of the methotrexate-loaded nanoemulsion was performed to determine the suitability of the composition and the effect on red blood cells. To perform the test, a slight modification of the method reported by Roka and colleagues was employed. A volume of 5 mL of blood was drawn from healthy volunteers and an anticoagulant was added to prevent blood clotting. The blood sample was further diluted with a phosphate buffer with a pH of 7.4. The different concentrations of methotrexate-loaded nanoemulsion (20, 40, 60, and 80 μg/mL) and marketed drug (80 μg/mL) were added separately to the diluted blood samples and incubated for 24 h. The sample was then centrifuged at 3500 rpm for 10 min at 4 °C. The supernatant containing blood cells was placed in a microtiter plate, and absorbance was measured at 540 nm using an Elisa plate reader [42]. The percentage of hemolysis was calculated by using the following formula:

$$\%\text{Hemolysis} = \frac{\text{Blood cells in the supernatant of the sample solutions}}{\text{Blood cells in the distilled water}} \times 100$$

3.2.8. Cytotoxic Assay

MTT assay on RAW 264.7 (macrophage) cell lines was used for the evaluation of the in vitro cytotoxicity effect of the methotrexate-loaded nanoemulsion. The RAW 264.7 cell lines were cultured in DMEM (Dulbecco's modified Eagle's medium) media which was supplemented with 10% FBS, L-glutamine, sodium bicarbonate, and antibiotics at 37 °C maintaining humidification of 5% CO_2. The different concentrations of 6.25 μg/L, 12.5 μg/L, 25 μg/L, 50 μg/L, and 100 μg/L of nanoemulsion were prepared by diluting with 5% DMEM. These samples were added to the cell culture and incubated at 37 °C in a humidity of 5% CO_2. The normal saline solution was considered as the control solution for the measurement of cell viability.

The samples were incubated for 24 h, after which 30μL of reconstituted MTT solution was added to all test and control wells. The plate was shaken slowly and then incubated at 37 °C for 4 h to maintain the humidification of 5% CO_2 in the incubator. After completion of incubation, the supernatant was taken out, and added to 100μL of MTT solubilization solution. To solubilize the formazan crystals, the wells were carefully agitated by pipetting up and down. A microplate reader was used to assess the absorbance of samples at a wavelength of 570 nm, which signifies the optical density (OD) [43].

The percentage of viable cells was measured by using the following equation:

$$\%\text{ of viability} = \frac{\text{Mean OD of samples}}{\text{Mean OD of control}} \times 100$$

3.2.9. In Vivo Anti-Arthritic Activity

Approval for the animal study was obtained from the Central Animal Facility of the Amrita Institute of Medical Sciences (IAEC/2017/3/9) for the animal experiments. Experiments were conducted on adult male Sprague Dawley rats weighing about 200–250 g in order to determine the anti-arthritic effect. The animals were segregated into three groups of six animals each and kept in polypropylene cages at a temperature of 23 ± 2 °C. Intra-articular injection of 1 mg/mL CFA was injected to all groups of animals to induce the Chronic Arthritis condition. After that, methotrexate-loaded nanoemulsion (10 mg/mL) and marketed methotrexate drug (10 mg/mL) were administered subcutaneously once a week for 30 days to groups II and III, respectively, while no treatment was given to group I. Using a digital micrometer, the knee circumference (in mm) of each group was determined every day for 30 days. After 30 days of testing, retro-orbital blood was drawn from the rats and the serum was separated to assess the functions of liver enzyme markers (SGPT, SGOT, and ALP). In addition, radiographs of the knee joints of rats were performed to evaluate the effect of marketed methotrexate and methotrexate-loaded nanoemulsion on the severity of arthritis in FCA-induced rats [44,45].

3.2.10. Stability Studies

For three months, methotrexate-loaded nanoemulsion was stored at three different temperatures: refrigerator temperature (4 ± 2 °C), room temperature (30 ± 2 °C), and accelerated temperature (40 ± 2 °C, 75 ± 5% RH). Particle size (PS) and entrapment efficiency (EE) were evaluated at 15-day intervals over 90 days to check the changes in physical and chemical stability of the nanoemulsion [41].

4. Conclusions

In the treatment of RA, methotrexate is a DMARD that is considered a first-line therapy. There are only a small number of drug-delivery systems on the market that provide significant pharmacotherapy for rheumatoid arthritis. Methotrexate formulations currently on the market have been associated with systemic side effects. Therefore, a methotrexate-loaded subcutaneous nanoemulsion that delivers the drug exclusively to the lymphatic system was developed and studied in vitro and in vivo in order to minimize adverse effects and improve therapeutic efficacy. The nanoemulsion was found to encapsulate a significant amount of methotrexate with a smaller particle size and shape suitable for subcutaneous injection. Physicochemical studies confirmed that the methotrexate-loaded nanoemulsion exhibited homogeneous nanosized droplets and a stable formulation. In vitro drug-release studies confirmed the sustained release of the drug from the nanoemulsion. The results of hemolysis and cytotoxicity studies showed that the nanoemulsion is safer for systemic circulation at lower concentrations. The anti-arthritic study of the nanoemulsion in CFA-induced animals showed improved anti-arthritic activity compared to the marketed drug. The preliminary studies of the methotrexate-loaded nanoemulsion suggest that the formulation may be able to improve the lymphatic transport of drugs after systemic administration. In addition, this formulation may reduce drug doses while minimizing adverse effects on cell integrity. It can be concluded that methotrexate-loaded nanoemulsion is a superior formulation for the treatment of RA as it mitigates systemic side effects and improves treatment efficacy compared to the conventional dosage form. In the future, the drug-uptake study in the lymph node will demonstrate the ability of the nanoemulsion to deliver drugs subcutaneously into the lymphatic circulation.

Author Contributions: Conceptualization, V.V. and P.S.; methodology, H.P.V.; validation, S.C.N., S.A., M.M.G. and V.V.; formal analysis, V.V. and I.E.; investigation, P.S.; writing—original draft preparation, V.V., S.I.B. and R.K.S.; writing—review and editing, R.K.S., E.I.T. and M.M.S.-B.; supervision, V.V. All authors have read and agreed to the published version of the manuscript.

Funding: This work was supported by the Researchers Supporting Project (number RSP2022R497) at King Saud University, Riyadh, Saudi Arabia.

Data Availability Statement: Not applicable.

Acknowledgments: The authors would like to thank our dean research Shantikumar and Sabitha M for providing the facilities for research. Authors are thankful to King Saud University, Riyadh, Saudi Arabia for financial support for this paper.

Conflicts of Interest: The authors declare no conflict of interest.

References

1. Wong, S.H.; Lord, J.M. Factors underlying chronic inflammation in rheumatoid arthritis. *Arch. Immunol. Ther. Exp.* **2004**, *52*, 379–388.
2. Qiang, G.; Yuxiang, W.; Dan, X.; Johannes, N.; Nathan, J.P.; Jiake, X. Rheumatoid arthritis: Pathological mechanisms and modern pharmacologic therapies. *Bone Res.* **2018**, *6*, 1–14.
3. Cross, M.; Smith, E.; Hoy, D.; Carmona, L.; Frederick, W.; Theo, W. The global burden of rheumatoid arthritis: Estimates from the Global Burden of Disease 2010 study. *Ann. Rheum. Dis.* **2014**, *73*, 1316–1322. [CrossRef] [PubMed]
4. Allan, G. Epidemiology, Pathophysiology, and Diagnosis of Rheumatoid Arthritis: A Synopsis. *Am. J. Manag. Care* **2014**, *20*, S128-35.
5. Jefferies, W.M. The etiology of rheumatoid arthritis. *Med. Hypothesis* **1998**, *51*, 111–114. [CrossRef]
6. Kahlenberg, J.M.; David, A.F. Advances in the medical treatment of rheumatoid arthritis. *Hand Clin.* **2011**, *27*, 11–20. [CrossRef]
7. Chauhan, K.; Jagmohan, S.J.; Mohammed, A.A. *Rheumatoid Arthritis, StatPearls [Internet]*; StatPearls Publishing: Treasure Island, FL, USA, 2019.
8. Jacqueline, B.; Syed, A.A.R.; Ayman, M.S. Rheumatoid Arthritis: A Brief Overview of the Treatment. *Med. Princ. Pract.* **2018**, *27*, 501–507.
9. Rajitha, P.; Biswas, R.; Sabitha, M.; Jayakumar, R. Methotrexate in the treatment of psoriasis and rheumatoid arthritis: Mechanistic insights, current issues and novel delivery approaches. *Curr. Pharm. Des.* **2017**, *23*, 3550–3566. [CrossRef]
10. Amy, M.W. Diagonosis and management of Rheumatoid Arthritis. *Am. Fam. Physician* **2011**, *84*, 1245–1252.

11. Ram, M.; Anita, M.; Sujithra, R.M.; Aramya, A.R.; Mathew, J.; Parvathy, M. Comparative Effectiveness Research on Different Treatment Options for Rheumatoid Arthritis in Ayurveda. *J. Altern. Complement. Med.* **2014**, *20*, A75–A76.
12. Christine, T.N.P. Nanotherapeutic approaches for the treatment of rheumatoid arthritis. *Wiley Interdiscip. Rev. Nanomed. Nanobiotechnol.* **2011**, *3*, 607–619.
13. Cutolo, M.; Sulli, A.; Pizzorni, C.; Seriolo, B.; Straub, R. Anti-inflammatory mechanisms of methotrexate in rheumatoid arthritis. *Ann. Rheum. Dis.* **2001**, *60*, 729–735. [CrossRef] [PubMed]
14. Simon, S.; Michael, D. Inflammation and Lymphatic Function. *Front. Immunol.* **2019**, *10*, 308.
15. Schwartz, N.; Chalasani, M.L.S.; Li, T.M.; Feng, Z.; Shipman, W.D.; Lu, T.T. Lymphatic Function in Autoimmune Diseases. *Front. Immunol.* **2019**, *10*, 519. [CrossRef] [PubMed]
16. Zhang, X.Y.; Lu, W.Y. Recent advances in lymphatic targeted drug delivery system for tumor metastasis. *Cancer Biol. Med.* **2014**, *11*, 247–254.
17. Yang, Q.; Laird, F. *Drug Delivery to the Lymphatic System, Drug Delivery: Principles and Applications*, 2nd ed.; Chapter 21; Wiley Publisher: Hoboken, NJ, USA, 2016.
18. Isabel, M.O.; Cristiana, G.; Rui, L.R.; Joaquim, M.O. Engineering nanoparticles for targeting rheumatoid arthritis: Past, present, and future trends. *Nano Res.* **2018**, *11*, 4489–4506.
19. Satheesh, J.; Shyam, S.R.; Jithan, A. Development of subcutaneous sustained release nanoparticles encapsulating low molecular weight heparin. *J. Adv. Pharm. Technol. Res.* **2015**, *6*, 58–64.
20. McLennan, D.N.; Porter, C.J.; Charman, S.A. Subcutaneous drug delivery and the role of the lymphatics. *Drug Discov. Today Technol.* **2005**, *2*, 89–96. [CrossRef]
21. Ghosh, V.; Saranya, S.; Mukherjee, A.; Chandrasekaran, N. Cinnamon oil nanoemulsion formulation by ultrasonic emulsification: Investigation of its bactericidal activity. *J. Nanosci. Nanotechnol.* **2013**, *13*, 114–122. [CrossRef]
22. Danaei, M.; Dehghankhold, M.; Ataei, S.; Hasanzadeh Davarani, F.; Javanmard, R.; Dokhani, A.; Khorasani, S.; Mozafari, M.R. Impact of Particle Size and Polydispersity Index on the Clinical Applications of Lipidic Nanocarrier Systems. *Pharmaceutics* **2018**, *10*, 57. [CrossRef]
23. Jang, J.H.; Jeong, S.H.; Lee, Y.B. Enhanced Lymphatic Delivery of Methotrexate Using W/O/W Nanoemulsion: In Vitro Characterization and Pharmacokinetic Study. *Pharmaceutics* **2020**, *12*, 978. [CrossRef] [PubMed]
24. Kaur, G.; Singh, P.; Sharma, S. Physical, morphological, and storage studies of cinnamon based nanoemulsions developed with Tween 80 and soy lecithin: A comparative study. *Food Meas.* **2021**, *15*, 2386–2398. [CrossRef]
25. Wik, J.; Bansal, K.K.; Assmuth, T.; Rosling, A.; Rosenholm, J.M. Facile methodology of nanoemulsion preparation using oily polymer for the delivery of poorly soluble drugs. *Drug Deliv. Transl. Res.* **2020**, *10*, 1228–1240. [CrossRef] [PubMed]
26. Zhou, H.; Yue, Y.; Liu, G.; Li, Y.; Zhang, J.; Gong, Q.; Yan, Z., Duan, M. Preparation and Characterization of a Lecithin Nanoemulsion as a Topical Delivery System. *Nanoscale Res. Lett.* **2010**, *5*, 224. [CrossRef] [PubMed]
27. El-Refai, A.A.; Rabie, M.M.; El-Gammal, R.E.; Al-Saban, W.A. Nanoemulsion of Sesame Seeds Oil: Preparation, Evaluation and Stability. *Asian J. Chem.* **2019**, *31*, 3004–3008. [CrossRef]
28. Antil, S.; Gupta, C. Formulation and Evaluation of Nanoemulsion for Bioavailability Enhancement of Metaxalone. *Int. J. Cur. Res. Rev.* **2021**, *13*, 47–53. [CrossRef]
29. Jangde, R.; Elhassan, G.O.; Khute, S.; Singh, D.; Singh, M.; Sahu, R.K.; Khan, J. Hesperidin-Loaded Lipid Polymer Hybrid Nanoparticles for Topical Delivery of Bioactive Drugs. *Pharmaceuticals* **2022**, *15*, 211. [CrossRef]
30. Aswathy, K.N.; Asdaq, S.M.B.; Saritha, C.K.; Thomas, L.; Haridas, N.; Viswanad, V.; Sahu, R.K.; Fattepur, S.; Alamri, A.S.; Alsanie, W.F.; et al. Formulation and In-vitro Characterization of Fast-Disintegrating Herbal Extract Sublingual Immunotherapy Tablet for Peanut-Induced Allergic Asthma. *Saudi J. Biol. Sci.* **2022**, *29*, 1283–1297. [CrossRef]
31. Rashid, S.A.; Bashir, S.; Naseem, F.; Farid, A.; Rather, I.A.; Hakeem, K.R. Olive Oil Based Methotrexate Loaded Topical Nanoemulsion Gel for the Treatment of Imiquimod Induced Psoriasis-like Skin Inflammation in an Animal Model. *Biology* **2021**, *10*, 1121. [CrossRef]
32. Rathee, J.; Kanwar, R.; Kumari, L.; Pawar, S.V.; Salunke, D.B.; Mehta, S.K. Preparation of α-Tocopherol based nanoemulsion for efficacious delivery of Methotrexate. *J. Dispers. Sci. Technol.* **2022**. [CrossRef]
33. Kelmann, R.G.; Kuminek, G.; Teixeira, H.F.; Koester, L.S. Carbamazepine parenteral nanoemulsions prepared by spontaneous emulsification process. *Int. J. Pharm.* **2007**, *342*, 231–239. [CrossRef]
34. Weber, M.; Steinle, H.; Golombek, S.; Hann, L.; Schlensak, C.; Wendel, H.P.; Avci-Adali, M. Blood-Contacting Biomaterials: In Vitro Evaluation of the Hemocompatibility. *Front. Bioeng. Biotechnol.* **2018**, *6*, 99. [CrossRef] [PubMed]
35. Akrawi, S.H.; Gorain, B.; Nair, A.B.; Choudhury, H.; Pandey, M.; Shah, J.N.; Venugopala, K.N. Development and Optimization of Naringenin-Loaded Chitosan-Coated Nanoemulsion for Topical Therapy in Wound Healing. *Pharmaceutics* **2020**, *12*, 893. [CrossRef]
36. Ali, H.H.; Hussein, A.A. Oral nanoemulsions of candesartan cilexetil: Formulation, characterization and in vitro drug release studies. *AAPS Open* **2017**, *3*, 4. [CrossRef]
37. Alhamdany, A.T.; Saeed, A.M.H.; Alaayedi, M. Nanoemulsion and Solid Nanoemulsion for Improving Oral Delivery of a Breast Cancer Drug: Formulation, Evaluation, and a Comparison Study. *Saudi Pharm. J.* **2021**, *29*, 1278–1288. [CrossRef] [PubMed]
38. Saani, S.M.; Abdolalizadeh, J.; Heris, S.Z. Ultrasonic/sonochemical synthesis and evaluation of nanostructured oil in water emulsions for topical delivery of protein drugs. *Ultrason. Sonochem.* **2019**, *55*, 86–95. [CrossRef]

39. Rosso, A.; Lollo, G.; Chevalier, Y.; Troung, N.; Bordes, C.; Bourgeois, S.; Maniti, O.; Granjon, T.; Dugas, P.; Urbaniak, S.; et al. Development and structural characterization of a novel nanoemulsion for oral drug delivery. *Colloids Surf. A Physicochem. Eng. Asp.* **2020**, *593*, 124614. [CrossRef]
40. Sarheed, O.; Dibi, M.; Ramesh, K.V.R.N.S. Studies on the Effect of Oil and Surfactant on the Formation of Alginate-Based O/W Lidocaine Nanocarriers Using Nanoemulsion Template. *Pharmaceutics* **2020**, *12*, 1223. [CrossRef]
41. Vijaya Rani, K.R.; Rajan, S.; Bhupathyraaj, M.; Priya, R.K.; Halligudi, N.; Al-Ghazali, M.A.; Sridhar, S.B.; Shareef, J.; Thomas, S.; Desai, S.M.; et al. The Effect of Polymers on Drug Release Kinetics in Nanoemulsion In Situ Gel Formulation. *Polymers* **2022**, *14*, 427. [CrossRef]
42. Róka, E.; Ujhelyi, Z.; Deli, M.; Bocsik, A.; Fenyvesi, É.; Szente, L.; Fenyvesi, F.; Vecsernyés, M.; Váradi, J.; Fehér, P.; et al. Evaluation of the Cytotoxicity of α-Cyclodextrin Derivatives on the Caco-2 Cell Line and Human Erythrocytes. *Molecules* **2015**, *20*, 20269–20285. [CrossRef]
43. Sahlan, M.; Mahira, K.F.; Pratami, D.K.; Rizal, R.; Ansari, M.J.; Al-Anazi, K.M.; Farah, M.A. The cytotoxic and anti-inflammatory potential of Tetragonula sapiens propolis from Sulawesi on raw 264.7 cell lines. *J. King Saud Univ.-Sci.* **2021**, *33*, 101314. [CrossRef]
44. Cui, X.; Wang, R.; Bian, P.; Wu, Q.; Seshadri, V.D.D.; Liu, L. Evaluation of antiarthritic activity of nimbolide against Freund's adjuvant induced arthritis in rats. *Artif. Cells Nanomed. Biotechnol.* **2019**, *47*, 3391–3398.
45. Singh, V.S.; Dhawale, S.C.; Shakeel, F.; Faiyazuddin, M.; Alshehri, S. Antiarthritic Potential of Calotropis procera Leaf Fractions in FCA-Induced Arthritic Rats: Involvement of Cellular Inflammatory Mediators and Other Biomarkers. *Agriculture* **2021**, *11*, 68. [CrossRef]

nanomaterials

MDPI

Article

Impact of Citrate and Lipid-Functionalized Magnetic Nanoparticles in Dehydropeptide Supramolecular Magnetogels: Properties, Design and Drug Release

Sérgio R. S. Veloso [1], Joana F. G. Silva [1], Loic Hilliou [2], Cacilda Moura [1], Paulo J. G. Coutinho [1], José A. Martins [3], Martín Testa-Anta [4,5], Verónica Salgueiriño [4,5], Miguel A. Correa-Duarte [5], Paula M. T. Ferreira [3] and Elisabete M. S. Castanheira [1,*]

1 Centro de Física (CFUM), University of Minho, Campus de Gualtar, 4710-057 Braga, Portugal; sergioveloso96@gmail.com (S.R.S.V.); joanasilva.tg@gmail.com (J.F.G.S.); cmoura@fisica.uminho.pt (C.M.); pcoutinho@fisica.uminho.pt (P.J.G.C.)

2 Institute for Polymers and Composites, Department of Polymer Engineering, University of Minho, Campus de Azurém, 4800-058 Guimarães, Portugal; loic@dep.uminho.pt

3 Centro de Química (CQUM), University of Minho, Campus de Gualtar, 4710-057 Braga, Portugal; jmartins@quimica.uminho.pt (J.A.M.); pmf@quimica.uminho.pt (P.M.T.F.)

4 Departamento de Física Aplicada, Universidade de Vigo, 36310 Vigo, Spain; mtesta@uvigo.es (M.T.-A.); vsalgue@uvigo.es (V.S.)

5 CINBIO, Universidade de Vigo, 36310 Vigo, Spain; macorrea@uvigo.es

* Correspondence: ecoutinho@fisica.uminho.pt; Tel.: +351-253-604321

Abstract: Currently, the nanoparticle functionalization effect on supramolecular peptide-based hydrogels remains undescribed, but is expected to affect the hydrogels' self-assembly and final magnetic gel properties. Herein, two different functionalized nanoparticles: citrate-stabilized (14.4 $\pm$ 2.6 nm) and lipid-coated (8.9 $\pm$ 2.1 nm) magnetic nanoparticles, were used for the formation of dehydropeptide-based supramolecular magnetogels consisting of the ultra-short hydrogelator Cbz-L-Met-Z-ΔPhe-OH, with an assessment of their effect over gel properties. The lipid-coated nanoparticles were distributed along the hydrogel fibers, while citrate-stabilized nanoparticles were aggregated upon gelation, which resulted into a heating efficiency improvement and decrease, respectively. Further, the lipid-coated nanoparticles did not affect drug encapsulation and displayed improved drug release reproducibility compared to citrate-stabilized nanoparticles, despite the latter attaining a stronger AMF-trigger. This report points out that adsorption of nanoparticles to hydrogel fibers, which display domains that improve or do not affect drug encapsulation, can be explored as a means to optimize the development of supramolecular magnetogels to advance theranostic applications.

Keywords: magnetic gels; drug release; magnetic hyperthermia; magnetic lipogels; supramolecular hydrogels; magnetic nanoparticles; self-assembly; nanoparticle functionalization

Citation: Veloso, S.R.S.; Silva, J.F.G.; Hilliou, L.; Moura, C.; Coutinho, P.J.G.; Martins, J.A.; Testa-Anta, M.; Salgueiriño, V.; Correa-Duarte, M.A.; Ferreira, P.M.T.; et al. Impact of Citrate and Lipid-Functionalized Magnetic Nanoparticles in Dehydropeptide Supramolecular Magnetogels: Properties, Design and Drug Release. *Nanomaterials* **2021**, *11*, 16. https://dx.doi.org/10.3390/nano11010016

Received: 29 November 2020
Accepted: 21 December 2020
Published: 23 December 2020

1. Introduction

Supramolecular magnetogels basically comprise two main components: the hydrogel and the magnetic nanoparticles. Following a stimulus, the self-assembly of the supramolecular hydrogelators is driven towards a kinetically-trapped intertwined fibrillar structure, such that solvent pocket microdomains are formed. This process takes place through the cooperative effect of different non-covalent intermolecular interactions: hydrogen bonding, van der Waals, electrostatic, and/or hydrophobic and aromatic interactions [1–7].

For example, dehydropeptide-based hydrogelators, such as the minimalist Cbz-L-Met-Z-ΔPhe-OH (see Structure S1 in Supplementary Materials), have been used due to the simplicity of producing biocompatible systems at low cost, and its promising properties for drug delivery [8–10]. The aforementioned dehydropeptide uses of a non-polar amino acid, methionine, to induce hydrophobic collapse, while the dehydrophenylalanine moiety provides a means for π-π interactions (like the *N*-capping group) and conformational constraints in the peptide backbone, meaning that it promotes the self-assembly into fibers and provides resistance to enzymatic degradation [8–10]. As assessed in this work, the hydrogel displays a moderate gelation kinetics and elastic modulus, which allows following the impact of the nanoparticles, so that diverse parameters can be optimized. Further, the gels can be formed in a skin pH range of 4-6 [11], thus being suitable for dermal application.

The retention or entrapment of nanoparticles during the gelation process allows for the tailoring and modulation of the matrix structure, along with the possibility of tuning different physical properties [3,12,13]. The modulation of the matrix structure also enables optimizing the loading of a wide variety of hydrophilic and hydrophobic drugs, reducing potential side effects, and enabling higher doses in therapy at the right location when using magnetogels for drug delivery. In this regard, taking advantage of the magnetic nanoparticles, the application of a magnetic field gradient allows the control and targeting of the nanosystem to a specific location [14–16], which can be further coupled with an alternating magnetic field (AMF), such that the nanoparticles can absorb energy and then release it as heat [17–20]. As a consequence, there is a synergistic effect that involves this magnetic hyperthermia and the subsequently associated enhancement of drug release [21–23], such that a much higher therapeutic efficiency can be attained. Along these lines, among transition metal ferrites, manganese ferrite nanoparticles exhibit very suitable magnetic properties, particularly in terms of the high magnetic susceptibility and in terms of magnetophoretic mobility, which render them as an appealing option to improve the supramolecular magnetogels for drug delivery applications [24,25]. Further, manganese ferrites display good biocompatibility and the large saturation magnetization reduces the required concentration of nanoparticles, thus averting side effects [8,25].

Supramolecular magnetogels endorsed with bare nanoparticles were demonstrated to be easily prone to aggregation, which led us to limit the concentration of nanoparticles that were encapsulated [8,9]. Alternatively, the use of a thick shell coating the nanoparticles was confirmed to hamper the gelation process, with the additional cost of requiring a larger concentration of hydrogelator to compensate for the bulkier nanocomposite [10]. On the other hand, forcing the supramolecular design strategies to conjugate the nanoparticles with the hydrogelator molecules has also been considered [25–27]. In general, these existing strategies for supramolecular magnetogels either imply complex fabrication steps and/or fail on the homogeneous incorporation of high concentrations of magnetic material and target drugs, which strongly limit their usability in theranostics. Furthermore, the current literature on supramolecular magnetogels lacks exploration of the composite functionalization and its effect on supramolecular magnetic gels properties. Consequently, the large amounts of hydrogelator employed and the aggregation of nanoparticles in most of the reported systems have demonstrated lower heating efficiency, hindering the sought on-demand controlled release of the payload.

Doxorubicin is one of the most commonly used chemotherapeutic drugs in a wide variety of cancers. However, its various side effects (e.g., neutropenia and heart failures) lead to a worsening of the patient's quality of life, which can be improved through incorporation with drug delivery system to increase the safety profile of the therapy. Considering this need, a useful strategy could be the encapsulation of doxorubicin in liposomal systems, such as the liposomal formulation Doxil®, or the use of citrate-stabilized nanoparticles, which strongly adsorb the positively charged drug molecules through electrostatic interactions [28].

Herein, supramolecular magnetogels endorsed with high nanoparticle concentration, employing a lipid coating (magnetoliposome-like structure) or citrate stabilization, are evaluated. Whereas the citrate stabilization provides electrostatic stabilization, the lipid coating assessment tackles the strategy, not only to ensure the steric stabilization of manganese ferrite nanoparticles, but also to provide the structure with enough hydrophobic cavities for an efficient drug loading and subsequent controlled drug diffusion. This option to obtain magnetogels systems, compared with the use of negatively-charged citrate-stabilized nanoparticles, demonstrates that the gelation becomes optimized, such that the concentration of stable magnetic nanoparticles within the gel is increased. As a consequence, besides improving the system stability and magnetic properties, the strategy demonstrates the role of the magnetic nanoparticles taking advantage of the hyperthermia they are responsible for, on drug release, using the antitumor drug doxorubicin.

2. Materials and Methods

2.1. Synthesis Procedure of Magnetic Nanoparticles

2.1.1. Preparation of Citrate-Stabilized Manganese Ferrite Nanoparticles

A modified synthesis of citrate-stabilized nanoparticles was adapted from reference [29]. Trisodium citrate dehydrate (1 mmol) and NaOH (4 mmol) were added to 19 mL of ultrapure water at 100 °C. A 1 mL aqueous solution of $FeSO_4{\cdot}7H_2O$ (1.33 mmol) and $MnSO_4{\cdot}H_2O$ (all reagents from Merck-Sigma, St. Louis, MO, USA) (0.66 mmol) was added, drop by drop, into the mixture under vigorous agitation. After 2 h, the solution was cooled down to room temperature, washed through magnetic decantation with water/ethanol 1:1, and dried at 100 °C.

2.1.2. Preparation of Lipid-Coated Manganese Ferrite Nanoparticles

Lauric acid (1 mmol) and NaOH (4 mmol) were added to 19 mL of ultrapure water at 100 °C. A 1 mL aqueous solution of $FeSO_4{\cdot}7H_2O$ (1.33 mmol) and $MnSO_4{\cdot}H_2O$ (0.66 mmol) was added, drop by drop, into the mixture under vigorous agitation. After 2 h, the solution was cooled down to room temperature, washed through repeated centrifugation with water, and dried at 100 °C. The stock solution was prepared by dispersion of the nanoparticles (4 mg) in 2 mL of 2 mM L-α-phosphatidylcholine (from egg yolk, egg-PC) (Merck-Sigma, St. Louis, MO, USA) solution through sonication at 190 W. The lipid-coated nanoparticles were then washed and purified with ultrapure water by magnetic decantation.

2.2. Self-Assembly of Magnetogels

2.2.1. Optimization of Hydrogel Gelation

Gelation optimization was carried out through turbidity measurements at 500 nm. The hydrogel and glucono-δ-lactone (GdL) concentrations were screened. The self-assembly was induced by dissolving the hydrogelator in basic pH through the addition of 2 v/v% NaOH (1 M) and, then, glucono-δ-lactone (GdL) was added to decrease the pH homogeneously. A solution of 0.05 wt% hydrogel was used to assess the fiber-catalyzed secondary nucleation. The aggregates fraction, $f(t)$, was defined as follows:

$$f(t) = \frac{T_{obs} - T_{free}}{T_{agg} - T_{free}} \quad (1)$$

where T_t, T_{free} and T_∞ stand for turbidity measured at 500 nm observed at time t, before the addition of GdL and when aggregated (turbidity at $t = 5\,h$), respectively. An empirical exponential decay function was fitted according to Equation (2) [30]:

$$T(t) = \frac{T(\infty)}{\sqrt[v]{\left(1 + ve^{-k_{emp}(t-t_m)}\right)}} \quad (2)$$

where k_{emp} is the rate constant (inverse of the relaxation time) of fibril formation and t_m is the point of the maximum elongation rate. This constant rate includes various steps and its interpretation between different systems is misleading. As a result, other models were fitted to understand the influence of the parameters on the nucleation and elongation rates. Saitô's fractional aggregation model has been successful in the aggregation studies of β-amyloid, calcitonin, prion, and α-sinuclein [31–33]. At a concentration larger than critical micellar concentration ($[M] \gg cmc$), the aggregation can be described by a two-step mechanism scheme [33]:

$$n_0 M(monomers) \rightleftharpoons M_{n_0}(micelle) \overset{k_n}{\rightarrow} P_n$$
$$M + P_n \overset{k_e}{\rightarrow} P_{n+1}.$$

In this mechanism, k_n and k_e represent the average nucleation and growth rate constants, M is the monomer, M_{n_0} represents the micelle (precatalytic form of the monomer), P_n is the nucleus of fibril with n monomer molecules, and P_{n+1} is the extended fibril with $n+1$ monomer molecules; thus, P is both a product and catalyst in the growth step reaction. The fibril formation can be described according to the equation:

$$f(t) = \frac{\rho\left(e^{(1+\rho)k_s t} - 1\right)}{1 + \rho e^{(1+\rho)k_s t}} \quad (3)$$

where $k_s = k_e[M]$ is the effective growth rate constant, $\rho = \frac{k_n}{k_s}$, and the initial condition is $f(0) = 0$. The secondary nucleation was evaluated through the aggregation models of Knowles et al. [30,34] and Cohen et al. [35]. The former model describes the concentration of monomer in the fibers according to the equation:

$$f(t) = 1 - e^{-C_+ e^{kt} + C_- e^{-kt} + k_n m_{tot}^{n_c - 1} k_-^{-1}} \quad (4)$$

where m_{tot} is the total hydrogelator concentration and $k = \sqrt{2m_{tot}k_+k_-}$. In the absence of fibrils, at the beginning of the aggregation,

$$C_\pm = \frac{\pm k_n m_{tot}^{n_c - 1}}{2k_-} \quad (5)$$

where k_+ is the polymerization rate and k_- is the secondary nucleation (fragmentation). The latter model describes the concentration of monomer in the fibers according to the equation:

$$f(t) = 1 - \left(\frac{B_+ + C_+}{B_+ + C_+ e^{\kappa t}} \frac{B_- + C_+ e^{\kappa t}}{B_- + C_+}\right)^{\frac{k_\infty^2}{\kappa k j_\infty}} - e^{-k_\infty t} \quad (6)$$

where $\kappa = \sqrt{2k_+k_2m(0)^{n_2+1}}$ is associated with the secondary pathways, k_2 is the fibril-catalyzed secondary nucleation, and $k_2 = k_-$ when $n_2 = 0$, $C_\pm = \pm\lambda^2/(2\kappa^2)$, $\lambda = \sqrt{2k_+k_nm(0)^{n_c}}$ is related with the rate of formation of new aggregates through primary pathways, $B_\pm = (k_\infty \pm kj_\infty)/(2\kappa)$, $k_\infty = \sqrt{2\kappa^2/[n_2(n_2+1)] + 2\lambda^2/n_c}$, and $kj_\infty = \sqrt{k_\infty^2 - 4C_+C_-\kappa^2}$. The parameters n_c and n_2 describe the dependencies of the primary and secondary pathways, and $m(0)$ is the initial concentration of soluble monomers.

2.2.2. Development of Magnetogels

The prepared nanoparticles were added to the hydrogel solution at a final volume of 200 μL and at the required concentration from a starting solution at 2 wt%. All hydrogel/magnetogel solutions were left standing at room temperature until the gel phase was attained. Here, the unit wt% stands for m/v%.

2.3. Spectroscopic Measurements

2.3.1. General Methods

Fluorescence measurements were carried out using a Horiba-Jobin Yvon Fluorolog 3 spectrofluorimeter (HORIBA Jobin Yvon IBH Ltd., Glasgow, UK), equipped with double excitation and emission monochromators, Glan-Thompson polarizers (HORIBA Jobin Yvon IBH Ltd., Glasgow, UK), and a temperature-controlled cuvette holder. Fluorescence emission spectra were corrected for the instrumental response of the system. The excitation of the hydrogelator was set at 280 nm, and the emission spectrum was collected between 290 nm and 600 nm with a slit of 6 nm in both excitation and emission. Absorption spectra were recorded in a Shimadzu UV-3600 Plus UV-Vis-NIR spectrophotometer (Shimadzu Corporation, Kyoto, Japan).

The fluorescence quantum yield, Φ_s, can be determined by Equation (7) (standard method) [36,37],

$$\Phi_s = \frac{(A_r F_s n_s^2)}{(A_s F_r n_r^2)} \Phi_r \tag{7}$$

where A is the absorbance at the excitation wavelength, F is the integrated emission area, and n is the refraction index of the solvents. Subscripts r and s refer to the reference and sample compound, respectively. The absorbance value at excitation wavelength was always less than 0.1 in order to avoid inner filter effects. L-Tryptophan in aqueous buffer solution (pH = 7.2) was used as a reference (Φ_r = 0.14 at 25 °C) [38].

2.3.2. Fluorescence Anisotropy Measurements

The steady-state fluorescence anisotropy values, r, provide information on the average microviscosity of the gel matrix where the fluorophore is localized and can be determined by Equation (8) [39],

$$r = \frac{I_{VV} - G I_{VH}}{I_{VV} + 2G I_{VH}} \tag{8}$$

where I_{VV} and I_{VH} are the intensities of the emission spectra obtained with vertical and horizontal polarization, respectively (for vertically polarized excitation light), I_{HV} and I_{HH} are the emission intensities obtained with vertical and horizontal polarization (for horizontally polarized excitation light), and $G = I_{HV}/I_{HH}$ is the instrumental correction factor.

2.3.3. FRET Measurements

The drug incorporation into the magnetogels network was investigated by Förster Resonance Energy Transfer (FRET). FRET efficiency, Φ_{FRET}, defined as the proportion of donor molecules that have transferred their excess energy to acceptor molecules, can be expressed by Equation (9) [39],

$$\Phi_{FRET} = 1 - \frac{I_{DA}}{I_D} \tag{9}$$

where I_{DA} and I_D are the donor integrated fluorescence intensities in the presence and absence of an acceptor, respectively. FRET efficiency can also be determined using the donor-acceptor intermolecular distance, R_{DA}, and the Förster radius (critical diameter), R_0, through Equation (10) [39],

$$\Phi_{FRET} = \frac{1}{1 + \left(\frac{R_{DA}}{R_0}\right)^6}. \tag{10}$$

The Förster radius, R_0, the distance at which FRET efficiency is 50%, can be determined by the spectral overlap, $J(\lambda)$ between the donor fluorescence emission and the acceptor absorption, according to Equations (11) and (12) (with R_0 in Å, λ in nm, $\varepsilon_A(\lambda)$ in M^{-1} cm^{-1}) [39],

$$R_0 = 0.2108 \times \left[\kappa^2 \Phi_D n^{-4} J(\lambda)\right]^{1/6} \tag{11}$$

$$J(\lambda) = \int_0^\infty I_D(\lambda)\varepsilon_A(\lambda)\lambda^4 d\lambda \tag{12}$$

where $\kappa^2 = 2/3$ is the orientational factor assuming random orientation of the dyes, Φ_D is the donor fluorescence quantum yield in the absence of energy transfer, n is the refraction index of the medium, $I_D(\lambda)$ is the fluorescence spectrum of the donor normalized so that $\int_0^\infty I_D(\lambda)\, d\lambda = 1$, and $\varepsilon_A(\lambda)$ is the molar absorption coefficient of the acceptor.

2.4. *Characterization Techniques*

2.4.1. Scanning Transmission Electron Microscopy (STEM)

STEM images were recorded using a NanoSEM—FEI Nova 200 (FEI Company, Hillsboro, OR, USA), operating at 15 kV, coupled to an Electron Dispersive Spectroscopic analyzer (EDS) and Electron Backscatter Diffraction EDAX—Pegasus X4M analyser (AMETEK Inc., Berwyn, PA, US) and detection system (EBSD) at SEMAT (Serviços de Caracterização de Materiais, Guimarães, Portugal). After preparation of the hydrogel, a small portion of each sample was placed onto a TEM 400 mesh copper grid with Formvar/Carbon (ref. S162-4 from Agar Scientific), held by tweezers and the excess solution was cleaned. The processing of STEM images was performed using ImageJ software (National Institutes of Health (NIH), version 1.52p, Bethesda, MD, USA),which consisted of enhancing local contrast and adjusting brightness followed by a manual selection of fibers.

2.4.2. X-ray Diffraction

A conventional PAN'alytical X'Pert PRO diffractometer (Malvern Panalytical Ltd., Malvern, UK) was used for X-ray diffraction (XRD) analyses, operating with Cu K_α radiation, in a Bragg-Brentano configuration.

2.4.3. Raman Spectroscopic Measurements

Raman spectroscopy was used to assess the effect of nanoparticles in the secondary structure of the hydrogel fibers. Measurements were performed at room temperature with a Jobin Yvon T64000 triple Raman Spectrometer (HORIBA Jobin Yvon IBH Ltd., Glasgow, UK), equipped with a liquid nitrogen cooled charge couple device (CCD) detector, with a resolution better than 1 cm^{-1}. The excitation line, 514.5 nm, of an argon ion laser was focused onto the sample using a $\times 50$ objective (focused to ~1.5 μm of diameter) of an Olympus Microscope BHSM (Olympus Corporation, Tokyo, Japan) in a backscattering geometry. The spectra were acquired with a measured power of about 350 μW on the sample, with a spectral acquisition time of 45 s averaged over 10 scans, over the range 770–1800 cm^{-1}.

2.4.4. Magnetic Properties

Magnetic measurements were performed using a SQUID magnetometer from Quantum Design (Quantum Design Inc., San Diego, CA, USA). The magnetization dependence with temperature in zero-field-cooling (ZFC) and field-cooling (FC) conditions was performed at 100 Oe in the 10–320 K range. Hysteresis loops were measured at different temperatures up to an external field of 50 kOe.

2.4.5. Rheology

The viscoelastic characterization of gels was performed with a stress-controlled rotational rheometer Anton Paar MCR300 (Anton Paar GmbH, Graz, Austria). Liquid samples were loaded into the Couette geometry of the rheometer and temperature was kept at 25 °C during testing. After a five hour rest period ensuring gel setting and structural equilibrium of samples, a sweep in the strain amplitude was performed from 0.001% to 500% to assess the linear regime of viscoelasticity and the large amplitude oscillatory strain (LAOS) regime.

2.5. *Drug Release Assays*

2.5.1. Incorporation of Doxorubicin

To study the incorporation and microenvironment of doxorubicin in gels through fluorescence spectroscopy, the drug was added to gel solutions prior to gelation, for a final concentration of 20 μM (to guarantee that fluorescence intensity is proportional to concentration). From the hydrogelator solution, 200 μL were transferred to a fluorescence microcuvette and left standing until the gel was formed.

2.5.2. Drug Release to pH = 7 Buffer

To assess doxorubicin release through fluorescence spectroscopy, gels (100 μL) loaded with 0.1 mM doxorubicin were prepared and left stabilizing overnight in Amicon® Ultra-0.5 mL centrifugal filters (MilliporeSigma, St. Louis, MO, USA) with a 0.1 μm pore size. Then, the filter tube was immersed in pH = 7 buffer (800 μL) to keep pH constant (besides neutralizing the gels), and left standing at room temperature, with or without an alternating magnetic field (AMF). The AMF was generated in a custom-designed solenoid device (800 turns per meter, length: 31 cm and internal diameter: 4.8 cm) by applying an alternating electric current. A magnetic field of 2.98 mT at 1000 kHz was used. Aliquots were taken and replaced with pH = 7 buffer, then fluorescence was measured to determine the concentration at each time point. Release profile assays were performed in triplicate.

3. Results

3.1. Optimization of Hydrogel Gelation Kinetics

Turbidity kinetic assays were carried out to optimize hydrogel gelation, which is required to ensure quasi-homogeneous encapsulation of nanocomposites. Initially, the hydrogelator molecules were majorly organized in a mixture of micelles/aggregates and free monomer as suggested by fluorescence emissions at 450 nm and 360 nm, respectively (Figure 1A).

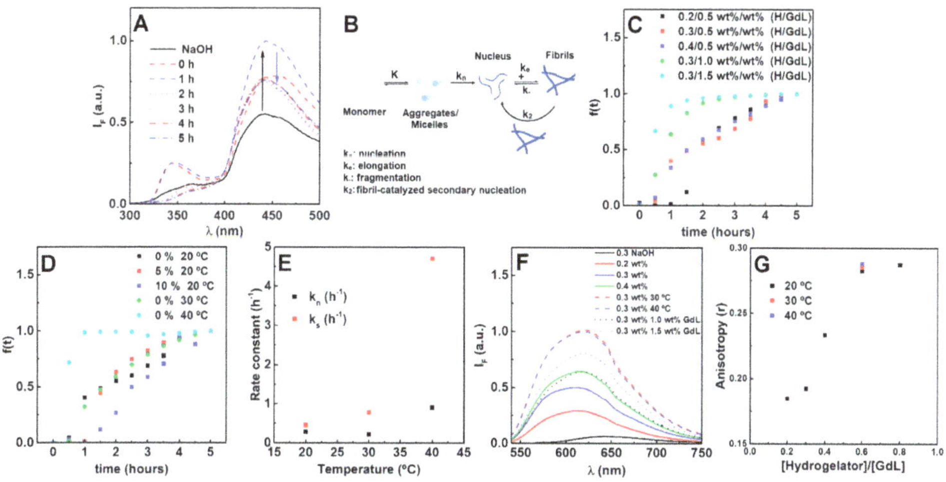

Figure 1. (**A**) Fluorescence emission spectra of the hydrogelator (0.3 wt%) over time after the gelation trigger (0.5 wt%). (**B**) Scheme of the evaluated pathways during the gelation process, where the fibril-catalyzed secondary nucleation contribution was found to be negligible. (**C**) Gelation kinetic profile dependence on hydrogelator and GdL concentration, (**D**) fibril concentration and temperature. (**E**) Average nucleation and elongation rate constants obtained from the Saitô's aggregation model fitting to turbidity profiles at increasing temperatures (hydrogel 0.3 wt%, GdL 0.5 wt%). Fluorescence emission (**F**) and anisotropy *r* (**G**) of Nile Red (2 μM) in hydrogels prepared at different hydrogelator-to-GdL ratios and temperatures (fixed hydrogel and GdL concentration).

Once gelation is initiated (after the addition of GdL), both emission bands increase, indicating the occurrence of a reorganization process. Here, the models of Knowles et al. [34] and Cohen et al. [35] (commonly used for β-amyloid aggregation) were also used as a strategy to assess the possible occurrence and influence of monomer independent (k_-;

fragmentation) and/or monomer dependent (k_2; fibril-catalyzed secondary nucleation) secondary pathways (Figure 1B).

The gelation is characterized by a sigmoidal profile, which is a common feature of fibrillation processes comprising a nucleation and elongation phase (Figure 1C,D) [30]. Increasing the hydrogelator concentration at the same GdL (0.5 wt%) concentration resulted in an increase of the nucleation rate, while the elongation rate decreased. The increase in GdL concentration promoted both the nucleation and elongation phases. Hereby, the primary nucleation process is dependent on both the hydrogelator and GdL, while the elongation process is majorly affected by the GdL. The elongation phase dependence on GdL concentration (for a fixed hydrogelator concentration) demonstrates that its rate can be increased if more protons are made available over time, considering that GdL proton dissociation is also a kinetically-dependent process [40]. Such results evidence that an increase of hydrogelator concentration has to be accompanied by an increase of GdL concentration to keep the molar equivalents, thus favoring both nucleation and elongation. Notably, by increasing both hydrogelator and GdL concentrations, no major differences were obtained for the final pH, while increasing GdL alone strongly decreased pH (Table S1 in Supplementary Materials).

A weak scaling of the half-time with the initial monomer concentration was obtained (-0.63), which is characteristic of monomer independent secondary pathways (such as fragmentation). Yet, the light scattering kinetic profiles cannot be satisfactorily matched using fixed k_n/k_- and k_+k_n parameters in the Knowles' aggregation model (Figure S1 in Supplementary Materials) [35]. The addition of pre-formed fibrils inhibited the average nucleation phase and enhanced the average elongation phase (Figure 1D), i.e., the secondary nucleation rate can be neglected. Further, increasing the hydrogelator decreased both secondary pathways rates, while GdL enhanced the monomer independent pathway and inhibited the monomer-dependent pathway (Table S2 in Supplementary Materials). Nonetheless, self-assembly can be majorly attributed to the primary pathways. Temperature exponentially affected the average nucleation and elongation phase, which is associated with a faster GdL proton dissociation (Figure 1E) [40]. As a result, overall gelation can be enhanced by increasing both GdL and hydrogelator concentrations and preparing the gel at 30 or 40 °C, which favors the primary pathways (nucleation and elongation).

The dye Nile Red was used to evaluate the effect of preparation conditions on the microenvironment, as it is a solvatochromic probe that has almost negligible emission in water, but intensely emits fluorescence in non-polar environments [41–44], as observed after hydrogel formation (Figure 1F). Furthermore, the emission is accompanied by a blue-shift with a reduction of polarity [41–44]. Here, Nile Red is localized in a microenvironment with a polarity between acetone and ethanol [41]. The higher Nile Red fluorescence emission intensity on the gels prepared at 30 °C and 40 °C suggests that more hydrophobic regions were made available. The fluorescence anisotropy values reveal similar fluidity compared to the hydrogel prepared at room temperature. Although increasing GdL (fixed hydrogelator concentration at 0.3 wt%) also contributed for more hydrophobic regions, the microfluidity was lower than that obtained by increasing the hydrogelator concentration (fixed GdL concentration at 0.5 wt%). As such, increasing both GdL and hydrogelator concentrations (0.4 hydrogelator-to-GdL ratio or higher) promotes more hydrophobic regions with higher microviscosity.

3.2. Nanoparticles Characterization

Nanoparticles of manganese ferrite with different coatings were prepared using different synthetic methods, and are named, from now on, as citrate-stabilized or lipid-coated manganese ferrite nanoparticles. The X-ray diffraction (XRD) patterns of both samples present well-defined peaks (Figure 2A,B) characteristic of a crystalline structure, which was obtained without calcination. The diffraction peaks of the $MnFe_2O_4$ crystalline spinel structure are observed at $2\theta = 29.7°$ (2 2 0), 34.9° (3 1 1), 36.5° (2 2 2), 42.5° (4 0 0), 52.7° (4 2 2), 56.2° (3 3 3) and (5 1 1), 61.8° (4 4 0), 65.0° (5 3 1), 70.1° (6 2 0), 73.1° (5 3 3), 74.0°

(6 2 2), 78° (4 4 4), 85.6° (6 4 2), 88.5° (7 3 1) and (5 5 3), according to CIF file 2300618 (space group Fd-3m). Rietveld analysis was performed using the FullProf software suite, confirming the spinel structure. As in previous works [45,46], we considered it to be important to use microabsorption correction [47], resulting in fits with good R_F values. The calculated parameters are presented in Table S3 in Supplementary Materials, offering a larger size of the crystalline domains in the nanoparticles that are citrate-stabilized. Additional diffraction peaks were observed for the lipid-coated nanoparticles, occurring at positions similar to those reported for layered manganese laurate [48], which can be ascribed to ordered lauric acid molecules at the surface of manganese ferrite.

The UV-visible absorption spectra of the prepared nanoparticles are represented in Figure 2C. From the Tauc plot (inset of Figure 2C), the optical band gap (E_g) between the citrate-stabilized and lipid-coated manganese ferrite nanoparticles was determined and a linear relation was obtained for an indirect semiconductor with a band gap of 1.13 eV and 1.19 eV, respectively, which are similar to the previous reported value of 1.08 eV [49].

The sedimentation profiles for bare, lipid-coated, and citrate-stabilized nanoparticles are displayed in Figure 2D. The dependence of the sedimentation rate on nanoparticle concentration (obtained through fitting of a Becquerel function or compressed hyperbola) [50] is included (values are reported in Table S4 in Supplementary Materials). The citrate-stabilized nanoparticles sedimentation profile suggests the occurrence of nanoparticles aggregation into stable agglomerates [51,52], which settle down at a faster rate than single nanoparticles. The lipid-coated nanoparticles show a sedimentation rate independent of the used concentration range (0.025–0.2 wt%). Hereby, the longer-term stability of the lipid-coated nanoparticles is expected to provide homogeneous gels at higher concentration of nanoparticles, compared to the ones that are citrate-stabilized.

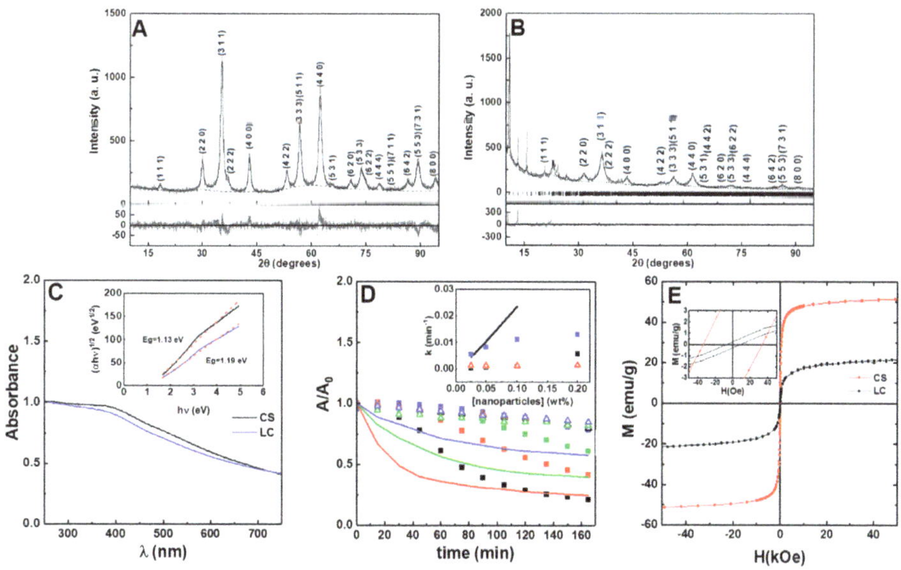

Figure 2. X-ray diffraction pattern of (**A**) citrate-stabilized and (**B**) lipid-coated manganese ferrite nanoparticles. Gray lines: experimental patterns; black lines: fitted patterns; dotted lines: fitted background; the vertical lines in (**B**) are the Bragg diffraction positions of manganese laurate. (**C**) Absorption spectra of citrate-stabilized (CS, black line) and lipid-coated (LC, blue line) $MnFe_2O_4$ nanoparticles. Inset: Tauc plot of citrate-stabilized and lipid-coated nanoparticles. (**D**) Sedimentation profiles of the citrate-stabilized (squares), lipid-coated (triangles), and bare (line) $MnFe_2O_4$ nanoparticles at 0.2 wt% (black), 0.1 wt% (red), 0.05 wt% (green), and 0.025 wt% (blue). Inset: sedimentation rate dependence on nanoparticle concentration. The citrate-stabilized aggregation rate is included (black squares). (**E**) Magnetization hysteresis loops of citrate-stabilized and lipid-coated manganese ferrite nanoparticles measured at room temperature (T = 300 K). Inset: Enlargement of the loops in the low field region.

The magnetic hysteresis loops show that the saturation magnetization (emu/g) is higher for the citrate-stabilized nanoparticles than for the lipid-coated nanoparticles (Figure 2E), which can be explained by taking a different stoichiometry into account in the manganese ferrite in the two samples, as pointed out by the different lattice constant obtained in the X-ray diffraction analysis. A smaller size and a higher wt% of organic matter present in the second sample can also influence the final value of saturation magnetization. Consequently, considering that SAR $\propto M_s^2$ (SAR—specific absorption rate), the lipid-coated nanoparticles are expected to have lower heating efficiency than the citrate-stabilized nanoparticles [53,54]. The low M_r/M_s ratio, of around 0.1 (see Table S5 in Supplementary Materials), is an indication that both types of nanoparticles display a superparamagnetic behavior [54].

3.3. Development of Magnetogels

An empirical equation was used to assist the estimation of the conditions required to maximize the homogeneity of the gel (see deduction, discussion and Figure S2 in Supplementary Materials). Figure 3 displays plots of the obtained k_{emp}/k_{sed} for various nanoparticle concentration and GdL-to-hydrogelator concentration ratio. The estimation implies that the gelation conditions have to guarantee a $k_{emp}/k_{sed} > 41.7$, so that when gel fraction $f(t)$ attains 0.9, the nanoparticles suspended fraction is also at 0.9 (if $v = 1$ and $a = 0.5$ is assumed).

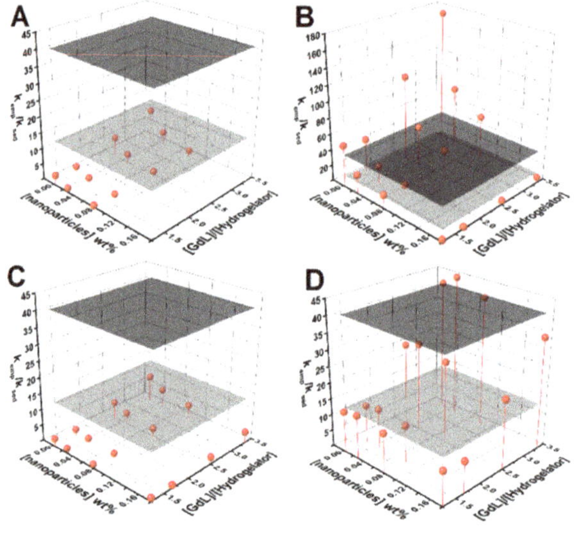

Figure 3. Surface plots of k_{emp}/k_{sed} dependence on nanoparticle concentration and the GdL-to-hydrogelator concentration ratio for: (**A**) bare nanoparticles, (**B**) the citrate-stabilized nanoparticles aggregation rate, (**C**) the sedimentation rate, and (**D**) lipid-coated nanoparticles. The gray planes define the estimated k_{emp}/k_{sed} required to guarantee that when the gel is about 10% of gelation completion, 90% of the nanoparticles remains in suspension ($\nu = 1$, a = 0.5, dark gray plane; $\nu = 0.5$, a = 0.5, gray plane).

Gels prepared with 1.5 wt% GdL and 0.3 wt% hydrogelator retained the nanoparticles, though the pH value was lower than 4. Increasing the hydrogelator content to 0.5 wt% and reducing GdL to 1 wt% (has higher microviscosity) allowed the preparation of homogeneous magnetogels at 0.1 wt% of nanoparticles, with a pH of ~5. The 0.2 wt% content of nanoparticles can also be prepared at 0.5 wt% of hydrogelator and 1 wt% of GdL but preparation at 30 °C or 40 °C is required (yields k_{emp}/k_{sed} larger than 200). Nonetheless, the parameter v obtained from curve fitting was around 0.5, thus decreasing the required k_{emp}/k_{sed} to 14, which allowed the possibility of obtaining quasi-homogeneous gels at 0.1 wt% of nanoparticles at a [GdL]/[hydrogelator] ratio of 2 (see magnetogels and respective pH values in Figure S3 in Supplementary Materials).

3.4. Gels Microviscosity

The effect of the nanoparticles in the gels matrix microenvironment was studied using Nile Red as a fluorescence probe, while considering its sensitivity to polarity and viscosity [55,56]. In the 20–40 °C temperature range, no major fluorescence emission decay changes were observed, which might be associated with the structure maintaining its integrity. A steep fluorescence emission decrease indicated that the phase transition occurs above 45 °C for the hydrogel and the citrate-stabilized nanoparticle-containing magnetogels, while in the lipid-coated nanoparticle-containing magnetogels, it was shifted to 50 °C (Figure 4A). The fluorescence anisotropy increases at the phase transition temperature, which might be associated with a fluorescence emission lifetime decrease (Figure 4B) [57].

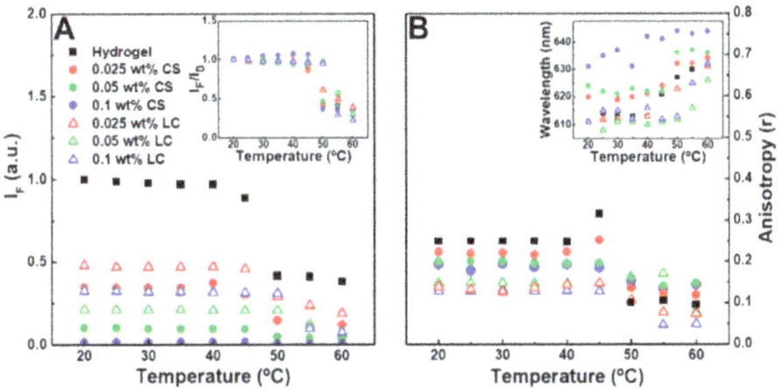

Figure 4. (**A**) Nile Red (2 μM) maximum fluorescence emission dependence on the temperature of hydrogel, citrate-stabilized (CS), and lipid-coated (LC) nanoparticles containing magnetogels. Inset: Normalized maximum fluorescence emission dependence on the temperature of the respective gels. (**B**) Nile Red fluorescence anisotropy dependence on temperature. Inset: Nile Red maximum emission wavelength dependence on temperature.

The results suggest that citrate-stabilized nanoparticles destabilized the hydrophobic domains, leading to a microviscosity reduction and polarity increase of the cavities where Nile Red is localized, which might occur through hydrogen bonding and ionic interaction between the nanoparticles and the fibers. The polarity in the hydrogel is similar to acetone, which was changed towards ethanol after addition of citrate-stabilized nanoparticles. Furthermore, after phase transition and at 0.1 wt% of citrate-stabilized nanoparticles, the Nile Red emission wavelength (640 nm) became close to the reported maximum wavelength in water (657 nm) [42]. The lipid-coated nanoparticles induced a lower fluorescence anisotropy than citrate-stabilized nanoparticles and one that is similar to the reported anisotropy values of Nile Red in mixed vesicles and micelle membranes [44], thus suggesting that lipid-fiber domains are formed with a polarity similar to the fibers and a viscosity near that of membranes.

3.5. Gels Secondary Structure

The Raman spectra of the hydrogels and magnetogels (0.1 wt% of nanoparticles) were obtained to assess influence of nanoparticles in the secondary structure, which are displayed in Figure 5. Reported Raman shifts of the phenylalanine phenyl ring and methionine side chain (CH_3 deformation at 1440 cm^{-1} and CH_2 wagging at 1320 cm^{-1}) are also displayed [58,59]. Structural changes upon the addition of lipid-coated nanoparticles were suggested by the appearance of a band at 982 cm^{-1}. Gaussian curves were fitted to the major phenyl ring signal at around 1003 cm^{-1} (see Figure S4 in Supplementary

Materials). A blue shift and decreasing cross-section were observed with an increasing nanoparticle concentration. The latter effect can be associated with an increasing exposure of the aromatic rings to a more hydrated environment [60].

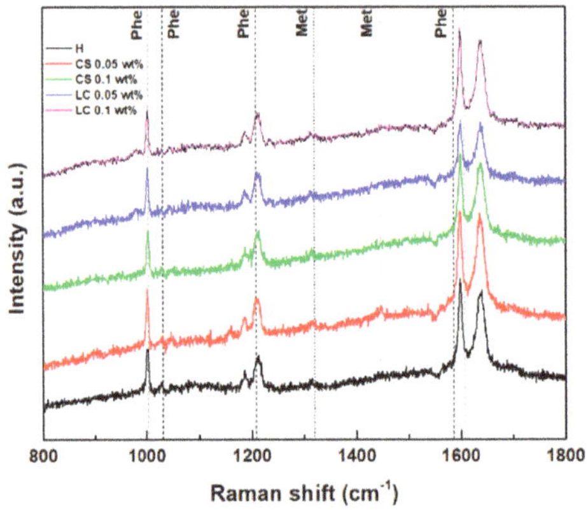

Figure 5. Raman scattering spectra of the hydrogel and magnetogels containing lipid-coated (LC) and citrate-stabilized (CS) nanoparticles. Vertical lines represent the reported Raman shifts of methionine (Met) and phenylalanine (Phe).

The Amide I region (1580–1700 cm^{-1}) arises from the C=O stretching vibration, which is sensitive to changes in backbone peptide conformation [61]. Deconvolution of the amide I band in its sub-bands is correlated with various secondary structure contributions (see Figure S5 in Supplementary Materials) [60–62]. A major contribution at 1637 cm^{-1} is common to all systems, which can be associated with a major β-sheet content [61,62]. The band at 1338 cm^{-1} also confirms the predominance of the β-sheet in all gels [61].

3.6. Rheological Properties

Large amplitude oscillatory shear strain sweeps (LAOSS) were carried out to assess the effect of the nanoparticles on the gels structure (Figure S6 in Supplementary Materials). Further, the effect of temperature at 37 °C on the hydrogel shear elastic and loss moduli was also assessed (Figure S6), the decrease of which evidences the phase transition behavior observed in the microviscosity studies. The addition of nanoparticles reduced the elasticity of gels, as previously observed for other systems [10,63]. Interestingly, the lipid-coated nanoparticles induced a lower elasticity than the citrate-stabilized nanoparticles, which was similar to their influence on the gels' microviscosity. Further, the strain at which G' crosses G'' increased in the presence of citrate-stabilized nanoparticles (at 0.05 wt%). As a result, the profiles suggest that different structures were obtained for each nanoparticle content, compared to the bare gel. The lipid-coated nanoparticle-containing magnetogel displays the Payne effect characteristic of viscoelastic matrices reinforced by solid fillers, i.e., a local maximum of G'' concomitant with a significant decrease of G', which can be associated with the breakage and recovery of weak interaction bonds linking adjacent clusters, aggregation/disaggregation of nanoparticles, or molecular disentanglement [64].

3.7. Electron Microscopy

Figure 6 displays the STEM images of the citrate-stabilized and lipid-coated nanoparticles in solution and incorporated into the hydrogel matrix (magnetogels) prepared at 0.5 wt% of hydrogelator. The negative charge of the citrate molecules stabilizing the nanoparticles ensured there were well dispersed nanoparticles with an average size of 14.4 ± 2.6 nm (see histograms in Figure S7 in Supplementary Materials), though some aggregates were observed (Figure 6A). Alternatively, the lipid-coated nanoparticles, with an average size of 8.9 ± 2.1 nm, displayed a tendency to form spherical aggregates, which can be associated with the dynamic membrane coating being prone to self-assembly (Figure 6B). The hydrogel network is also displayed for a matter of comparison with the magnetogels. The preparation conditions used here afforded a network comprised of thin and thick short fibers with a cross-section of 21.3 ± 3.4 nm and 48.4 ± 13.8 nm, and with an average length of 1030 ± 389 nm.

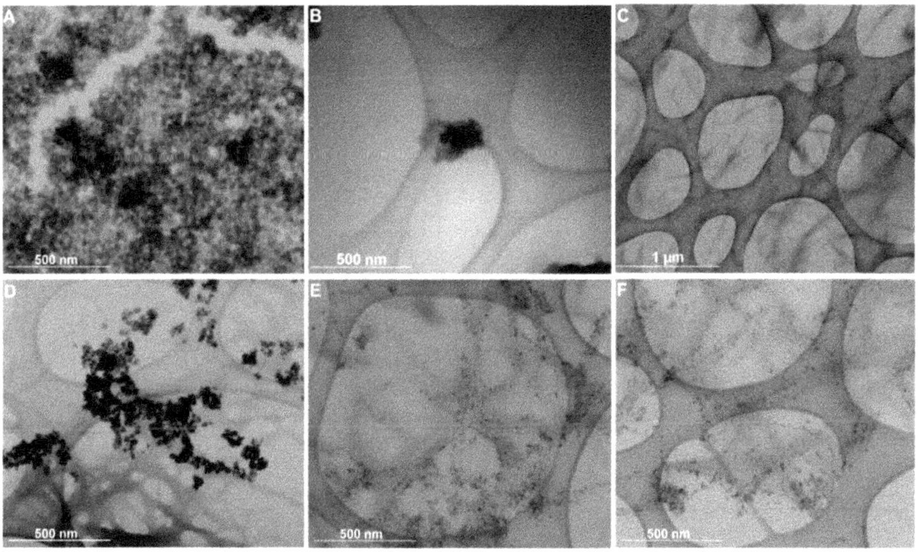

Figure 6. Scanning transmission electron microscopy (STEM) images of (**A**) citrate-stabilized nanoparticles, (**B**) lipid-coated nanoparticles, and (**C**) the hydrogel structure. Magnetogels containing (**D**) citrate-stabilized nanoparticles and (**E**,**F**) lipid-coated nanoparticles prepared at 0.5 wt% of hydrogelator, 1 wt% of GdL, and 0.05 wt% of nanoparticles.

Furthermore, different effects on the magnetogels microstructure were observed when using the two types of nanoparticles. The citrate-stabilized nanoparticles are randomly distributed within the hydrogel matrix as aggregates. This stems from the fact that the magnetogels based on N-protected peptides lacked a cationic group in the hydrogelator structure electrostatically interacting with these negatively-charged stabilized nanoparticles and fixing them in the matrix. Nevertheless, the lipid-coated nanoparticles displayed an affinity towards the fibers surface, rendering them more adequate to avoid any potential leaking of nanoparticles. Furthermore, upon gelation, the lipid-coated nanoparticle aggregates become destabilized, as no aggregates were observed (see Figure 6E,F).

3.8. Hyperthermia Studies

The calorimetric approach was carried out to assess the magnetic nanoparticle hyperthermia effect in gels, while considering the medical threshold limit of $H_0 f \leq 5 \times 10^9$ A m^{-1} s^{-1} [65,66], or $H_0 f \leq 4.85 \times 10^8$ A m^{-1} s^{-1} [66], depending on the area exposed. The increase in temperature over time, when the nanoparticles are dispersed in aqueous solution and in gels while applying an alternating magnetic field, is displayed in Figure 7.

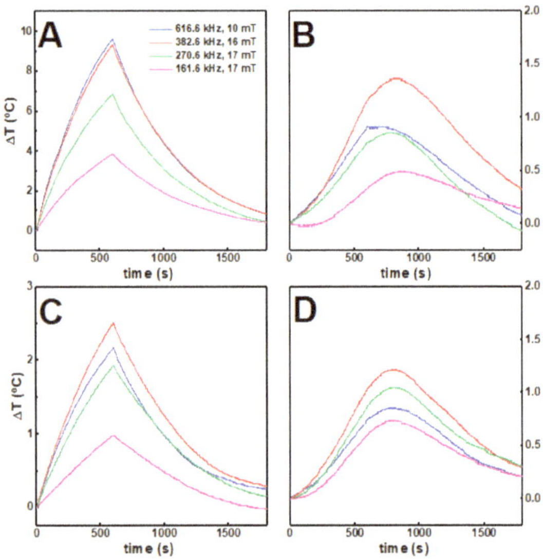

Figure 7. The temperature variation over time of (**A**) citrate-stabilized and (**B**) lipid-coated nanoparticles at 1 mg/mL (0.1 wt%) in water and in gels (**C**,**D**) respectively) under different magnetic field strengths and frequencies.

In water solution, the citrate-stabilized nanoparticles attained higher temperatures than the lipid-coated nanoparticles, which could be associated with the different stoichiometry in the spinel ferrites belonging to the two samples, with the different average sizes, and with the fact that the lipid-coated nanoparticles are more likely to aggregate in aqueous solution. In both cases, the increments in temperature decreased when the magnetic nanoparticles were incorporated in the gels, likely because some of the heat generated was employed for inducing local changes in the gel structure.

Furthermore, nanoparticles at high concentration within the gels underwent stronger magnetic dipolar interactions, which has a detrimental effect in the heat delivery capacity [67,68]. The heating efficiency was quantitatively evaluated through the intrinsic loss power (ILP) (see Table S6 in Supplementary Materials). While the ILP decreases for the citrate-stabilized nanoparticles around 80% when incorporated into the gels, for the lipid-coated nanoparticles no major changes were obtained. Accordingly, despite being less efficient than citrate-stabilized nanoparticles, the lipid-coated nanoparticles keep a similar heat delivery capacity when they are incorporated in the gels.

3.9. Drug Release Assays

3.9.1. Incorporation of Doxorubicin

Previously, supramolecular dehydropeptide-based hydrogels have shown promising results as drug delivery nanosystems [8]. Here, the nanoparticles concentration effect over doxorubicin incorporation is assessed. FRET (Förster Resonance Energy Transfer) process from the emissive moieties of the hydrogel aromatic moieties (acting as the energy donors) to doxorubicin (acting as the energy acceptor) allowed us to follow the encapsulation of doxorubicin owing to the overlap between the drug absorption band and hydrogel fluorescence emission (see Figure S8 in Supplementary Materials).

Fluorescence spectra of hydrogel and magnetogels at 0.025 wt% of citrate-stabilized and lipid-coated nanoparticles, with and without doxorubicin, are displayed in Figure 8. The absence of doxorubicin results in a strong fluorescence emission (Figure 8A) of the aggregates associated with the stacking of the aromatic rings (λ_{max} ~ 450 nm).

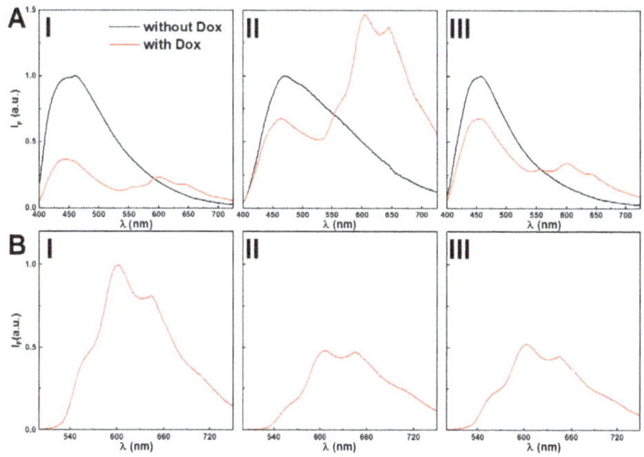

Figure 8. (**A**) Fluorescence emission spectra (λ_{exc} = 375 nm) of hydrogel (I) and magnetogels containing citrate-stabilized (II) and lipid-coated (III) manganese ferrite nanoparticles (0.025 wt%) incorporating the doxorubicin and comparison with the plain gels. (**B**) Fluorescence spectra (λ_{exc} = 480 nm) of directly-excited doxorubicin in gels, (I) and magnetogels con-taining citrate-stabilized (II) and lipid-coated (III) manganese ferrite nanoparticles (0.025 wt%) incorporating the doxorubicin and comparison with the plain gels

The addition of citrate-stabilized nanoparticles induced a red-shift (Figure 8B), which was modulated by increasing the nanoparticle concentration (see Figure S9 in Supplementary Materials). The lipid-coated nanoparticles induced a thinning of the emission band shape, suggesting that the variety of microenvironments in the vicinity of the fibers was restrained.

Doxorubicin displays reverse solvatochromism, making the assignment of its location ambiguous [69]. Further, the assays with Nile Red demonstrated that a higher concentration of citrate-stabilized nanoparticles was observed to increase the polarity of the gel hydrophobic cavities. Thus, the decreasing fluorescence emission ratio of doxorubicin between the peaks at λ = 560 nm and λ = 600 nm upon addition of citrate-stabilized nanoparticles might be associated with an increasingly acidic region and its location in cavities, which are destabilized by the increasing concentration of nanoparticles (see Figure S10 in Supplementary Materials) [69]. Nonetheless, the possibility of aggregation is not excluded, considering that at 10 μM, a fraction of 47% is dimerized [70,71], and the obtained spectra in gels resemble the fluorescence emission spectrum of a doxorubicin concentrated solution (0.1 mM) at pH = 7 (see Figure S11 in Supplementary Materials). The lipid-coated nanoparticles do not show the same effects, as the wavelength and emission ratio remain mostly unchanged, i.e., no major changes are induced in the doxorubicin location microenvironment. The fluorescence quenching by increasing nanoparticle concentration further suggests its proximity to the nanoparticles.

The hydrogelator fluorescence quantum yield at a nanoparticle concentration of 0.025 wt% was determined through Equation (7). The calculated FRET efficiencies (Φ_{FRET}), Förster radius (R_0), and donor-acceptor distances (R_{DA}) are presented in Table 1. As reported in other systems [8,63], the distances between the fiber aromatic groups and doxorubicin remain similar with or without nanoparticles, and suggests a host-guest type interaction [8,63].

Here, similarly to the obtained results with Nile Red, the lipid-coated nanoparticles induced a lower anisotropy, while the fluorescence emission spectrum remains similar, suggesting the occurrence of lipid-fiber domains. Overall, the high anisotropy values suggest that doxorubicin has affinity towards the gel network fibers and the nanoparticles affect the arrangement of the fibers.

Table 1. Förster Resonance Energy Transfer (FRET) efficiencies (Φ_{FRET}), fluorescence quantum yields of the donor (hydrogel aggregates) (Φ_D^0), donor-acceptor distances (R_{DA}) and steady-state fluorescence anisotropy (r) values for gels with an incorporated drug. Anisotropy value of doxorubicin in glycerol (at 25 °C): r = 0.285.

System	Content (wt%)	Φ_{FRET}	Φ_D^0 [a]	R_0 (nm)	R_{DA} (nm)	r
H	-	0.66	0.012	2.1	1.9	0.17
CS	0.025	0.33	0.001	1.7	1.4	0.16
	0.05	-	-	-	-	0.15
	0.1	-	-	-	-	0.14
LC	0.025	0.33	0.010	2.3	2.0	0.13
	0.05	-	-	-	-	0.11
	0.1	-	-	-	-	0.11

[a] Relative to L-tryptophan in aqueous buffer solution, pH = 7.2 (Φ_r = 0.14 at 25 °C) [39]. The error rate is about 10%.

3.9.2. Doxorubicin Release Assays

In a previous drug release assay of hydrogels, a mixed behavior of gel erosion and drug diffusion was observed [8]. Doxorubicin release from the hydrogel and magnetogels to pH = 7 buffer (to keep pH conditions constant and neutralize gels) was assessed.

The release profiles of the hydrogel and both limiting conditions (0.1 wt% of nanoparticles) of magnetogels are displayed in Figure 9A. Similarly to previous results for dehydropeptide hydrogels (without nanoparticles) and other magnetogels [62], an initial burst release occurs, followed by a slow release phase. Moreover, both systems display a low drug release profile that can be associated with the strong interactions established between the drug and gels components. Such profiles are useful for therapeutic applications, as the systems can be loaded with high amounts of chemotherapeutic drugs and ensure a prolonged and controlled release of a therapeutically relevant dose in the target site.

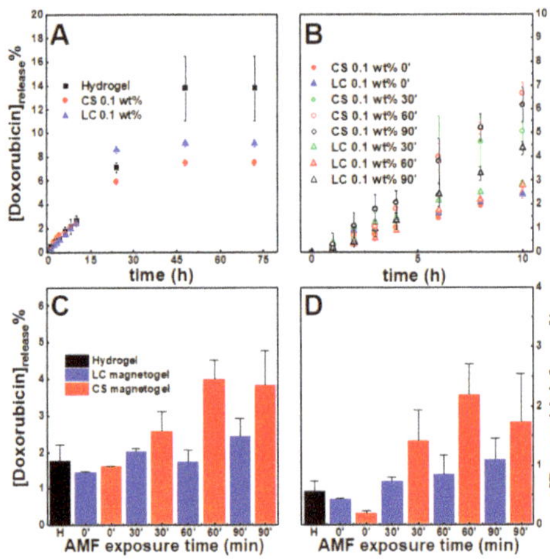

Figure 9. Cumulative doxorubicin percentage release profiles from the hydrogel and gels containing citrate-stabilized (CS) and lipid-coated (LC) nanoparticles (0.1 wt%) (**A**) without an externally applied alternating magnetic field (AMF) and (**B**) comparison of the magnetogels release when an AMF is applied for 30, 60, and 90 min between t = 4 h and t = 6 h. (**C**) Cumulative doxorubicin released at t = 6 h. (**D**) Variation of doxorubicin percentage released between t=4 h and t = 6 h, that is, for a period of 2 h.

Quantitative analysis of the cumulative drug release profiles without a magnetic field was carried out through the fitting of various mathematical models (see Tables S7 and S8 in Supplementary Materials) [72–75].

Overall, the obtained Gompertz model fitting suggests that both nanoparticles limited the maximum drug quantity released, while the Korsmeyer-Peppas mechanistic model ($0.45 < n < 0.89$) points to a mechanism that combines diffusion and erosion drug release (non-Fickian release) [72–74]. The latter contribution was also evidenced by the good fitting of the first-order model (associated with the diffusion of water-soluble drugs in porous matrices) and the Hixson-Crowell model (describes systems where changes in the surface area and diameter occur over time, but the initial geometrical shape is kept constant).

An alternating magnetic field (AMF) was applied for 30, 60, and 90 min, 2.98 mT at 1000 kHz after 4 h, which resulted into an increase of cumulative doxorubicin release comparatively to the systems that were not subjected to the AMF (Figure 9B,C). To remove the contribution from preparation anomalies, the enhancement of drug release was further evidenced by comparing the amount of drug release in the period between t = 4 h and t = 6 h (Figure 9D), that is, during two hours. The enhancement is larger in the citrate-stabilized nanoparticles than lipid-coated nanoparticles, which might be associated with the higher heating efficiency. However, the enhanced drug release displays a non-linear relation with the AMF exposure time. This can be associated with the nanoparticles being distributed as aggregates, as opposed to the lipid-coated nanoparticles that incrementally enhanced drug release, as the sample was exposed to longer periods. Further, gels subjected to the AMF for 30 min revealed a decrease of fluorescence anisotropy to 0.12 for the citrate-stabilized nanoparticles, while in the lipid-coated ones, it remained close to 0.15, i.e., the citrate-stabilized nanoparticles heating induced an irreversible collapse of the gel network that led to an increased variance of drug released, while the other nanoparticles might have not affected the network structure.

As a result, the incremental drug release enhancement and retention of the lipid-coated nanoparticles developed here indicate that this supramolecular system architecture is a suitable approach for controlling drug release, since the heating-induced effect displayed improved reproducibility. However, lipid-coated nanoparticles with a higher heating efficiency and gels that can homogeneously include more nanoparticles are required in future developments to improve the percentage of drug released upon the AMF-trigger, as suggested from the higher triggered release from citrate-stabilized nanoparticle-containing gels.

4. Conclusions

Pursuing the effect of citrate and lipid-functionalized nanoparticles in the development of supramolecular magnetogels, the gelation of the hydrogel Cbz-L-Met-Z-ΔPhe-OH was systematically optimized by using kinetic models to prepare homogeneous magnetogels, while considering both the kinetics of gelation and sedimentation of nanoparticles. Lipid-coated nanoparticles formed lipid-fiber domains and increased the gel irreversible phase transition temperature. The heating efficiency of lipid-coated nanoparticles was improved (maximum heat and reproducibility) when they were incorporated in the gels, while a detrimental effect was obtained for citrate-stabilized nanoparticles. Further, the former did not produce major changes in doxorubicin encapsulation, while the latter increased the micropolarity of its location and induced aggregation. The magnetogels revealed similar doxorubicin release profiles and AMF-trigger was stronger in the citrate-stabilized nanoparticles, though the triggered release was more reproducible in the lipid-coated nanoparticle-containing gels.

Overall, the lipid-coated nanoparticles displayed promising results for future developments of supramolecular magnetogels aiming at the control of drug release. This was mainly associated with the improved nanoparticle distribution (along the hydrogel fibers), unaffected heating efficiency upon gelation, and reproducible triggered drug release. On the other hand, despite the higher heating efficiency of the negatively-charged citrate-stabilized nanoparticles in solution, they are prone to aggregation upon gelation, which is reflected in a decreased heating efficiency, and local inhomogeneous distribution consequently leading to less reproducibility in drug release.

Hereby, this work reveals that negatively-charged stabilized and lipid-coated nanoparticles affect the final gel architecture differently and, thus, also affect its properties and the encapsulation of drugs in different ways. Further, comparison of both systems points out that on-demand drug release in dehydropeptide-based supramolecular magnetogels can be optimized by developing nanoparticles that can adsorb onto hydrogel fibers, while providing domains that improve or do not affect drug encapsulation (lipid-coated nanoparticles). In particular, the unaffected drug encapsulation and reproducible release from the lipid-coated nanoparticle-containing gels, upon application of AMF, is anticipated to potentiate the supramolecular magnetic gels in drug delivery towards on-demand drug release.

Future developments will be focused on improving heating efficiency of the lipid-coated nanoparticles and the synergy between hyperthermia and triggered drug release, without inducing a major collapse of the hydrogel.

Supplementary Materials: The following are available online at https://www.mdpi.com/2079-4991/11/1/16/s1, Structure S1: Hydrogelator structure, Table S1: Final hydrogel pH, Figure S1: Knowles' aggregation model fitting to experimental aggregation profiles, Table S2: Calculated curve-fitting parameters of the turbidity profiles, Table S3: Calculated X-ray diffraction parameters, Table S4: Calculated curve-fitting parameters of the sedimentation profiles, Table S5: Obtained parameters from the SQUID hysteresis curves, Figure S2: dependence of $\frac{k_{emp}}{k_{sed}}$ on $s(t)$ and $f(t)$ for fixed v and a values, Figure S3: Image of the magnetogels containing citrate-stabilized and lipid-coated nanoparticles, Figure S4: Curve-fitting of the Raman scattering spectra in the wavelength range 960–1040 cm^{-1} and dependence of the Raman shift and full width at half maximum of the major phenyl ring vibration, Figure S5: Curve-fitting of the Raman scattering spectra of gels in the amide I region, Figure S6: Strain dependence of the shear elastic and loss moduli for gels, Figure S7: STEM images and histograms of gels, Table S6: Calculated magnetic hyperthermia parameters, Figure S8: Normalized fluorescence emission spectrum of the hydrogel and doxorubicin absorption spectrum, Figure S9: Normalized fluorescence emission spectra of hydrogel and magnetogels, Figure S10: Fluorescence emission spectra of doxorubicin in hydrogel and magnetogels at various nanoparticle concentration, Figure S11: Normalized fluorescence emission spectra of doxorubicin in pH = 7 buffer at various concentrations, Table S7: Coefficients of determination for doxorubicin release profiles, Table S8: Release coefficients of the models fitted to the experimental release profiles.

Author Contributions: Conceptualization, S.R.S.V., P.M.T.F., and E.M.S.C.; methodology, S.R.S.V., L.H., C.M., P.J.G.C., J.A.M., V.S., M.A.C.-D., P.M.T.F., and E.M.S.C.; formal analysis, S.R.S.V., J.F.G.S., L.H., C.M., P.J.G.C., J.A.M., M.T.-A., V.S., M.A.C.-D., P.M.T.F., and E.M.S.C.; investigation, S.R.S.V., J.F.G.S., L.H., C.M., M.T.-A., V.S., M.A.C.-D., P.M.T.F., and E.M.S.C.; writing—original draft preparation, S.R.S.V., P.M.T.F., and E.M.S.C.; writing—review and editing, M.A.C.-D. and E.M.S.C.; supervision, M.A.C.-D., P.M.T.F., and E.M.S.C.; project administration, P.J.G.C. and P.M.T.F. All authors have read and agreed to the published version of the manuscript.

Funding: This work was supported by the Portuguese Foundation for Science and Technology (FCT) in the framework of the Strategic Funding of CF-UM-UP (UIDB/04650/2020) and CQUM (UIDB/00686/2020). FCT, FEDER, PORTUGAL2020 and COMPETE2020 are also acknowledged for funding under research projects PTDC/QUI-QFI/28020/2017 (POCI-01-0145-FEDER-028020), PTDC/QUI-QOR/29015/2017 (POCI-01-0145-FEDER-029015), UIDB/05256/2020 and UIDP/05256/2020. L.H. acknowledges the funding from the "FCT Researcher" Programme through grant IF/00606/2014. S. R. S. Veloso acknowledges FCT for a PhD grant (SFRH/BD/144017/2019). Support from MAP-Fis Doctoral Programme is also acknowledged. Ministerio de Economía y Competitividad (MINECO, Spain) under the Grant (CTM2017-84050-R), Xunta de Galicia/FEDER (IN607A 2018/5 and Centro Singular de Investigación de Galicia accreditation 2019-2022, ED431G 2019/06), 0712_ACUINANO_1_E, 0624_2IQBIONEURO_6_E and NANOCULTURE cofounded by FEDER through the program Interreg V-A España-Portugal (POCTEP), NANOCULTURE (ERDF: 1.102.531) Interreg Atlantic Area, the European Union (European Regional Development Fund-ERDF).

Conflicts of Interest: The authors declare no conflict of interest.

References

1. Ma, M.; Kuang, Y.; Gao, Y.; Zhang, Y.; Gao, P.; Xu, B. Aromatic-aromatic interactions induce the self-assembly of pentapeptidic derivatives in water to form nanofibers and supramolecular hydrogels. *J. Am. Chem. Soc.* **2010**, *132*, 2719–2728. [CrossRef] [PubMed]
2. Sutton, S.; Campbell, N.L.; Cooper, A.I.; Kirkland, M.; Frith, W.J.; Adams, D.J. Controlled release from modified amino acid hydrogels governed by molecular size or network dynamics. *Langmuir* **2009**, *25*, 10285–10291. [CrossRef] [PubMed]
3. Bhattacharya, S.; Samanta, S. Soft-nanocomposites of nanoparticles and nanocarbons with supramolecular and polymer gels and their applications. *Chem. Rev.* **2016**, *116*, 11967–12028. [CrossRef] [PubMed]
4. Jian, H.; Wand, M.; Dong, Q.; Li, J.; Wang, A.; Li, X.; Ren, P.; Bai, S. Dipeptide self-assembled hydrogels with tunable mechanical properties and degradability for 3D bioprinting. *ACS Appl. Mater. Interfaces* **2019**, *11*, 46419–46426. [CrossRef] [PubMed]
5. Fichman, G.; Gazit, E. Self-assembly of short peptides to form hydrogels: Design of building blocks, physical properties and technological applications. *Acta Biomater.* **2014**, *10*, 1671–1682. [CrossRef] [PubMed]
6. Yuan, D.; Bing, X. Heterotypic supramolecular hydrogels. *J. Mater. Chem. B* **2016**, *4*, 5638–5649. [CrossRef]
7. Draper, E.R.; Adams, D.J. Low-molecular-weight gels: The state of the art. *Chem* **2017**, *2*, 716–731. [CrossRef]
8. Veloso, S.R.S.; Magalhães, C.A.B.; Rodrigues, A.R.O.; Vilaça, H.; Queiroz, M.J.R.P.; Martins, J.A.; Coutinho, P.J.G.; Ferreira, P.M.T.; Castanheira, E.M.S. Novel dehydropeptide-based magnetogels containing manganese ferrite nanoparticles as antitumor drug nanocarriers. *Phys. Chem. Chem. Phys.* **2019**, *21*, 10377–10390. [CrossRef]
9. Veloso, S.R.S.; Ferreira, P.M.T.; Martins, J.A.; Coutinho, P.J.G.; Castanheira, E.M.S. Core-shell magnetic-plasmonic nanoparticles enclosed in a biocompatible dehydropeptide-based hydrogel containing lysine. In Proceedings of the Fourth International Conference on Applications of Optics and Photonics, Lisbon, Portugal, 3 October 2019.
10. Carvalho, A.; Gallo, J.; Pereira, D.M.; Valentão, P.; Andrade, P.B.; Hilliou, L.; Ferreira, P.M.T.; Bañobre-Lopez, M.; Martins, J.A. Magnetic dehydrodipeptide-based self-assembled hydrogels for theragnostic applications. *Nanomaterials* **2019**, *9*, 541. [CrossRef]
11. Ali, S.; Yosipovitch, G. Skin pH: From basic science to basic skin care. *Acta Derm. Venereol.* **2013**, *93*, 261–267. [CrossRef]
12. Bonhome-Espinosa, A.; Campos, F.; Rodriguez, I.; Carriel, V.; Marins, J.; Zubarev, A.; Duran, J.; Lopez-Lopez, M. Effect of particle concentration on the microstructural and macromechanical properties of biocompatible magnetic hydrogels. *Soft Matter* **2017**, *13*, 2928–2941. [CrossRef] [PubMed]
13. Antman-Passig, M.; Shefi, O. Remote magnetic orientation of 3D collagen hydrogels for directed neuronal regeneration. *Nano Lett.* **2016**, *16*, 2567–2573. [CrossRef] [PubMed]
14. Kim, H.; Jo, A.; Baek, A.; Lim, D.; Park, S.; Cho, S.; Chung, J.; Yoon, J. Synergistically enhanced selective intracellular uptake of anticancer drug carrier comprising folic acid-conjugated hydrogels containing magnetite nanoparticles. *Sci. Rep.* **2017**, *7*, 1–10. [CrossRef] [PubMed]
15. Testa-Anta, M.; Liébana-Viñas, S.; Rivas-Murias, B.; Rodríguez-González, B.; Farle, M.; Salgueiriño, V. Shaping iron oxide nanoparticles for magnetic separation applications. *Nanoscale* **2018**, *10*, 20462–20467. [CrossRef]
16. Munaweera, I.; Shi, Y.; Koneru, B.; Saez, R.; Aliev, A.; Di Pasqua, A.; Balkus, K. Chemoradiotherapeutic magnetic nanoparticles for targeted treatment of nonsmall cell lung cancer. *Mol. Pharm.* **2015**, *12*, 3588–3596. [CrossRef]
17. Hervault, A.; Thanh, N. Magnetic nanoparticle-based therapeutic agents for thermo-chemotherapy treatment of cancer. *Nanoscale* **2014**, *6*, 11553–11573. [CrossRef]
18. Akbarzadeh, A.; Samiei, M.; Davaran, S. Magnetic nanoparticles: Preparation, physical properties, and applications in biomedicine. *Nanoscale Res. Lett.* **2012**, *7*, 144. [CrossRef]
19. Veloso, S.R.S.; Ferreira, P.M.T.; Martins, J.A.; Coutinho, P.J.G.; Castanheira, E.M.S. Magnetogels: Prospects and main challenges in biomedical applications. *Pharmaceutics* **2018**, *10*, 145. [CrossRef]
20. Zhang, Z.Q.; Song, S.C. Thermosensitive/superparamagnetic iron oxide nanoparticle-loaded nanocapsule hydrogels for multiple cancer hyperthermia. *Biomaterials* **2016**, *106*, 13–23. [CrossRef]
21. Ramos-Docampo, M.; Fernández-Medina, M.; Taipaleenmäki, E.; Hovorka, O.; Salgueiriño, V.; Stadler, B. Microswimmers with heat delivery capacity for 3D cell spheroid penetration. *Acs Nano* **2019**, *13*, 12192–12205. [CrossRef]
22. Bhardwaj, A.; Parekh, K.; Jain, N. In vitro hyperthermic effect of magnetic fluid on cervical and breast cancer cells. *Sci. Rep.* **2020**, *10*, 15249. [CrossRef] [PubMed]
23. Guisasola, E.; Asín, L.; Beola, L.; de la Fuente, J.; Baeza, A.; Vallet-Regí, M. Beyond Traditional Hyperthermia: In Vivo Cancer Treatment with Magnetic-Responsive Mesoporous Silica Nanocarriers. *Acs Appl. Mater. Interfaces* **2018**, *10*, 12518–12525. [CrossRef] [PubMed]
24. Amiri, M.; Gholami, T.; Amiri, O.; Pardakhti, A.; Ahmadi, M.; Akbari, A.; Amanatfard, A.; Salavati-Niasari, M. The magnetic inorganic-organic nanocomposite based on $ZnFe_2O_4$-Imatinib-liposome for biomedical applications, in vivo and in vitro study. *J. Alloy. Compd.* **2020**, *849*, 156604. [CrossRef]
25. Colombo, M.; Carregal-Romero, S.; Casula, M.F.; Gutiérrez, L.; Morales, M.P.; Böhm, I.B.; Heverhagen, J.T.; Prosperi, D.; Parak, W.J. Biological applications of magnetic nanoparticles. *Chem. Soc. Rev.* **2012**, *41*, 4306–4334. [CrossRef]
26. Gila-Vilchez, C.; Mañas-Torres, M.; Contreras-Montoya, R.; Alaminos, M.; Duran, J.; de Cienfuegos, L.; Lopez-Lopez, M. Anisotropic magnetic hydrogels: Design, structure and mechanical properties. *Philos. Trans. R. Soc. A* **2019**, *377*, 20180217. [CrossRef]

27. Yang, Z.; Gu, H.; Du, J.; Gao, J.; Zhang, B.; Zhang, X.; Xu, B. Self-assembled hybrid nanofibers confer a magnetorheological supramolecular hydrogel. *Tetrahedron* **2007**, *63*, 7349–7357. [CrossRef]
28. Nawara, K.; Romiszewski, J.; Kijewska, K.; Szczytko, J.; Twardowski, A.; Mazur, M.; Krysinski, P. Adsorption of doxorubicin onto citrate-stabilized magnetic nanoparticles. *J. Phys. Chem. C* **2012**, *116*, 5598–5609. [CrossRef]
29. Hui, C.; Shen, C.; Yang, T.; Bao, L.; Tian, J.; Ding, H.; Li, C.; Gao, H.J. Large-scale Fe_3O_4 nanoparticles soluble in water synthesized by a facile method. *J. Phys. Chem. C* **2008**, *112*, 11336–11339. [CrossRef]
30. Pagano, R.; López Medus, M.; Gómez, G.; Couto, P.; Labanda, M.; Landolfo, L.; D'Alessio, C.; Caramelo, J. Protein fibrillation lag times during kinetic inhibition. *Biophys. J.* **2014**, *107*, 711–720. [CrossRef]
31. Wang, S.; Chen, Y.; Chen, P.; Liu, K. A kinetic study on the aggregation behavior of β-amyloid peptides in different initial solvent environments. *Biochem. Eng. J.* **2006**, *29*, 129–138. [CrossRef]
32. Morris, A.; Watzky, M.; Finke, R. Protein aggregation kinetics, mechanism, and curve-fitting: A review of the literature. *Biochim. Biophys. Actaproteins Proteom.* **2009**, *1794*, 375–397. [CrossRef] [PubMed]
33. Kamihira, M.; Naito, A.; Tuzi, S.; Saitô, H.; Nosaka, A. Conformational transitions and fibrillation mechanism of human calcitonin as studied by high-resolution solid-state 13C NMR. *Protein Sci.* **2000**, *9*, 867–877. [CrossRef] [PubMed]
34. Knowles, T.; Waudby, C.; Devlin, G.; Cohen, S.; Aguzzi, A.; Vendruscolo, M.; Terentjev, E.; Welland, M.; Dobson, C. An analytical solution to the kinetics of breakable filament assembly. *Science* **2009**, *326*, 1533–1537. [CrossRef] [PubMed]
35. Cohen, S.; Linse, S.; Luheshi, L.; Hellstrand, E.; White, D.; Rajah, L.; Otzen, D.; Vendruscolo, M.; Dobson, C.; Knowles, T. Proliferation of amyloid-β42 aggregates occurs through a secondary nucleation mechanism. *Proc. Natl. Acad. Sci. USA* **2013**, *110*, 9758–9763. [CrossRef]
36. Crosby, G.A.; Demas, J.N. The measurement of photoluminescence quantum yields—Review. *J. Phys. Chem.* **1971**, *75*, 991–1024. [CrossRef]
37. Fery-Forgues, S.; Lavabre, D. Are fluorescence quantum yields so tricky to measure? A demonstration using familiar stationery products. *J. Chem. Educ.* **1999**, *76*, 1260–1264. [CrossRef]
38. Kirby, E.P.; Steiner, R.F. Influence of solvent and temperature upon the fluorescence of indole derivatives. *J. Phys. Chem.* **1970**, *74*, 4480–4490. [CrossRef]
39. Valeur, B. *Molecular Fluorescence—Principles and Applications*, 2nd ed.; Wiley-VCH: Weinheim, Germany, 2001.
40. Cardoso, A.; Alvarez, A.; Cattoz, B.; Griffiths, P.; King, S.; Frith, W.; Adams, D. The influence of the kinetics of self-assembly on the properties of dipeptide hydrogels. *Faraday Discuss.* **2013**, *166*, 101–116. [CrossRef]
41. Greenspan, P.; Fowler, S. Spectrofluorometric studies of the lipid probe Nile Red. *J. Lipid Res.* **1985**, *26*, 781–789.
42. Ira; Krishnamoorthy, G. Probing the link between proton transport and water content in lipid membranes. *J. Phys. Chem. B* **2001**, *105*, 1484–1488. [CrossRef]
43. Coutinho, P.J.G.; Castanheira, E.M.S.; Rei, M.C.; Oliveira, M.E.C.D.R. Nile Red and DCM fluorescence anisotropy studies in $C_{12}E_7$/DPPC Mixed Systems. *J. Phys. Chem. B* **2002**, *106*, 12841–12846. [CrossRef]
44. Hungerford, G.; Castanheira, E.M.S.; Real Oliveira, M.E.C.D.; Miguel, M.G.; Burrows, H. Monitoring ternary systems of $C_{12}E_5$/water/tetradecane via the fluorescence of solvato-chromic probes. *J. Phys. Chem. B* **2002**, *106*, 4061–4069. [CrossRef]
45. Rodrigues, A.R.O.; Matos, J.O.G.; Nova Dias, A.M.; Almeida, B.G.; Pires, A.; Pereira, A.M.; Araújo, J.P.; Queiroz, M.J.R.P.; Castanheira, E.M.S.; Coutinho, P.J.G. Development of multifunctional liposomes containing magnetic/plasmonic $MnFe_2O_4$/Au core/shell nanoparticles. *Pharmaceutics* **2019**, *11*, 10. [CrossRef] [PubMed]
46. Cardoso, B.D.; Rodrigues, A.R.O.; Almeida, B.G.; Amorim, C.O.; Amaral, V.S.; Castanheira, E.M.S.; Coutinho, P.J.G. Stealth magnetoliposomes based on calcium-substituted magnesium ferrite nanoparticles for curcumin transport and release. *Int. J. Mol. Sci.* **2020**, *21*, 3641. [CrossRef] [PubMed]
47. Pitschke, W.; Hermann, H.; Mattern, N. The influence of surface roughness on diffracted X-ray intensities in Bragg-Brentano geometry and its effect on the structure determination by means of Rietveld analysis. *Powder Diffr.* **1993**, *8*, 74–83. [CrossRef]
48. Lisboa, F.S.; Gardolinski, J.E.F.C.; Cordeiro, C.S.; Wypych, F. Layered metal laurates as active catalysts in the methyl/ethyl esterification reactions of lauric acid. *J. Braz. Chem. Soc.* **2012**, *23*, 46. [CrossRef]
49. Rodrigues, A.R.O.; Ramos, J.M.F.; Gomes, I.T.; Almeida, B.G.; Araújo, J.P.; Queiroz, M.J.R.P.; Coutinho, P.J.G.; Castanheira, E.M.S. Magnetoliposomes based on manganese ferrite nanoparticles as nanocarriers for antitumor drugs. *RSC Adv.* **2016**, *6*, 17302–17313. [CrossRef]
50. Foderà, V.; Donald, A. Tracking the heterogeneous distribution of amyloid spherulites and their population balance with free fibrils. *Eur. Phys. J. E* **2010**, *33*, 273–282. [CrossRef]
51. Hsiung, C.; Lien, H.; Galliano, A.; Yeh, C.; Shih, Y. Effects of water chemistry on the destabilization and sedimentation of commercial TiO_2 nanoparticles: Role of double-layer compression and charge neutralization. *Chemosphere* **2016**, *151*, 145–151. [CrossRef]
52. Phenrat, T.; Saleh, N.; Sirk, K.; Tilton, R.; Lowry, G. Aggregation and sedimentation of aqueous nanoscale zerovalent iron dispersions. *Environ. Sci. Technol.* **2007**, *41*, 284–290. [CrossRef]
53. Obaidat, I.; Issa, B.; Haik, Y. Magnetic properties of magnetic nanoparticles for efficient hyperthermia. *Nanomaterials* **2015**, *5*, 63–89. [CrossRef] [PubMed]
54. Smit, J. *Magnetic Properties of Materials*; McGraw Hill: New York, NY, USA, 1971.

55. Swain, J.; Mishra, A. Nile red fluorescence for quantitative monitoring of micropolarity and microviscosity of pluronic F127 in aqueous media. *Photochem. Photobiol. Sci.* **2016**, *15*, 1400–1407. [CrossRef] [PubMed]
56. Swain, J.; Mishra, J.; Ghosh, G.; Mishra, A. Quantification of micropolarity and microviscosity of aggregation and salt-induced gelation of sodium deoxycholate (NaDC) using Nile red fluorescence. *Photochem. Photobiol. Sci.* **2019**, *18*, 2773–2781. [CrossRef] [PubMed]
57. Marchi, M.; Bilmes, S.; Negri, R. Microviscosity in the cavities of titania gels studied by steady-state fluorescence anisotropy. *Langmuir* **1997**, *13*, 3665–3673. [CrossRef]
58. Hernández, B.; Pflüger, F.; Kruglik, S.; Ghomi, M. Characteristic Raman lines of phenylalanine analyzed by a multiconformational approach. *J. Raman Spectrosc.* **2013**, *44*, 827–833. [CrossRef]
59. Noothalapati, H.; Iwasaki, K.; Yoshimoto, C.; Yoshikiyo, K.; Nishikawa, T.; Ando, M.; Hamaguchi, H.; Yamamoto, T. Imaging phospholipid conformational disorder and packing in giant multilamellar liposome by confocal Raman microspectroscopy. *Spectrochim. Acta A Mol. Biomol. Spectrosc.* **2017**, *187*, 186–190. [CrossRef]
60. Kurouski, D.; Van Duyne, R.; Lednev, I. Exploring the structure and formation mechanism of amyloid fibrils by Raman spectroscopy: A review. *Analyst* **2015**, *140*, 4967–4980. [CrossRef]
61. Kuhar, N.; Sil, S.; Verma, T.; Umapathy, S. Challenges in application of Raman spectroscopy to biology and materials. *RSC Adv.* **2018**, *8*, 25888–25908. [CrossRef]
62. Maiti, N.; Apetri, M.; Zagorski, M.; Carey, P.; Anderson, V. Raman spectroscopic characterization of secondary structure in natively unfolded proteins: α-Synuclein. *J. Am. Chem. Soc.* **2004**, *126*, 2399–2408. [CrossRef]
63. Veloso, S.R.S.; Martins, J.A.; Hilliou, L.; Amorim, C.O.; Amaral, V.S.; Almeida, B.G.; Jervis, P.J.; Moreira, R.; Pereira, D.M.; Coutinho, P.J.G.; et al. Dehydropeptide-based plasmonic magnetogels: A supramolecular composite nanosystem for multimodal cancer therapy. *J. Mater. Chem. B* **2020**, *8*, 45–64. [CrossRef]
64. Fan, X.; Xu, H.; Zhang, Q.; Xiao, D.; Song, Y.; Zheng, Q. Insight into the weak strain overshoot of carbon black filled natural rubber. *Polymer* **2019**, *167*, 109–117. [CrossRef]
65. Mehdaoui, B.; Meffre, A.; Carrey, J.; Lachaize, S.; Lacroix, L.; Gougeon, M.; Chaudret, B.; Respaud, M. Optimal size of nanoparticles for magnetic hyperthermia: A combined theoretical and experimental study. *Adv. Funct. Mater.* **2011**, *21*, 4573–4581. [CrossRef]
66. Umut, E.; Coşkun, M.; Pineider, F.; Berti, D.; Güngüneş, H. Nickel ferrite nanoparticles for simultaneous use in magnetic resonance imaging and magnetic fluid hyperthermia. *J. Colloid Interface Sci.* **2019**, *550*, 199–209. [CrossRef] [PubMed]
67. Mehdaoui, B.; Tan, R.; Meffre, A.; Carrey, J.; Lachaize, S.; Chaudret, B.; Respaud, M. Increase of magnetic hyperthermia efficiency due to dipolar interactions in low-anisotropy magnetic nanoparticles: Theoretical and experimental results. *Phys. Rev. B* **2013**, *87*, 174419. [CrossRef]
68. Abenojar, E.; Wickramasinghe, S.; Bas-Concepcion, J.; Samia, A. Structural effects on the magnetic hyperthermia properties of iron oxide nanoparticles. *Prog. Nat. Sci-Mater.* **2016**, *26*, 440–448. [CrossRef]
69. Karukstis, K.; Thompson, E.; Whiles, J.; Rosenfeld, R. Deciphering the fluorescence signature of daunomycin and doxorubicin. *Biophys. Chem.* **1998**, *73*, 249–263. [CrossRef]
70. Tasca, E.; Alba, J.; Galantini, L.; D'Abramo, M.; Giuliani, A.; Amadei, A.; Palazzo, G.; Giustini, M. The self-association equilibria of doxorubicin at high concentration and ionic strength characterized by fluorescence spectroscopy and molecular dynamics simulations. *Colloids Surf. A* **2019**, *577*, 517–522. [CrossRef]
71. Changenet-Barret, P.; Gustavsson, T.; Markovitsi, D.; Manet, I.; Monti, S. Unravelling molecular mechanisms in the fluorescence spectra of doxorubicin in aqueous solution by femtosecond fluorescence spectroscopy. *Phys. Chem. Chem. Phys.* **2013**, *15*, 2937. [CrossRef]
72. Dash, S.; Murthy, P.N.; Nath, L.; Chowdhury, P. Kinetic modelling on drug release from controlled drug delivery systems. *Acta Pol. Pharm.* **2010**, *67*, 217–223.
73. Ghitman, J.; Stan, R.; Ghebaur, A.; Cecoltan, S.; Vasile, E.; Iovu, H. Novel PEG-modified hybrid PLGA-vegetable oils nanostructured carriers for improving performances of indomethacin delivery. *Polymers* **2018**, *10*, 579. [CrossRef]
74. Kini, S.; Bahadur, D.; Panda, D. Mechanism of anti-cancer activity of benomyl loaded nanoparticles in multidrug resistant cancer cells. *J. Biomed. Nanotechnol.* **2015**, *11*, 877–889. [CrossRef] [PubMed]
75. Joy, M.; Iyengar, S.; Chakraborty, J.; Ghosh, S. Layered double hydroxide using hydrothermal treatment: Morphology evolution, intercalation and release kinetics of diclofenac sodium. *Front. Mater. Sci.* **2017**, *11*, 395–408. [CrossRef]

nanomaterials

Review

Emerging Nano-Formulations and Nanomedicines Applications for Ocular Drug Delivery

Dawin Khiev [1], Zeinab A. Mohamed [1], Riddhi Vichare [1], Ryan Paulson [2], Sofia Bhatia [2], Subhra Mohapatra [3,4], Glenn P. Lobo [5], Mallika Valapala [6], Nagaraj Kerur [7], Christopher L. Passaglia [8], Shyam S. Mohapatra [1,4,9] and Manas R. Biswal [1,2,9,10,*]

1 MSPN Graduate Programs, Taneja College of Pharmacy, University of South Florida, Tampa, FL 33612, USA; dawinkhiev@usf.edu (D.K.); zmohamed@usf.edu (Z.A.M.); vicharer@usf.edu (R.V.); smohapat@usf.edu (S.S.M.)
2 Department of Pharmaceutical Sciences, Taneja College of Pharmacy, University of South Florida, Tampa, FL 33612, USA; rjpaulso@usf.edu (R.P.); bhatias@usf.edu (S.B.)
3 Department of Molecular Medicine, Morsani College of Medicine, University of South Florida, Tampa, FL 33612, USA; smohapa2@usf.edu
4 James A. Haley Veterans' Hospital, Tampa, FL 33612, USA
5 Department of Medicine, Medical University of South Carolina, Charleston, SC 29425, USA; lobo@musc.edu
6 School of Optometry, Indiana University, Bloomington, IN 47401, USA; mvalapal@iu.edu
7 Department of Ophthalmology, University of Virginia School of Medicine, Charlottesville, VA 22908, USA; nk8m@virginia.edu
8 Department of Medical Engineering, College of Engineering, Tampa, FL 33620, USA; passaglia@usf.edu
9 Department of Internal Medicine, Morsani College of Medicine, University of South Florida, Tampa, FL 33612, USA
10 Department of Ophthalmology, Morsani College of Medicine, University of South Florida, Tampa, FL 33612, USA
* Correspondence: biswal@usf.edu; Tel.: +1-813-974-8333

Citation: Khiev, D.; Mohamed, Z.A.; Vichare, R.; Paulson, R.; Bhatia, S.; Mohapatra, S.; Lobo, G.P.; Valapala, M.; Kerur, N.; Passaglia, C.L.; et al. Emerging Nano-Formulations and Nanomedicines Applications for Ocular Drug Delivery. *Nanomaterials* **2021**, *11*, 173. https://doi.org/10.3390/nano11010173

Received: 16 December 2020
Accepted: 7 January 2021
Published: 12 January 2021

Publisher's Note: MDPI stays neutral with regard to jurisdictional claims in published maps and institutional affiliations.

Abstract: Ocular diseases can deteriorate vision to the point of blindness and thus can have a major impact on the daily life of an individual. Conventional therapies are unable to provide absolute therapy for all ocular diseases due to the several limitations during drug delivery across the blood-retinal barrier, making it a major clinical challenge. With recent developments, the vast number of publications undergird the need for nanotechnology-based drug delivery systems in treating ocular diseases. The tool of nanotechnology provides several essential advantages, including sustained drug release and specific tissue targeting. Additionally, comprehensive in vitro and in vivo studies have suggested a better uptake of nanoparticles across ocular barriers. Nanoparticles can overcome the blood-retinal barrier and consequently increase ocular penetration and improve the bioavailability of the drug. In this review, we aim to summarize the development of organic and inorganic nanoparticles for ophthalmic applications. We highlight the potential nanoformulations in clinical trials as well as the products that have become a commercial reality.

Keywords: organic and inorganic nanoparticles; ophthalmic applications; clinical trials

1. Introduction

Vision stems from integrated and intricate communication between the eyes and brain. Visual impairment has a deteriorating effect on the life of an individual both physically and mentally. The global prevalence of ocular diseases based on 2015 statistics is an estimated 216.6 million people suffering from moderate to severe visual impairment and 188.5 million people affected by mild visual impairment. In the upcoming years, growth in the aging population will cause a spike in the number of cases of visual impairment, making it a major public health concern [1]. Common eye diseases include glaucoma, age-related macular degeneration (AMD), and diabetic retinopathy [2]. The delivery of therapeutics to the anterior segment of the eye is challenging, owing to the presence of anatomical and physiological barriers. Briefly, anatomical barriers are divided into static and dynamic barriers. Static barriers include corneal epithelium and blood-aqueous barrier, whereas dynamic barriers comprise tear drainage and conjunctival blood and lymph flow. On the other hand, physiological barriers include blinking action, tear film turnover, and nasolacrimal drainage. For drug delivery in the posterior segment, the major limitation is due to the presence of a blood-retinal barrier. Topical administration, like eye drops, is the most common and simple route for drug delivery to the anterior segment. However, the aforementioned problems result in poor bioavailability of topical applications, which has urged researchers to shift focus to novel nanotherapeutics for ophthalmic diseases [1,3]. Topical applications do not reach the posterior segment of the eyes, which makes intravitreal administration (IVT) of a drug the common route of administration. Although IVT injection efficiently delivers drugs to the posterior segment, it can be difficult to achieve for small and large bioorganic compounds like proteins and DNA, and some drugs need repeated injections to reach a therapeutic dose. Multiple injections can increase the risk of complications and possibly damage the eyes. Therefore, it is not a preferable option in treating ocular disease for most patients due to the lack of patient compliance [4].

In the past decade, clinical pharmacologists have focused on developing nanomedicine that can overcome barriers to provide a sustained and targeted release with minimal toxicity. The development of lipid-based nanocarriers has emerged into pharmaceutical industries and attracted attention from formulation scientists for its high drug solubility, substance release, and specific targeting. Organic nanoparticles, such as liposomes, polymeric nanoparticles, and dendrimers, are fabricated and developed to improve therapeutic drug delivery for the ocular disease [5,6]. Inorganic nanoparticles have garnered significant attention for their theranostic properties, substance release, and specific targeting. In this review, we aim to provide an insight into current treatment modalities based on nanotechnology for drug delivery in the eye. Finally, we will summarize the current nanoformulations under clinical trials and the potential formulations available on the market.

2. Nanotechnology in Ocular Diseases Therapy

Nanotechnology combines science and technology with the ability to manipulate structures and properties at the nanoscale size range, which lies between 1 and 100 nm. The ability to manipulate molecules on such a fine scale has application in many areas of basic science and in clinical treatment, diagnosis, and managing disease, which is known as "nanomedicine" [6]. Nanotechnology has been introduced since the 1980s in drug delivery systems to treat ocular disease [7]. The development of nanoformulation produces an ability to overcome ocular barriers, improve drug residence time on the cornea surface, increase permeability and bioavailability of the drug, reduce degradation of unstable drug and be well tolerated by the patient compared to the conventional drug [8]. Organic and inorganic nanoparticles (NPs) provide a new tool to address the unmet clinical needs, especially in the ophthalmic fields, with a significant improvement in drug delivery [5,6]. Therefore, various nanoparticle formulations have been fabricated, including lipid-based nanoparticles, nanosuspension, nanoemulsion, and metal-based nanoparticles.

3. Organic Nanoparticles

The group of organic nanocarriers comprises liposomes, niosomes, dendrimers, solid-lipid NPs, polymers and protein/peptide-based NPs, which are fabricated from proteins, lipids, carbohydrates or other organic compounds (Figure 1). The use of organic nanoparticles seems to offer several advantages to facilitate drug or gene delivery to the ocular surfaces.

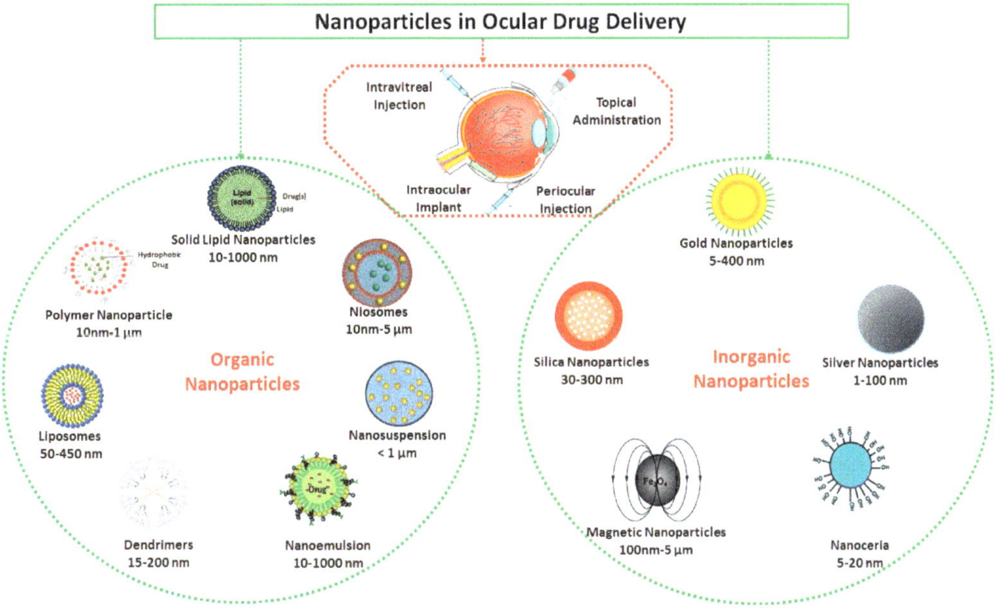

Figure 1. Schematic drawings for potential organic and inorganic nanoparticles for ophthalmic drug delivery.

3.1. Liposomes

The liposome is a spherical-shaped nanoparticle with an aqueous core composed of one or more phospholipid bilayers, which resemble a cell membrane. Base on size and lamellarity, liposomes can be classified into three types: multilamellar vesicles, small unilamellar vesicles, and large unilamellar vesicles. Liposomes are widely used as nanocarriers in drug delivery systems and provide a significant advantage due to their size, biocompatibility, biodegradable, low toxicity, and potential to encapsulate hydrophilic, lipophilic, and amphiphilic drugs [9]. Due to their physicochemical and biophysical properties, liposomes can deliver drugs at different concentrations [10]. The liposome surface can be modified for targeted delivery, improving drug absorption to the posterior segment of the eyes, and the positive charge of the liposome can increase drug permeation to the cornea, prolonging drug release, which results in reducing multiple drug administration [7,10]. Law et al. [11] investigated acyclovir (ACV) embedded positively and negatively charged liposomes to target corneal penetration and absorption improvement. In an in vivo study, ACV-loaded liposomes were administered topically to the cornea surface, and the result suggested greater corneal permeation ability of ACV-loaded positively charged liposomes in comparison to negatively charged and plain ACV solution. The drug Edaravone loaded in submicron-sized liposomes was developed by Shimazaki et al. [12] for protection from light-induced retinal damage in mice. Their outcome demonstrated the eyedrop application of Edaravone encapsulated liposomes successfully inhibited light-induced reactive species oxygen production without any toxicity, making it a promising therapeutic potential in dry age-related macular degeneration management. Natarajan et al. [13] explored the therapeutic effect of liposome formulation to deliver latanoprost to treat glaucoma through subconjunctival injection administration. The liposomal formulation increased sustainability and drug targeting in the subconjunctival space. Along with this, latanoprost-encapsulated liposomes displayed an extended reduction of intraocular pressure (IOP) up to 50 days in rabbit eyes as compared to the conventional topical applications of eyedrops [13]. These promising effects have made liposomes an essential formulation in ophthalmic drug delivery. Bochot et al. [14] studied a novel approach to design an ocular drug delivery system that is based on the dispersion of PEGylated liposomes into a thermosensitive gel, which is made of a copolymer of ethylene oxide and propylene oxide (poloxamer 407). This approach offers new potentialities for ocular drug delivery.

3.2. Niosomes

Niosomes were first widely used in the cosmetic industry, but later their potential in drug delivery was studied for different routes of administration [15,16]. Niosomes are bilayer structured vesicles with a submicron size of 10 nm to over 5 μm composed of amphiphilic nonionic surfactants in the aqueous phase [7,17,18]. The essential properties of these vesicle nanocarriers offer the ability to entrap both lipophilic and hydrophilic drugs like liposomes but with the additional advantages of chemical stability, minimum toxicity due to its nonionic surface, and improved permeation of drug in the corneal cells [19]. The small size of niosomes offers an essential property to overcome the ocular drainage mechanism and produce better drug retention on the ocular surface [20]. Furthermore, compared to liposomes, niosomes are convenient for handling and storage and cost less to make this vesical more preferable to produce large-scale in the pharmaceutical industry [15]. Gugleva et al. [21] reported that there was a high encapsulation efficiency of doxycycline hyclate in niosomes prepared by sorbitan monostearate (span60) and cholesterol formulation, which produced a prolonged drug-release rate and was well tolerated in mice eye. Abdelbary and El-Gendy [22] achieved an encapsulation efficiency of gentamycin sulfate in niosomes that consisted of polyoxyethylene sorbitan monostearate, also called Tween-60 dicetyl phosphate (DCP), and cholesterol with an effect of controlled drug release from the ophthalmic delivery system comparable to the drug solution. Zeng et al. [23] prepared hyaluronic acid-coated niosomes for improving the ocular bioavailability of the tacrolimus drug. The combination of niosomes with hyaluronic acid coating on tacrolimus improved ocular adhesion increased precorneal drug retention, and enhanced permeation of the transcorneal area compared to noncoated niosomes.

3.3. Solid Lipid Nanoparticles

Recently, solid lipid nanoparticles (SLNs) have been explored as a novel delivery system that includes liposome nanoparticles as a colloidal lipid nanocarrier [24]. SLNs offer essential characteristics such as small size, large surface area, controlled drug release, excellent physical stability, and prevention of drug degradation. SLNs typically range from 50 to 1000 nm in size, are made of biodegradable, biocompatible materials, and are stabilized in an aqueous surfactant solution [25,26]. Recently, SLNs have attracted great attention as potential candidates for drug delivery to ocular surfaces due to their capability of increasing corneal absorption and thus improving bioavailability with sustained and controlled drug release properties for hydrophilic and lipophilic drugs [27,28]. A previous study incorporated cyclosporine-A into solid lipid nanoparticles for ocular delivery and showed that the nanoparticle achieved higher concentration in aqueous and vitreous humor due to the controlled and substance release effects of SNLs as compared to a cyclosporine-A solution [29]. Cavalli et al. [30] evaluated solid lipid nanoparticles for the delivery of the tobramycin (TOB) drug. The formulation was applied topically to New Zealand albino rabbits, and the results indicated prolonged preocular drug retention, which is associated with increased ocular bioavailability. Attama et al. [31] investigated the ocular permeation and sustained release effect of diclofenac sodium by synthesized SNL combined with phospholipid and a homolipid obtained from goat fat. It was found that drug-loaded with SNL produced sustained release, and permeation studies of diclofenac from SNL through human cornea construct in cell culture revealed the improvement of cornea permeation potential [31].

3.4. Polymeric Nanoparticles

Polymeric nanoparticles are spherical-shaped solid colloidal particles composed of biocompatible and biodegradable polymers in a size range from 10 nm to 1000 nm [32].

Some polymeric NPs that are obtained from natural polymers include chitosan, alginate, albumin, gelatin, and dextran, while some are obtained from synthetic sources such as polylactic acid (PLA), polyglycolide (PGA), poly(L-lysine), polyaspartic acid, poly-alkyl cyanoacrylate, and polyethyleneimine (PEI) which are biodegradable synthetic polymers. Based on preparation methods, polymeric nanoparticles can be fabricated into nanospheres and nanocapsules [25,33]. Nanosphere polymer has a polymeric matrix core structure, in which drugs can either be entrapped inside a matrix or attached to the polymer surface. Nanocapsules are formed with a liquid phase in the core, surrounded by a solid polymeric wall [6]. These polymer characteristics make them potential nanocarriers as it is possible to conjugate a drug molecule at the surface or encapsulate it inside and deliver it to a specific target site of action at a high concentration with low systemic toxicity [32]. Poly lactic-co-glycolic acid (PLGA) nanoparticles, a biodegradable copolymer, have been used in ophthalmic drug delivery due to their high potential to entrap hydrophilic/hydrophobic drug molecules with prolonged drug release [34]. Glycol chitosan cerium oxide nanoparticles have been studied in treating dry eye diseases in which cerium oxide was entrapped in glycol chitosan. These studies have shown the ability of the drug to enter the corneal cells and conjunctiva cells and promoting SOD2 expression [35]. Sharma et al. [36] found that chloro trimethyl-ammonium methyl methacrylate (Eudragit RS 100/RL 100) polymeric nanoparticles containing amikacin sulfate formulation produced high corneal adherence, which led to an increase in drug retention in the cornea, and drug controlled release efficiency was observed during in vivo studies. Another interesting research conducted by Li et al. [37] reported the efficiency of the combination of hydroxypropyl-beta-cyclodextrin (HP-β-CD) and PLGA (PLGA) nanoparticles for triamcinolone acetonide encapsulation in ophthalmic delivery. The in vitro drug release study demonstrated a higher rate of drug release from hydroxypropyl-beta-cyclodextrin (HP-β-CD) /PLGA nanoparticles compared to the suspension of drug-loaded PLGA without β-CD. Moreover, the significantly improved drug penetration of HP-β-CD /PLGA in the transcorneal area led to increased drug concentration in the aqueous humor, which makes an HP-β-CD/PLGA nanoparticle-loaded triamcinolone acetonide a potential formulation in eye drops [37].

3.5. Dendrimers

Dendrimers, a name derived from the Greek etymology meaning "tree", are monodisperse macromolecules with well-defined size and a molecular structure composed of three major parts: (i) an inner core located inside the central molecule of the dendrimer, (ii) a highly branching unit forming a multilayer that links to the core called generation, and (iii) an outer surface with numerous valent functional groups [38]. Dendrimers have gained a prominent interest in the pharmaceutical application for drug delivery due to their tree-like branched structure with many covalent bonds allowing them to attach multiple functional groups, which can incorporate drugs inside dendrimers and conjugate them to the surface through covalent bonds. Their advantages include hydrophilic or lipophilic properties that offer control insolubility, and with their global shape, dendrimers can entrap small drug molecules inside their branches [39,40]. This unique architecture of dendrimers makes them an ideal nanocarrier for ophthalmic drug delivery with several purposes, such as improving permeability, water-solubility, biocompatibility, and bioavailability [41]. The most commonly used dendrimers-based nanocarriers are polyamidoamine (PAMAM) dendrimers that are hyperbranched [40]. It has been shown that polycationic PAMAM dendrimers are cytotoxic, and their toxicity increases as the generation increases, but hydroxy-terminated G4 PAMAM dendrimers are considered nontoxic due to their near-neutral surface charge that significantly reduces nonspecific retention and interactions in the tissues [42]. In other studies, PEGylation of PAMAM dendrimers has been shown to increase circulation time and biodistribution as well as reduce cytotoxicity of these dendrimers [43]. Mishra and Jain [44] have evaluated the intraocular pressure-lowering potential of acetazolamide (ACZ)-loaded poly-(propylene imine) dendrimer nanostructures for topical ocular administration. Acetazolamide-loaded dendrimer formulation was topically applied to adult male New Zealand albino rabbit eyes. The results showed that the drug-loaded dendrimer formulation caused a reduction in IOP for an extended time, approximately 4 h, compared to ACZ solution, which only showed a reduction effect up to 2 h [44]. In another study, PAMAM dendrimer was synthesized with primary amine and carboxylate surface groups to obtain better drug corneal permeation and properties during in vitro release. A dendrimer-based formulation was used for puerarin drug delivery and successfully tested on the albino rabbit model for ocular hypertension and cataracts glaucoma treatment [45]. Marano and coworkers [43] have designed lipophilic amino-acid dendrimer for anti-vascular endothelial growth factor (VEGF) oligonucleotide (ODN-1) retina delivery to treat choroidal neovascularization. It was reported that dendrimer conjugated ODN-1 significantly inhibits choroidal neovascularization (CNV) progression, and no side effects were observed during examination [46]. Given all the benefits of dendrimers, these formulations have the potential to improve the efficacy of drugs in ophthalmology.

3.6. *Nanosuspensions*

Poor solubility and low availability are the main concerns for the formulation of drugs in pharmaceutical industries as conventional formulations cannot overcome these problems, and over 40% of newly discovered drugs are reported to have low water solubility. Many strategies have been employed to improve poor drug solubility, but due to limitations in these strategies, scientists have found nanosuspensions to be a promising additional approach in drug delivery. Nanosuspensions are colloidal dispersions in nanoparticle size which have a surfactant as a stabilizer. Administration of nanosuspension offers benefits in sustained release of poorly soluble drug in the ocular surface, which leads to the increase of retention time [47–49]. Nanosuspensions have been reported in oral and topical formulations to enhance ocular drug bioavailability. In previous studies, a nanosuspension of ibuprofen sodium salt (IBU) sodium salt-coated chloro trimethyl-ammonium methyl methacrylate (Eudragit RS 100) polymer tested on rabbit eye by topical administration demonstrated penetration improvement to the anterior segment of the eye, extended drug release, and increased level of drug in the aqueous humor [50]. Some glucocorticoids used topically to treat eye inflammation, such as dexamethasone, hydrocortisone, and prednisolone, have been formulated in nanosuspension, showing an effect in prolonged drug absorption with higher drug bioavailability in ocular drug delivery, thus reducing frequent drug administration [51,52].

3.7. *Nanoemulsion*

Nanoemulsion is a colloidal system composed of two liquid phases, oil dispersed in water or water dispersed in oil, in which the nanoemulsion droplets with size 20–500 nm are stabilized by surfactants such as cetyltrimethylammonium bromide and sodium dodecyl sulfate [53,54]. The small size of droplets offers a high surface area with a significant potential to provide better efficacy in ophthalmic drug delivery with its benefits of improving ocular permeability and high bioavailability [55]. Nanoemulsion is considered to have low surface tension and greater drug spreading on the cornea, which can mix properly with the precorneal constituents. This enhances the drug contact time in the corneal epithelium [56]. The study by Ismail et al. [57] on antiglaucoma drug dorzolamide hydrochloride nanoemulsion showed promising effects on ocular therapy. The drug formulation offered prolonged effect with a quick onset of action and thus reduced frequent administration of eye drops. Tayel et al. [58] developed terbinafine hydrochloride-loaded nanoemulsion gels. The formulation that was developed indicated extended drug residence time and bioavailability improvement. Clinical research conducted by Mahboobian et al. [59] evaluated the uptake of acyclovir nanoemulsion in the bovine cornea following transcorneal permeation. The results showed increased cornea permeation of acyclovir through the corneal membrane cells, and no irritation occurred on the rabbit eye studied, which indicates that the formulation can be considered safe for treating ocular infection [59].

Table 1. Summarizes the applications of organic nanoparticles for drug delivery to ocular surfaces.

Table 1. Applications of organic nanoparticles for ophthalmic drug delivery.

Nanomaterials	Drug	Application	Animal Model	Function	Ref.
Liposomes	Acyclovir	Topical	Rabbit	Prolong drug penetration	[11]
	Edaravone	Topical	Rabbit	Decrease progression of dry AMD	[12]
	Latanoprost	Subconjunctival injection	Mice	Sustained drug release Reduce intraocular pressure	[13]
Niosomes	Doxycycline hyclate		Rabbit	Prolong drug release rate	[21]
	Tacrolimus	Topical	Rabbit	Increased precorneal drug retention	[23]
Solid lipid nanoparticle	Cyclosporine A	Topical	Sheep	Prolonged drug release	[29]
	Tobramycin	Topical	Rabbit	Increased drug retention	[30]
Polymeric NPs (Eudragit®)	Amikacin	Topical	Rabbit	Improved ocular penetration Controlled release	[36]
Polymeric NPs (chitosan)	Cerium oxide		Mice	Improved corneal penetration	[35]
Dendrimers	Acetazolamide	Topical	Rabbit	Enhanced drug residence time	[44]
	Puerarin		Rabbit	Improved corneal permeation	[45]
	Anti-VEGF			Reduced CNV progression	[46]
Nanosuspension	IBU sodium salt	Topical	Rabbit	Increased penetration Prolonged drug release	[50]
	Glucocorticoid			Prolonged drug absorption	[51]
Nanoemulsion	Terbinafine hydrochloride			Improved drug residence time Increased bioavailability	[58]
	Acyclovir			Increased corneal permeation	[59]

4. Inorganic Nanoparticles

Inorganic nanoparticles comprise mainly metallic nanoparticles (Figure 1) and quantum dots. Metallic nanoparticles have gained much attention over the past decade since Faraday proved that they could exist in solution [35]. They can be categorized into four different groups: metal nanoparticles, metal oxide nanoparticles, doped metal–metal/oxide–metal nanoparticles, and metal sulfide, and metal–organic frameworks. Metal NPs such as silver (Ag), gold (Au), copper (Cu), magnesium (Mg), titanium (Ti), platinum (Pt), zinc (Zn), and iron (Fe) nanoparticle have been investigated in different research areas and have succeeded as efficient, stable drug delivery platforms with fewer side effects and powerful imaging probes. Doped metallic nanoparticles, like ZnO doped with Co, were reported to have lower toxicities than single metals with better properties [60,61]. In the ocular drug and gene delivery field, metals such as cerium oxide (CeO2) NPs and gold NPs have demonstrated their potential antioxidant capabilities with a high safety profile after delivery to the eye and high chemical stability.

4.1. Gold NPs (AuNPs)

AuNPs are one of the noble metals that are known for their unique optical properties induced by the popular phenomenon of localized surface plasmon resonance (LSPR). This phenomenon is strongly influenced by the AuNP shape and is the main reason for its penetrative ability to the biological tissues [62]. AuNPs have other advantages that make them popular in the nanomedicine field, like their chemical stability, surface functionalization, biocompatibility, and unique surface and size characteristics [63]. The biocompatibility and internalization capability of different AuNP morphologies (rods, spheres, and cubes) have been assessed in retinal pigment epithelium (RPE) cells. Spheres and cubes of 50 and 100 nm showed no cytotoxicity and good internalization properties; however, rod-shaped particles were less biocompatible [64]. Dong et al. [65] showed that resveratrol coated AuNPs with a median size of 20 nm and doses of 200–300 mg/kg injected in streptozotocin-induced diabetic rats for 3 months could provide a protective effect against diabetic retinopathy. This protective effect of AuNPs could help to regain the balance of the stimulators and inhibitors of the angiogenesis process through the inhibitory effects of the ERK1/2 pathway and nuclear factor kappa B (NF-kb) expression, which could reduce inflammation and permeability of the blood-retinal barrier in the diabetic rats. Furthermore, there was a significant reduction of all the retinal mRNA expressions of vascular endothelial growth factor (VEGF-1), tumor necrosis factor-alpha (TNFα), and interleukin 6 (IL-6) [65]. In another study of the topical application of AuNPs, Pereira et al. [66] showed that AuNPs could decrease the intraocular oxidative damage and inflammation in endotoxin-induced uveitis rat models.

The treatment of choroid-retina endothelial (RF/6A) cells with AuNPs of 3–5 nm successfully inhibited the VEGF-induced RF/6A cell migration through the Akt/eNOS pathways. In their cell viability and cell adhesion studies, they found neither cytotoxic effects on AuNPS on RF/6A nor detrimental effects on the normal physiological adhesion of cells to fibronectin [62]. Hayashi et al. [67] examined the feasibility, biodistribution, and effects of the subretinal injections of immunoglobulin G (IgG) adsorbed on gold nanoparticles (GNPs) in rabbits' eyes for three months and in ARPE-19 cultured cells, and they found that GNPs was successfully delivered into photoreceptor cells and RPE in rabbits and reported no cytotoxicity in ARPE-19 cells. Moreover, in inherited retinal dystrophies gene delivery therapies, AuNPs have been examined as a safer and more efficient alternative to the viral-based approaches in RPE cells. The preliminary results of Trigueros et al. [68] have shown that plasmid DNA-wrapped AuNPs with a size of 40 nm could be successfully internalized in differentiated ARPE-19 cells with good transfection efficiency. These results were supported by the early expression of a reporter gene that was noted at 16 h post-transfection. Their results show a potentially successful gene delivery route to RPE cells using AuNPs; nevertheless, they propose that the interaction mechanism of cell to AuNPs must be better understood to increase the transfection efficiency of these particles and escape the autophagic pathways of the particles to ensure sustained gene expression effect of their system [68]. Generally, AuNPs are considered as a core-shell system, at which gold is the inert inner core, and the drug can be conjugated to either its core or outer active layer through thiol linkages. As mentioned earlier, gold is biocompatible, and it is approved for internal medicine; however, the main challenge for AuNPs to be used as drug delivery vehicles in the clinical practice is their non-degradable gold core that elongates their excretion from the body [69]. Interestingly, the successful passage of AuNPs of the size 20 nm through the blood-retinal barrier (BRB) and their distribution in all retinal layers after their intravenous injection into C57B1/6 mice without showing any structural damage or increased cell death compared to cells without NPs by Kim et al. [70] raise their potential existence in future clinical trials.

4.2. Silver Nanoparticles (AgNPs)

Silver nanoparticles (AgNPs) have been used widely in the nanomedicine field due to their advantages of the unique chemical and physical characteristics, large surface area to volume ratio, low production cost, and biocompatibility that make them good candidates as drug delivery carriers [71]. Various physical and chemical methods are being used to synthesize and stabilize the AgNPs. Out of these, the chemical method is most common, which includes chemical reduction, electrochemical techniques, reducing agents, physiochemical reduction, and radiolysis [72]. AgNPs synthesis using these chemical methods is experiencing problems such as stability and agglomeration of NPs, shape, and size of particles. The synthesis of AgNPs requires (a) silver precursors, (b) reducing agents that include organic (e.g., ethylene glycol, ascorbic acid, glucose) and inorganic (e.g., hydrazine, sodium borohydride), (c) capping agents or stabilizing agents such as polyvinylpyrrolidone (PVP) [72]. Capping agents offer electrostatic and steric stabilization effects for AgNPs dispersion in suspensions [73].

AgNPs have been reported as effective nanoparticles in the treatment of cancer and infections due to their powerful tumor-killing and bactericidal effects. In the ocular field, AgNPs have shown successful delivery of therapeutic agents to eyes and have been used as a coating agent for contact lenses. In addition, AgNPs alone or in conjugation with natural plant extracts demonstrated significant anti-angiogenic, antioxidant, anti-glycation end products, and anti-cataractogenic effects in various cell culture systems and animal models for ocular diseases [74]. Anbukkarasi et al. [75] investigated the antioxidant and anti-cataractogenic effect of AgNPs with a size range of 15–50 nm formulated with the ethanolic extract of T. divaricate leaves, which reportedly have antioxidant and anti-cataractogenic pharmacological effects. The study designed an in vitro selenite-induced cataractogenic model to evaluate the effects of AgNPs against the dense opacification of Wistar rat lenses. The results suggested that the formulated AgNPs acted as a potent antioxidant by dose-dependent scavenging of DPPH and H_2O_2 free radicals, which are the main triggers of cataract formation [75]. Kalishwaralal et al. [76] tested the anti-angiogenic properties of formulated AgNPs in a size range of 40–50 nm, acting as an inhibitor of VEGF, which induces cell proliferation and migration in bovine retinal endothelial cells (BRECs). The findings of Gurunathan et al. [77] suggested that AgNPs could inhibit angiogenesis via the inhibition of the PI3K/Akt signaling pathways. Sheikpranbabu et al. [78] studied the effects of AgNPs with a size of 50 nm on advanced glycation end-products (AGEs)-induced endothelial cell permeability in porcine retinal endothelial cells (PRECs), and the studies demonstrated that AGE-induced permeability was inhibited by AgNPs through the Src kinase pathway.

Since the nineteenth century, silver nitrate has been known for its potential antimicrobial effects as it was widely used against ophthalmia neonatorum. Xu et al. [79] investigated the antifungal effects of polymeric AgNPs of sizes ranging from 20 to 30 nm against three different types of filamentous fungi isolates that cause fungal keratitis. The antifungal effects of AgNPs in the fungal cultures were compared to those of the antifungal natamycin, and the AgNPs demonstrated a significantly superior antifungal activity that was 4–32 times higher than natamycin.

Despite the advantages that AgNPs offers as antifungal, antioxidant, anti-angiogenic, and anti-inflammatory agents, their limited advancement in the ocular drug delivery field goes back to their reported toxicity in several studies. The widely reported toxicity is related to the release of silver ions from the nanoparticle surface that induces irritation to different ocular parts. Jun et al. [80] investigated the toxicity of silver NPs at a dose of 0.4 mg/L and reported downregulation of several lens crystalline genes by the possible cell death or the nuclear DNA or RNA export blockage. Kim et al. [81] also observed edema and conjunctival redness after a one-hour exposure of New Zealand white rabbits' eyes to AgNPs. In a study, AgNPs with sizes of 22.4 nm and 42.5 nm, respectively, induced cytotoxic effects and activation of the reactive oxygen species (ROS) system in bovine retinal endothelial cells (BRECs) [82].

4.3. Cerium Oxide Nanoparticles (Nanoceria-CeO_2-NPs)

Cerium is a rare earth element in the lanthanide series of the periodic table and is considered a potent antioxidant. The mechanism by which CeO_2-NPs of 3–5 nm exhibits its catalytic antioxidant reaction involves two oxidation states, a reduction of Ce4+ to Ce3+ and a loss of an oxygen atom [83]. This reaction creates vacancies for reactive oxygen species (ROS) in cells to replace the loss of oxygen, which mimics the antioxidant enzymatic activities of superoxide dismutase (SOD) and catalase [84]. Nanoceria has proved their efficacy as antioxidants to provide protection in many diseases and act as neuroprotective, radioprotective, cardioprotective, as well as anti-inflammatory agents [85].

In the ophthalmic field, CeO_2-NPs have been investigated in several animal models of age-related macular degeneration (AMD) and showed promising antioxidant, anti-inflammatory, and anti-angiogenic effects. Moreover, they were reported to provide long-term neuroprotection for photoreceptors in retinitis pigmentosa (RP) models without collateral adverse effects. They were also used as an effective anti-cataractogenic material and were reported to be safe on lens cells [86]. Nanoceria is mainly fabricated by simple wet chemistry methods, as explained by Karakoti et al. [87], and their PEGylation and liposomal encapsulation enable them to permeate the cornea [86]. Chen et al. [88] reported that fabricated CeO_2-NPs of 20 nm particle size could prevent the increased concentration of free radicals in primary cell cultures of rat retinas. These findings prompted them to test the NPs in vivo, where a group of albino rats was exposed to 2700 lux of light for 6 h. The light exposure normally results in damage to 50–60% of photoreceptors, but their electroretinogram (ERG) data showed that a significant number of photoreceptor cells were rescued after posttreatment with CeO_2-NPs [88]. Kong et al. [85] testing of nanoceria on P34 mice including ROS reduction, increase in the expression of neuroprotection associated genes, and inhibition of the apoptosis signaling pathways, all of which delayed the degenerative condition of photoreceptor cells.

In a study by Fiorani et al. [89], pretreatment of albino Sprague–Dawley rats with intravenous and intravitreal injections of CeO_2-NPs for 3 weeks could prevent neuronal death induced by 1000 lux light for 24 h and helped preserve the integrity of the outer nuclear layer. The study also demonstrated that nanoceria remained stable in the outer layer and could inhibit the microglial activation and migration toward the outer nuclear layer for an extended time [89]. Intravitreal injections of nanoceria were also tested by Zhou et al. [90], who found inhibition of neovascularization in Vldlr knockout mice through their inhibitory effects of the VEGF cascade and their initiation of antioxidant genes expression pathways; they reported these results as promising findings that could make nanoceria good candidates in the treatment of AMD and diabetic retinopathy diseases. In a study by Bhargava et al. [91], nanoceria of 5–10 nm could increase the survival of rod and cone photoreceptors in culture by helping the maintenance of the cell line, which allows for improved drug screening experiments on these cell lines.

4.4. Mesoporous Silica Nanoparticles (MSNs)

Mesoporous silica nanoparticles (MSNs), one of the most well-studied inorganic nanoparticles for drug delivery. These are mesoporous systems made up of silica with particle sizes of 30–300 nm. They can adsorb exceptionally large quantities of drugs in their pores. Unique properties of MSNs include large specific surface area and pore volume, controllable particle size and better biocompatibility make them excellent nanoplatform for biomedical applications and drug delivery [92]. Park et al. [93] conducted a study to evaluate the cytotoxicity of three different sizes (50 nm, 100 nm and 150 nm) of silica nanoparticles (SiNPs) on ocular surface cells such as human corneal epithelial cells (HCECs). Experimental results suggested that cellular autophagy and mammalian target of rapamycin (mTOR) pathways activated with the addition of SiNPs without inducing any significant cytotoxicity in cultured HCECs.

An eye drop, brimonidine showed limited effectiveness in the treatment of glaucoma because of its rapid clearance from preocular space. To resolve this issue, a study was conducted by Kim et al. [94] to deliver brimonidine using amino-functionalized mesoporous silica (AMS) particles so that AMS particles form adhesion with the mucous layer and allow high preocular retention time. Because of the presence of mesopores, AMS encapsulated brimonidine released the drug in a sustained manner over 8 h. BMD-AMS stayed in preocular space for up to 12 h when topically administered into the eyes of a rabbit. To measure the in vivo efficacy, the variance in IOP and brimonidine concentration was examined after administering BMD-AMS and compared it with that induced by the marketed brimonidine eye drops, i.e., Alphagan P. The results showed that in BMD-AMS, the duration in the decrease in IOP and the area under the drug concentration in the aqueous humor-time curve was found to be 12 h and 2.68 μg/mL, respectively, which is twice as compared to Alphagan P. These finding indicated enhancement in ocular bioavailability of brimonidine with BMD-AMS.

4.5. Magnetic Nanoparticles (MNPs)

Magnetic nanoparticles (MNPs) belong to the group of nanotechnology-based materials consisting of magnetic elements such as iron, cobalt, chromium, manganese. It can be used for various biomedical applications such as drug delivery, magnetic resonance imaging, tissue repair, transfection and tissue targeting [95]. It forms a strong drug delivery system as their reactive surface can be functionalized with biocompatible coatings or bioactive molecules, which increase their specificity toward cellular targets and avoid their interaction with healthy cells [96]. In addition to this, MNPs also provides two other advantages; first, it can be controlled by noncontact forces, second, it has potential to be used in MRI tracking and could be used in targeted nano-drug delivery system [96].

Most used MNPs include iron-oxide magnetic particles. The coating of biocompatible material on iron-oxide magnetic particles avoids aggregation, biodegradation, and alterations of particles from their original state and allows entrapping of the bioactive agent on the particle via covalent attachment or adsorption [97]. It has a magnetite or maghemite core surrounded by a shell that contains a layer of polymer or functional group (such as antibodies, biotin, amines and streptavidin) [69]. The drug molecules are attached to the shell of magnetic nanoparticles. The main advantage of using magnetic iron oxide nanoparticles is that they can be visualized by magnetic resonance imaging, and the drug-loaded nanoparticles can be held in place by using a magnetic field. In a study done by Yanai et al. [98], superparamagnetic iron oxide nanoparticles (SPIONs) were used to magnetized rat mesenchymal stem cells (MSCs) in order to deliver cells to the diseased area in the dystrophic retina. Mesenchymal stem cells (MSCs) labeled with fluidMAG-D were injected intravitreally in a retinal degenerative transgenic rat. In vitro studies revealed that cells remain viable and retained differentiation ability even after magnetizing them using fluidMAG-D. The results showed that magnetic MSC delivery to the retina increased tenfold as compared to the normal intravitreal injected cells. Cryosection imaging revealed that magnetic MSC cells had migrated into the inner as well as outer retina. Moreover, magnetic nanoparticle treatment along with orbital magnet resulted in significantly higher concentrations of anti-inflammatory molecule IL-10 and hepatocyte growth factor in the retina. These findings suggested that this approach may provide optimal benefit in outer retinal diseases like AMD where controlled delivery to focal cells is required because it can deliver a higher drug load to the site of interest and resulted in therapeutically useful biochemical changes in the dystrophic retina.

In a study on 44 Sprague–Dawley rats, polymer-coated 50 nm or 4 μm MNPs were injected in the left eye, and the same amount of PBS into the right eye and their effect can be evaluated on IOP, corneal endothelial cell count, the morphology of retina and functioning of photoreceptors at different times [99]. Measurement of IOP, ERG and histology showed no toxicity after injecting magnetic nano- and microparticles. However, microparticles showed small toxicity in corneal endothelial cell counts and iron deposition in tissues. In another study, polymer-coated MNPs were found to be nontoxic to the photoreceptors at histologic as well as electrophysiologic levels [100]. Table 2. Summarizes the applications of inorganic nanoparticles for drug delivery to ocular surfaces.

Table 2. Applications of inorganic nanoparticles for drug delivery to ocular surfaces.

Inorganic Nanomaterial	Drug	Application	Animal Model	Function	Ref.
AuNPs	Resveratrol	Injection		Reduced retinal inflammation	[65]
		Topical	Rat	Decrease the intraocular oxidative damage	[66]
	Immunoglobulin-G	Subretinal injection	Rabbit	Improved biodistribution	[67]
	DNA			Treat retinal dystrophies	[68]
AgNPs			Rat	Produced antioxidants Exhibit anticataractogenic	[75]
				Anti-vasopermeability	[76]
				Antifungal activity	[79]
Cerium oxide NPs			Rat	Reduced photoreceptor damaged	[88]
			Mice	Decreased reactive oxygen species	[85]
		Intravenous injection	Rat	Decrease neurodegenerative	[89]
		Intravitreal injection	Mice	Inhibit neovascularization	[90]
MSNs	Brimonidine	Topical	Rabbit	Enhancement in ocular bioavailability	[91]
MNPs	Mesenchymal stem cells (MSCs)	Intravitreal injection	Rat	Increase tenfold in delivery of drug load to the site of interest	[98]
	Polymer-coated MNPs	Intravitreal injection	Rat	polymer-coated MNPs found to be nontoxic to the photoreceptors	[99]

4.6. *Implant Devices*

Implant devices made from polymeric materials can be used as a mode of drug delivery. They showed controllable release of the drug over a long period of time [69]. Based on the material, implants can be classified into two types; biodegradable and non-biodegradable devices. Non-biodegradable implants have shown more accurate control of drug release and longer release periods as compared to biodegradable implants. However, surgical removal of non-biodegradable devices may cause problems for a patient. Based on the places of implantation, the release of drugs in the retina may range from five weeks to six months. Currently, there are several types of commercial implants that are available for ocular disease treatment, such as Ozurdex (dexamethasone biodegradable implant; Allergan, Inc., Irvine, CA, USA), Trivaris (triamcinolone acetonide suspension; Allergan, Inc.), Kenalog (triamcinolone acetonide suspension; Bristol-Myers Squibb, Princeton, NJ, USA), Triesence (triamcinolone acetonide suspension; Alcon, Fort Worth, TX, USA), Retisert (fluocinolone acetonide non-biodegradable implant; Bausch and Lomb, Inc., Rochester, NY, USA), Iluvien (fluocinolone acetonide non-biodegradable implant; Alimera Sciences, Inc., Alpharetta, GA, USA) [69].

5. Potential Ocular Nanomedicine in Clinical Trials and on the Market

The astounding potential of nanomedicine in the pharmaceutical field to improve health care has captured scientists' attention, with extensive research globally to acquire a market position [101,102]. The first product of nanomedicine was placed on the market almost two decades ago. As of 2012, 33 nanotherapies were available on the market with more than 100 therapeutics in development, and around 50% of nanomedicine companies were based in the United States [103,104]. Numerous nanomedicine products have been under investigation in clinical trials, and some have been approved by the US Food and Drug Administration (FDA) and are available to use in clinical settings for the treatment of conditions such as cancer, autoimmune disease, infectious disease, age-related macular degeneration and others [105]. The great evolution of nanotechnology has dominated the drug delivery field, leading to the progression of formulations for ocular drug delivery, with many in clinical trials and some already introduced in the market.

Many nanoformulations for ocular therapies have been developed and commercialized in the market, but some formulations such as drug-loading emulsions are still slow in progressing for therapy even though drug-free nanoemulsion has been approved to treat dry eye. Restasis® is the first product on the market developed as a nanoemulsion containing cyclosporin A for chronic dry eye treatment, and Durezol®, a nanoemulsion containing the drug difluprednate, is approved to treat inflammation of the eye [7]. Another ocular nanomedicine is Visudyne®, a liposomal formulation containing verteporfin marketed by Novartis Pharma Ag that was approved by the FDA in 2000 for intravenous administration in the treatment of choroidal neovascularization due to age-related macular generation, pathological myopia, and ocular histoplasmosis syndrome. Macugen® is a PEG anti-vascular endothelial growth factor aptamer administered by intravitreal injection that was approved by the FDA in 2004 for wet age-related macular degeneration treatment [106]. According to Grumezescu, many ophthalmic nanoformulations are under development in clinical evaluation, including TLC399 (ProDex), which contains dexamethasone sodium phosphate and is currently in phase II for treatment of macular edema due to retinal vein occlusion. In addition, latanoprost coated liposome (POLAT-001) has completed phase II clinical trials for ocular hypertension and primary open-angle glaucoma treatment [107]. Based on clinicaltrials.gov, a website developed by the U.S. National Institutes of Health, Department of Health and Human Services, one current ocular nanomedicine under clinical trial in 2020 is SYSTANE®, a propylene glycol-based eye drops nanoemulsion, which has completed phase IV for dry eye disease treatment. The development of nanotechnology in ophthalmic therapies seems to be a promising strategy with an increasing number of nano-based formulations in clinical trials and on the market, although more research and studies are necessary regarding nanostructure delivery to the eye [108]. Table 3. Nanomedicine for ocular diseases under clinical trial and approved by FDA.

Table 3. Nanomedicine for ocular diseases under clinical trial and approved by FDA.

Product	Nanoformulation	Indication	FDA Approval Status	Ref
Restasis®	Nanoemulsion	Dry eye	Approved	[7]
Durezol®	Nanoemulsion	Eye inflammation	Approved	[7]
Ozurdex	Dexamethasone biodegradable implant	Macular edema, Non-infectious uveitis	Approved	[69]
Trivaris	Triamcinolone acetonide suspension	Uveitis	Approved	
Kenalog	Triamcinolone acetonide suspension	Macular edema	Approved	
Retisert	Fluocinolone acetonide non-biodegradable implant	Non-infectious uveitis	Approved	
Iluvien	Fluocinolone acetonide nonbiodegradable implant	Diabetic macular edema	Approved	
Triesence	Triamcinolone acetonide suspension	Macular edema	Approved	
Visudyne®	Liposome	AMD	Approved	[106]
Macugen®	Aptamer–polymer nanoparticle	Wet AMD	Approved	[106]
TLC399 (ProDex)	Lipid-based nanoparticle	Macular edema	Phase II	[107]
POLAT-001	Liposome	Glaucoma	Phase II	[107]
SYSTANE®	Propylene glycol-based nanoemulsion	Dry eye	Phase IV	[108]

6. Conclusions

The field of nanotechnology has boomed in recent years, accounting for half of the patent filings and one-third of all publications in the US alone. The nanotechnology market has a forecasted compound annual growth rate (CGAR) of 22%, making it a sector with significant investment potential. Around 200 companies and several potential startups have been identified as active in nanomedicine development worldwide. Nanomedicines accounts for approximately 5% of research publications in the field of nanotechnology worldwide. Several nanoparticles, either organic or inorganic, have provided tools that can overcome the problems associated with the conventional drug delivery of ophthalmic formulations. The nanomedicine market is in the early stages of development, and several drug delivery devices are in the clinical testing stage. Various Intravitreal, polymeric drug delivery implants for ocular diseases have been approved by the FDA.

The application of nanotech has shown great potential in ocular nanomedicine research. The nanomaterials could also be used in increasing the bioavailability of various therapeutic agents. Examples of such nanomaterials include liposomes, dendrimers, niosomes, metal NPs. Nanomedicine also has started to showcase its applied potential in improving the pharmacokinetic properties of drug delivery systems; many studies are focusing on integrating hybrid systems with hydrogel using micelles, dendrimers, and cyclo-dextrins [109]. Other aspects that are associated with ocular nanomedicine development have made uniform NPs with reproducible features on an enormous scale. To stimulate clinical translations of nanomedicine, it is critical to investigate the safety and toxicity profiles of such nanomedicine formulations. There could be many other factors that can influence toxicity in nanomedicine, such as administration dose, shape, and size of the particle, surface charge, and functional groups.

Advanced nanofabrication technologies like particle replication in non-wetting templates (PRINT) [110] and the hydrogel template method [111] have been introduced to create ocular nanomedicine. PRINT technology can be used to create monodispersed NPs and microparticles with controlled shape, size, and surface modification at a large-scale. Another applied example that demonstrated the use of PRINT technology is the AR13503 implant, which was manufactured to provide sustained release of API for more than two months in vitro [112]. This implant was prepared using biodegradable polymers PLGA/PDLA/poly(ester amide) (PEA) with rod-shaped and size suitable for injection through a 27 gauge needle. Clinical trials for AR 13,503 are underway and are expected to be used for treating diabetic macular edema (DME) and wet AMD. The hydrogel template method is another advanced nanofabrication technology that can be used to fabricate large amounts of homogenous nanoparticles and microparticles [111]. For example, nanowafer, an ultra-thin transparent lens prepared by filling drug solutions into the dissoluble PVA template. The amount of the drug can be controlled by adjusting the size of the particle.

With several advantages like drug targeting, sustainability, and increased bioavailability, nanomedicine promises to revolutionize the medical markets. Considering the numerous benefits of nanomedicine in ocular drug delivery, aggregation, toxicity, and clearance of nanoparticles remain a concern. Further study of nanodrug carrier development is necessary to provide a more inside understanding of the safety issue and progress in ocular drug delivery.

Author Contributions: Conceptualization, M.R.B., writing—original draft preparation, D.K., Z.A.M., R.V., M.R.B., S.B.; writing—review editing, S.B., R.P., G.P.L., M.V., N.K., S.M., S.S.M., C.L.P., M.R.B.; supervision, M.R.B.; funding acquisition, M.R.B. All authors have read and agreed to the published version of the manuscript.

Funding: This research was funded by NIH/NEI award EY027013-02 and USF Taneja College of Pharmacy New Faculty Development award to Manas R. Biswal.

Data Availability Statement: Not applicable.

Conflicts of Interest: The authors declare no conflict of interest.

References

1. Bourne, R.R.A.; Flaxman, S.R.; Braithwaite, T.; Cicinelli, M.V.; Das, A.; Jonas, J.B.; Keeffe, J.; Kempen, J.H.; Leasher, J.; Limburg, H.; et al. Vision Loss Expert Group. Magnitude, temporal trends, and projections of the global prevalence of blindness and distance and near vision impairment: A systematic review and meta-analysis. *Lancet Glob. Health* **2017**, *5*, e888–e897. [CrossRef]
2. Kim, Y.C.; Chiang, B.; Wu, X.; Prausnitz, M.R. Ocular delivery of macromolecules. *J. Control. Release* **2014**, *190*, 172–181. [CrossRef] [PubMed]
3. Agrahari, V.; Mandal, A.; Agrahari, V.; Trinh, H.M.; Joseph, M.; Ray, A.; Hadji, H.; Mitra, R.; Pal, D.; Mitra, A.K. A comprehensive insight on ocular pharmacokinetics. *Drug Deliv. Transl. Res.* **2016**, *6*, 735–754. [CrossRef]
4. Urtti, A. Challenges and obstacles of ocular pharmacokinetics and drug delivery. *Adv. Drug Deliv. Rev.* **2006**, *58*, 1131–1135. [CrossRef] [PubMed]
5. Weng, Y.; Liu, J.; Jin, S.; Guo, W.; Liang, X.; Hu, Z. Nanotechnology-based strategies for treatment of ocular disease. *Acta Pharm. Sin.* **2017**, *7*, 281–291. [CrossRef] [PubMed]
6. Nasimi, P.; Haidari, M. Medical use of nanoparticles: Drug delivery and diagnosis diseases. *Int. J. Green Nanotechnol.* **2013**, *1*. [CrossRef]
7. Reimondez-Troitiño, S.; Csaba, N.; Alonso, M.J.; Fuente, M. Nanotherapies for the treatment of ocular diseases. *Eur. J. Pharm. Biopharm.* **2015**, *95*, 279–293. [CrossRef]
8. Das, S.; Suresh, P.K. Drug delivery to eye: Special reference to nanoparticles. *Int. J. drug Deliv.* **2010**, *2*, 12–21. [CrossRef]
9. Rehman, A.U.; Akram, S.; Seralin, A.; Vandamme, T.; Anton, N. Lipid nanocarriers: Formulation, properties, and applications. In *Smart Nanocontainers*; Elsevier: Amsterdam, The Netherlands, 2020; pp. 355–382.
10. Asasutjarit, R.; Managit, C.; Phanaksri, T.; Treesuppharat, W.; Fuongfuchat, A. Formulation development and in vitro evaluation of transferrin-conjugated liposomes as a carrier of ganciclovir targeting the retina. *Int. J. Pharm.* **2020**, *577*, 119084. [CrossRef]
11. Law, S.L.; Huang, K.J.; Chiang, C.H. Acyclovir-containing liposomes for potential ocular delivery. Corneal penetration and absorption. *J. Control. Release* **2000**, *63*, 135–140. [CrossRef]
12. Shimazaki, H.; Hironaka, K.; Fujisawa, T.; Tsuruma, K.; Tozuka, Y.; Shimazawa, M.; Takeuchi, H.; Hara, H. Edaravone-loaded liposome eyedrops protect against light-induced retinal damage in mice. *Investig. Ophthalmol. Vis. Sci.* **2011**, *52*, 7289–7297. [CrossRef] [PubMed]
13. Natarajan, J.V.; Chattopaddyay, S.; Ang, M.; Darwitan, A.; Foo, S.; Zhen, M.; Koo, M.; Wong, T.T.; Venkatramam, S.S. Sustained release of an anti-glaucoma drug: Demonstration of efficacy of a liposomal formulation in the rabbit eye. *PLoS ONE* **2011**, *6*, e24513. [CrossRef] [PubMed]
14. Bochot, A.; Fattal, E.; Gulik, A.; Couarraze, G.; Couvreur, P. Liposomes dispersed within a thermosensitive gel: A new dosage form for ocular delivery of oligonucleotides. *Pharm. Res.* **1998**, *15*, 1364–1369. [CrossRef] [PubMed]
15. Moghassemi, S.; Hadjizadeh, A. Nano-niosomes as nanoscale drug delivery systems: An illustrated review. *J. Control. Release* **2014**, *185*, 22–36. [CrossRef]
16. Pardakhty, A.; Moazeni, E. Nano-niosomes in drug, vaccine and gene delivery: A rapid overview. *Nanomed. J.* **2013**, *1*, 1–12.
17. Herrero-Vanrell, R.; Torre, V.D.L., Andres Guerrero, V.; Barbosa-Alfaro, D.; Molina-Martinez, I.T.; Bravo-Osuna, I. Nano and microtechnologies for ophthalmic administration, an overview. *J. Drug Deliv. Sci. Technol.* **2013**, *23*, 75–102. [CrossRef]
18. Ge, X.; Wei, M.; He, S.; Yuan, W.E. Advances of Non-Ionic Surfactant Vesicles (Niosomes) and Their Application in Drug Delivery. *Pharmaceutics* **2019**, *11*, 55. [CrossRef]
19. Kaur, I.P.; Garg, A.; Singla, A.K.; Aggarwal, D. Vesicular systems in ocular drug delivery: An overview. *Int. J. Pharm.* **2004**, *269*, 1–14. [CrossRef]
20. Chen, S.; Hanning, S.; Falconer, J.; Locke, M.; Wen, J. Recent advances in non-ionic surfactant vesicles (niosomes): Fabrication, characterization, pharmaceutical and cosmetic applications. *Eur. J. Pharm. Biopharm.* **2019**, *144*, 18–39. [CrossRef]
21. Gugleva, V.; Titeva, S.; Rangelov, S.; Momekova, D. Design and in vitro evaluation of doxycycline hyclate niosomes as a potential ocular delivery system. *Int. J. Pharm.* **2019**, *567*, 118431. [CrossRef]
22. Abdelbary, G.; El-Gendy, N. Niosome-encapsulated gentamicin for ophthalmic controlled delivery. *AAPS PharmSciTech* **2008**, *9*, 740–747. [CrossRef] [PubMed]
23. Zeng, W.; Li, Q.; Wan, T.; Liu, C.; Pan, W.; Wu, Z.; Zhang, G.; Pan, J.; Qin, M.; Lin, Y.; et al. Hyaluronic acid-coated niosomes facilitate tacrolimus ocular delivery: Mucoadhesion, precorneal retention, aqueous humor pharmacokinetics, and transcorneal permeability. *Colloids Surf B Biointerfaces* **2016**, *141*, 28–35. [CrossRef] [PubMed]
24. Ekambaram, P.; Sathali, A.H.; Priyanka, K. Solid lipid nanoparticles: A review. *Sci. Rev. Chem. Commun.* **2012**, *2*, 80–102.
25. Wilczewska, A.Z.; Niemirowicz, K.; Markiewicz, K.H.; Car, H. Nanoparticles as drug delivery systems. *Pharmacol. Rep.* **2012**, *64*, 1020–1037. [CrossRef]
26. Mukherjee, S.; Ray, S.; Thakur, R.S. Solid lipid nanoparticles: A modern formulation approach in drug delivery system. *Indian J. Pharm. Sci.* **2009**, *71*, 349–358. [CrossRef] [PubMed]
27. Seyfoddin, A.; Shaw, J.; Al-Kassas, R. Solid lipid nanoparticles for ocular drug delivery. *Drug Deliv.* **2010**, *17*, 467–489. [CrossRef]
28. Mohanty, B.; Majumdar, D.K.; Mishra, S.K.; Panda, A.K.; Patnaik, S. Development and characterization of itraconazole-loaded solid lipid nanoparticles for ocular delivery. *Pharm. Dev. Technol.* **2015**, *20*, 458–464. [CrossRef]
29. Basaran, E.; Demirel, M.; Sirmagul, B.; Yazan, Y. Cyclosporine—A incorporated cationic solid lipid nanoparticles for ocular delivery. *J. Microencapsul.* **2010**, *27*, 37–47. [CrossRef]

30. Cavalli, R.; Gasco, M.R.; Chetoni, P.; Burgalassi, S.; Saettone, M.F. Solid lipid nanoparticles (SLN) as ocular delivery system for tobramycin. *Int. J. Pharm.* **2002**, *238*, 241–245. [CrossRef]
31. Attama, A.A.; Reichl, S.; Müller-Goymann, C.C. Diclofenac sodium delivery to the eye: In vitro evaluation of novel solid lipid nanoparticle formulation using human cornea construct. *Int. J. Pharm.* **2008**, *355*, 307–313. [CrossRef]
32. Yadav, H.K.; Almokdad, A.A.; Shaluf, S.I.M.; Debe, M.S. Polymer-based nanomaterials for drug-delivery carriers. In *Nanocarriers for Drug Delivery*; Elsevier: Amsterdam, The Netherlands, 2019; pp. 531–556.
33. Mishra, D.K.; Shandilya, R.R.; Mishra, P.K. Lipid based nanocarriers: A translational perspective. *Nanomedicine* **2018**, *14*, 2023–2050. [CrossRef] [PubMed]
34. Agarwal, P.; Huang, D.; Thakur, S.S.; Rupenthal, I.D. Nanotechnology for ocular drug delivery. In *Design of Nanostructures for Versatile Therapeutic Applications*; Elsevier: Amsterdam, The Netherlands, 2018; pp. 137–188.
35. Yu, F.; Zheng, M.; Zhang, A.Y.; Han, Z. A cerium oxide loaded glycol chitosan nano-system for the treatment of dry eye disease. *J. Control. Release* **2019**, *315*, 40–54. [CrossRef] [PubMed]
36. Sharma, U.K.; Verma, A.; Prajapati, S.K.; Panday, H.; Panday, A.C. In vitro, in vivo and pharmacokinetic assessment of amikacin sulphate laden polymeric nanoparticles meant for controlled ocular drug delivery. *Appl. Nanosci.* **2015**, *5*, 143–155. [CrossRef]
37. Li, F.; Wen, Y.; Zhang, Y.; Zheng, K.; Ban, J.; Xie, Q.; Wen, Y.; Liu, Q.; Chen, F.; Mo, Z.; et al. Characterisation of 2-HP-β-cyclodextrin-PLGA nanoparticle complexes for potential use as ocular drug delivery vehicles. *Artif. Cells Nanomed. Biotechnol.* **2019**, *47*, 4097–4108. [CrossRef] [PubMed]
38. Kalomiraki, M.; Thermos, K.; Chaniotakis, N.A. Dendrimers as tunable vectors of drug delivery systems and biomedical and ocular applications. *Int. J. Nanomed.* **2016**, *11*, 1–12.
39. Chaplot, S.P.; Rupenthal, I.D. Dendrimers for gene delivery—A potential approach for ocular therapy? *J. Pharm. Pharmacol.* **2014**, *66*, 542–556. [CrossRef] [PubMed]
40. Villanueva, J.R.; Navarro, M.G.; Villanueva, L.R. Dendrimers as a promising tool in ocular therapeutics: Latest advances and perspectives. *Int. J. Pharm.* **2016**, *511*, 359–366. [CrossRef]
41. Yavuz, B.; Pehlivan, B.S.; Ünlü, N. Dendrimeric systems and their applications in ocular drug delivery. *Sci. World J.* **2013**, 732340. [CrossRef]
42. Nance, E.; Zang, F.; Mishra, M.K.; Zhang, Z.; Kambhampati, S.P.; Kannan, R.M.; Kannan, S. Nanoscale effects in dendrimer-mediated targeting of neuroinflammatio. *Biomaterials* **2016**, *101*, 96–107. [CrossRef]
43. Jevprasesphant, R.; Penny, J.; Jalal, R.; Attwood, D.; Mckeown, N.B.; D'Emanuele, A. The influence of surface modification on the cytotoxicity of PAMAM dendrimers. *Int. J. Pharm.* **2003**, *252*, 263–266. [CrossRef]
44. Mishra, V.; Jain, N.K. Acetazolamide encapsulated dendritic nano-architectures for effective glaucoma management in rabbits. *Int. J. Pharm.* **2014**, *461*, 380–390. [CrossRef] [PubMed]
45. Yao, W.J.; Sun, K.X.; Liu, Y.; Liang, N.; Mu, H.J.; Yao, C.; Liang, R.C.; Wang, A.P. Effect of poly(amidoamine) dendrimers on corneal penetration of puerarin. *Biol. Pharm. Bull.* **2010**, *33*, 1371–1377. [CrossRef] [PubMed]
46. Marano, R.; Toth, I.; Wimmer, N.; Brankov, M.; Rakoczy, P.E. Dendrimer delivery of an anti-VEGF oligonucleotide into the eye: A long-term study into inhibition of laser-induced CNV, distribution, uptake and toxicity. *Gene Ther.* **2005**, *12*, 1544–1550. [CrossRef] [PubMed]
47. Patravale, V.B.; Date, A.A.; Kulkarni, R.M. Nanosuspensions: A promising drug delivery strategy. *J. Pharm. Pharmacol.* **2004**, *56*, 827–840. [CrossRef] [PubMed]
48. Jiraporn, C.; Walailak, J. Nanosuspension technology for drug delivery. *Walailak J. Sci. Technol.* **2007**, *4*, 139–153.
49. Jacob, S.; Nair, A.B.; Shah, J. Emerging role of nanosuspensions in drug delivery systems. *Biomater. Res.* **2020**, *24*, 3. [CrossRef]
50. Pignatello, R.; Bucolo, C.; Ferrara, P.; Maltese, A.; Puleo, A.; Puglisi, G. Eudragit RS100®nanosuspensions for the ophthalmic controlled delivery of ibuprofen. *Eur. J. Pharm. Sci.* **2002**, *16*, 53–61. [CrossRef]
51. Kassem, M.; Rahman, A.A.; Ghorab, M.M.; Ahmed, M.B.; Khalil, R.M. Nanosuspension as an ophthalmic delivery system for certain glucocorticoid drugs. *Int. J. Pharm.* **2007**, *340*, 126–133. [CrossRef]
52. Kamaleddin, M.A. Nano-ophthalmology: Applications and considerations. *Nanomedicine* **2017**, *13*, 1459–1472. [CrossRef]
53. Gupta, A. Nanoemulsions. In *Nanoparticles for Biomedical Applications*; Elsevier: Amsterdam, The Netherlands, 2020; pp. 371–384.
54. Gupta, A.; Eral, H.B.; Hatton, T.A.; Doyle, P.S. Nanoemulsions: Formation, properties and applications. *Soft Matter* **2016**, *12*, 826–841. [CrossRef]
55. Ismail, A.; Nasr, M.; Sammour, O. Nanoemulsion as a feasible and biocompatible carrier for ocular delivery of travoprost: Improved pharmacokinetic/pharmacodynamic properties. *Int. J. Pharm.* **2020**, *583*, 119402. [CrossRef] [PubMed]
56. Morsi, N.; Ibrahim, M.; Refai, H.; Sorogy, H.E. Nanoemulsion-based electrolyte triggered in situ gel for ocular delivery of acetazolamide. *Eur. J. Pharm. Sci.* **2017**, *104*, 302–314. [CrossRef] [PubMed]
57. Ammar, H.O.; Salama, H.A.; Ghorab, M.; Mahmoud, A.A. Nanoemulsion as a potential ophthalmic delivery system for dorzolamide hydrochloride. *AAPS PharmSciTech* **2009**, *10*, 808. [CrossRef] [PubMed]
58. Tayel, S.A.; El-Nabarawi, M.A.; Tadros, M.I. Promising ion-sensitive in situ ocular nanoemulsion gels of terbinafine hydrochloride: Design, in vitro characterization and in vivo estimation of the ocular irritation and drug pharmacokinetics in the aqueous humor of rabbits. *Int. J. Pharm.* **2013**, *443*, 293–305. [CrossRef] [PubMed]
59. Mahboobian, M.M.; Mohammadi, M.; Mansouri, Z. Development of thermosensitive in situ gel nanoemulsions for ocular delivery of acyclovir. *J. Drug Deliv. Sci. Technol.* **2020**, *55*, 101400. [CrossRef]

60. Yaqoob, A.A.; Ahmad, H.; Praveen, T.; Ahmad, A.; Oves, M.; Ismail, I.M.; Qari, H.A.; Umar, K.; Ibrahim, M.N.M. Recent Advances in Metal Decorated Nanomaterials and Their Various Biological Applications: A Review. *Front. Chem.* **2020**, *8*, 341. [CrossRef]
61. Guo, B.L.; Han, P.; Guo, L.C.; Cao, Y.Q.; Li, A.D.; Kong, J.Z.; Zhai, H.F.; Wu, D. The Antibacterial Activity of Ta-doped ZnO Nanoparticles. *Nanoscale Res. Lett.* **2015**, *10*, 1047. [CrossRef]
62. Boken, J.; Khurana, P.; Thatai, S.; Kumar, D.; Prasad, S. Plasmonic nanoparticles and their analytical applications: A review. *Appl. Spectrosc. Rev.* **2017**, *52*, 74–820. [CrossRef]
63. De Matteis, V.; Rizzello, L. Noble Metals and Soft Bio-Inspired Nanoparticles in Retinal Diseases Treatment: A Perspective. *Cells* **2020**, *9*, 679. [CrossRef]
64. Karakocak, B.B.; Raliya, R.; Davis, J.T.; Chavalmane, S.; Wang, W.N.; Ravi, N.; Biswas, P. Biocompatibility of gold nanoparticles in retinal pigment epithelial cell line. *Toxicol. In Vitro* **2016**, *37*, 61–69. [CrossRef]
65. Dong, Y.; Wan, G.; Yan, P.; Qian, C.; Li, F.; Peng, G. Fabrication of resveratrol coated gold nanoparticles and investigation of their effect on diabetic retinopathy in streptozotocin induced diabetic rats. *J. Photochem. Photobiol. B* **2019**, *195*, 51–57. [CrossRef] [PubMed]
66. Pereira, D.V.; Petronilho, F.; Pereira, H.R.; Vuolo, F.; Mina, F.; Possato, J.C.; Vitto, M.F.; de Souza, D.R.; da Silva, L.; da Silva Paula, M.M.; et al. Effects of gold nanoparticles on endotoxin-induced uveitis in rats. *Investig. Ophthalmol. Vis. Sci.* **2012**, *53*, 8036–8041. [CrossRef] [PubMed]
67. Hayashi, A.; Naseri, A.; Pennesi, M.E.; de Juan, E., Jr. Subretinal delivery of immunoglobulin G with gold nanoparticles in the rabbit eye. *Jpn. J. Ophthalmol.* **2009**, *53*, 249–256. [CrossRef] [PubMed]
68. Trigueros, S.; Domenech, E.B.; Toulis, V.; Marfany, G. In Vitro Gene Delivery in Retinal Pigment Epithelium Cells by Plasmid DNA-Wrapped Gold Nanoparticles. *Genes* **2019**, *10*, 289. [CrossRef] [PubMed]
69. Shen, H.H.; Chan, E.C.; Lee, J.H.; Bee, Y.S.; Lin, T.W.; Dusting, G.J.; Liu, G.S. Nanocarriers for treatment of ocular neovascularization in the back of the eye: New vehicles for ophthalmic drug delivery. *Nanomedicine* **2015**, *10*, 2093–2107. [CrossRef]
70. Kim, J.H.; Kim, J.H.; Kim, K.W.; Kim, M.H.; Yu, Y.S. Intravenously administered gold nanoparticles pass through the blood-retinal barrier depending on the particle size, and induce no retinal toxicity. *Nanotechnology* **2009**, *20*, 505101. [CrossRef]
71. Mathew, T.V.; Kuriakose, S. Photochemical and antimicrobial properties of silver nanoparticle-encapsulated chitosan functionalized with photoactive groups. *Mater. Sci. Eng. C Mater. Biol. Appl.* **2013**, *33*, 4409–4415. [CrossRef]
72. Iravani, S.; Korbekandi, H.; Mirmohammadi, S.V.; Zolfaghari, B. Syntheisi of silver nanoparticles: Chemial, physical, and biological methods. *Res. Pharm. Sci.* **2014**, *9*, 385–406.
73. Ajitha, B.; Reddy, Y.A.K.; Reddy, P.S.; Jeon, H.J.; Ahn, C.W. Role of capping agents in controlling silver nanoparticles size, antibacterial activity and potential application as optical hydrogen peroxide sensor. *RSC Adv.* **2016**, *6*, 36171–36179. [CrossRef]
74. Ge, L.; Li, Q.; Wang, M.; Ouyang, J.; Li, X.; Xing, M.M.Q. Nanosilver particles in medical applications: Synthesis, performance, and toxicity. *Int. J. Nanomed.* **2014**, *9*, 2399–2407.
75. Anbukkarasi, M.; Thomas, P.A.; Sheu, J.R.; Gerladine, P. In vitro antioxidant and anticataractogenic potential of silver nanoparticles biosynthesized using an ethanolic extract of Tabernaemontana divaricata leaves. *Biomed. Pharmacother.* **2017**, *91*, 467–475. [CrossRef] [PubMed]
76. Kalishwaralal, K.; Banumathi, E.; Pandian, S.R.K.; Deepak, V.; Muniyandi, J.; Eom, S.H.; Gurunathan, S. Silver nanoparticles inhibit VEGF induced cell proliferation and migration in bovine retinal endothelial cells. *Colloids Surf. B Biointerfaces* **2009**, *73*, 51–57. [CrossRef] [PubMed]
77. Gurunathan, S.; Lee, K.J.; Kalishwaralal, K.; Sheikpranbabu, S.; Vaidyanathan, R.; Eom, S.H. Antiangiogenic properties of silver nanoparticles. *Biomaterials* **2009**, *30*, 6341–6350. [CrossRef] [PubMed]
78. Sheikpranbabu, S.; Kalishwaralal, K.; Lee, K.J.; Vaidyanathan, R.; Eom, S.H.; Gurunathan, S. The inhibition of advanced glycation end-products-induced retinal vascular permeability by silver nanoparticles. *Biomaterials* **2010**, *31*, 2260–2271. [CrossRef] [PubMed]
79. Xu, Y.; Gao, C.; Li, X.; He, Y.; Zhou, L.; Pang, G.; Sun, S. In vitro antifungal activity of silver nanoparticles against ocular pathogenic filamentous fungi. *J. Ocul. Pharmacol. Ther.* **2013**, *29*, 270–274. [CrossRef]
80. Jun, Y.; Zi, Z.; Wanga, Y.; Zhaoa, G.; Liu, J.X. Silver nanoparticles affect lens rather than retina development in zebrafish embryos. *Ecotoxicol Environ Saf* **2018**, *163*, 279–288.
81. Kim, J.S.; Song, K.S.; Sung, J.H.; Ryu, H.R.; Choi, B.G.; Cho, H.S.; Lee, J.K.; Yu, I.J. Genotoxicity, acute oral and dermal toxicity, eye and dermal irritation and corrosion and skin sensitisation evaluation of silver nanoparticles. *Nanotoxicology* **2013**, *7*, 953–960. [CrossRef]
82. Maneewattanapinyo, P.; Banlunara, W.; Thammacharoen, C.; Ekgasit, S.; Kaewamatawong, T. An evaluation of acute toxicity of colloidal silver nanoparticles. *J. Vet. Med. Sci.* **2011**, *73*, 1417–1423. [CrossRef]
83. Patil, S.; Seal, S. Role of trivalent La and Nd dopants in lattice distortion and oxygen vacancy generation in cerium oxide nanoparticles. *Appl. Phys. Lett.* **2006**, *88*, 243110. [CrossRef]
84. Korsvik, C.; Patil, S.; Seal, S.; Self, W.T. Superoxide dismutase mimetic properties exhibited by vacancy engineered ceria nanoparticles. *Chem. Commun.* **2007**, *10*, 1056–1058. [CrossRef]
85. Kong, L.; Cai, X.; Zhou, X.; Wong, L.L.; Karakoti, A.S.; Seal, S.; McGinnis, J.F. Nanoceria extend photoreceptor cell lifespan in tubby mice by modulation of apoptosis/survival signaling pathways. *Neurobiol. Dis.* **2011**, *42*, 514–523. [CrossRef] [PubMed]
86. Maccarone, R.; Tisi, A.; Passacantando, M.; Ciancaglini, M. Ophthalmic Applications of Cerium Oxide Nanoparticles. *J. Ocul. Pharmacol. Ther.* **2019**, *36*, 376–383. [CrossRef] [PubMed]

87. Karakoti, A.; Monteiro-Riviere, N.A.; Aggarwal, R.; Davis, J.P.; Narayan, R.J.; Self, W.T.; McGinnis, J.; Seal, S. Nanoceria as antioxidant: Synthesis and biomedical applications. *Jom* **2008**, *60*, 33–37. [CrossRef] [PubMed]
88. Chen, J.; Patil, S.; Seal, S.; McGinnis, J.F. Rare earth nanoparticles prevent retinal degeneration induced by intracellular peroxides. *Nat. Nanotechnol.* **2006**, *1*, 142–150. [CrossRef]
89. Fiorani, L.; Passacantando, M.; Santucci, S.; Marco, S.D.; Bisti, S.; Maccarone, R. Cerium Oxide Nanoparticles Reduce Microglial Activation and Neurodegenerative Events in Light Damaged Retina. *PLoS ONE* **2015**, *10*, e0140387. [CrossRef]
90. Zhou, X.; Wong, L.L.; Karakoti, A.S.; Seal, S.; McGinnis, J.F. Nanoceria inhibit the development and promote the regression of pathologic retinal neovascularization in the Vldlr knockout mouse. *PLoS ONE* **2011**, *6*, e16733. [CrossRef]
91. Bhargava, N.; Shanmugaiah, V.; Saxena, M.; Sharma, M.; Sethy, N.K.; Singh, S.K.; Balakrishnan, K.; Bhargava, K.; Das, M. Nanocerium oxide increases the survival of adult rod and cone photoreceptor in culture by abrogating hydrogen peroxide-induced oxidative stress. *Biointerphases* **2016**, *11*, 031016. [CrossRef]
92. Pednekar, P.P.; Godiyal, S.C.; Jadhav, K.R.; Kadam, V.J. Mesoporous silica nanoparticles: A promising multifunctional drug delivery system. In *Nanostructures for Cancer Therapy*; Elsevier: Amsterdam, The Netherlands, 2017; pp. 593–621.
93. Park, J.H.; Jeong, H.; Hong, J.; Chang, M.; Kim, M.; Chuck, R.S.; Lee, J.K.; Park, C.Y. The effect of silica nanoparticles on human corneal epithelial cells. *Sci. Rep.* **2016**, *6*, 37762. [CrossRef]
94. Kim, S.N.; Ko, S.A.; Park, C.G.; Lee, S.H.; Huh, B.K.; Park, Y.H.; Kim, Y.K.; Ha, A.; Park, K.H.; Choy, Y.B. Amino-functionalized mesoporous silica particles for ocular delivery of brimonidine. *Mol. Pharm.* **2018**, *15*, 3143–3152. [CrossRef]
95. Gupta, A.K.; Naregalkar, R.R.; Vaidya, V.D.; Gupta, M. Recent advances on surface engineering of magnetic iron oxide nanoparticles and their biomedical applications. *Future Med.* **2007**. [CrossRef]
96. Giannaccini, M.; Giannini, M.; Calatayud, M.P.; Goya, G.F.; Cuschieri, A.; Dente, L.; Raffa, V. Magnetic nanoparticles as intraocular drug delivery system to target retinal pigmented epithelium (RPE). *Int. J. Mol. Sci.* **2014**, *15*, 1590–1605. [CrossRef] [PubMed]
97. Bruschi, M.L.; de Toledo, L.D.A.S. Pharmaceutical applications of iron-oxide magnetic nanoparticles. *Magnetochemistry* **2019**, *5*, 50. [CrossRef]
98. Yanai, A.; Häfeli, U.O.; Metcalfe, A.L.; Soema, P.; Addo, L.; Gregory-Evans, C.Y.; Po, K.; Shan, X.; Moritz, O.L.; Gregory-Evans, K. Focused magnetic stem cell targeting to the retina using superparamagnetic iron oxide nanoparticles. *Cell Transplant.* **2012**, *21*, 1137–1148. [CrossRef] [PubMed]
99. Raju, H.B.; Hu, Y.; Vedula, A.; Dubovy, S.R.; Goldberg, J.L. Evaluation of magnetic micro-and nanoparticle toxicity to ocular tissues. *PLoS ONE* **2011**, *6*, e17452. [CrossRef] [PubMed]
100. Häfeli, U.O.; Riffle, J.S.; Harris-Shekhawat, L.; Carmichael-Baranauskas, A.; Mark, F.; Dailey, J.P.; Bardenstein, D. Cell uptake and in vitro toxicity of magnetic nanoparticles suitable for drug delivery. *Mol. Pharm.* **2009**, *6*, 1417–1428. [CrossRef] [PubMed]
101. Morigi, V.; Tocchio, A.; Pellegrini, C.B.; Sakamoto, J.H.; Arnone, M.; Tasciotti, E. Nanotechnology in medicine: From inception to market domination. *J. Drug Deliv.* **2012**, *2012*, 389485. [CrossRef] [PubMed]
102. Nijhara, R.; Balakrishnan, K. Bringing nanomedicines to market: Regulatory challenges, opportunities, and uncertainties. *Nanomedicine* **2006**, *2*, 127–136. [CrossRef] [PubMed]
103. Weissig, V.; Pettinger, T.K.; Murdock, N. Nanopharmaceuticals (part 1): Products on the market. *Int. J. Nanomed.* **2014**, *9*, 4357. [CrossRef]
104. Wagner, V.; Dullaart, A.; Bock, A.K.; Zweck, A. The emerging nanomedicine landscape. *Nat. Biotechnol.* **2006**, *24*, 1211–1217. [CrossRef]
105. Ventola, C.L. The nanomedicine revolution: Part 2: Current and future clinical applications. *Pharm. Ther.* **2012**, *37*, 582–591.
106. Pooja, D.; Kadari, A.; Kulhari, H.; Sistla, R. Lipid-based nanomedicines: Current clinical status and future perspectives. In *Lipid Nanocarriers for Drug Targeting*, 1st ed.; William Andrew Publishing: Cambridge, MA, USA, 2018; pp. 509–528.
107. Grumezescu, A.M. *Design of Nanostructures for Versatile Therapeutic Applications*, 1st ed.; William Andrew Publishing: Cambridge, MA, USA, 2018.
108. Medicine USNLo. Study of Efficacy and Tolerability of SYSTANE Complete in Patients with Dry Eye Disease. Available online: https://clinicaltrials.gov/ct2/show/NCT03492541?term=nano&cond=ocular&cntry=US&draw=2&rank=1 (accessed on 27 June 2020).
109. Liu, S.; Jones, L.W.; Gu, F. Nanotechnology and naomaterials in ophthalmic drug delivery. In *Nano-Biomaterials for Ophthalmic Drug Delivery*; Springer: Cham, Switzerland, 2016.
110. Perry, J.L.; Herlihy, K.P.; Napier, M.E.; DeSimone, J.M. Print: A novel platform toward shape and size specific nanoparticle theranostics. *Acc. Chem. Res.* **2011**, *44*, 990–998. [CrossRef] [PubMed]
111. Acharya, G.; Shin, C.S.; McDermott, M.; Mishra, H.; Park, H.; Kwon, I.C.; Park, K. The hydrogel template method for fabrication of homogeneous nano/microparticles. *J. Control. Release* **2010**, *141*, 314–319. [CrossRef] [PubMed]
112. Glendenning, A.; Crews, K.; Sturdivant, J.; DeLong, M.A.; Kopczynski, C.; Lin, C.W. Sustained release, biodegradable PEA implants for intravitreal delivery of the ROCK/PKC in hibitor AR-13503. *ARVO Ann. Meet. Abstr.* **2018**, *59*, 5672.

nanomaterials

MDPI

Review

Nanotechnology Integration for SARS-CoV-2 Diagnosis and Treatment: An Approach to Preventing Pandemic

Syed Mohammed Basheeruddin Asdaq [1,*], Abu Md Ashif Ikbal [2], Ram Kumar Sahu [3,4,*], Bedanta Bhattacharjee [5], Tirna Paul [5], Bhargab Deka [5], Santosh Fattepur [6,*], Retno Widyowati [3], Joshi Vijaya [7], Mohammed Al mohaini [8,9], Abdulkhaliq J. Alsalman [10], Mohd. Imran [11], Sreeharsha Nagaraja [12,13], Anroop B. Nair [12], Mahesh Attimarad [12] and Katharigatta N. Venugopala [12,14]

1. Department of Pharmacy Practice, College of Pharmacy, AlMaarefa University, Dariyah, Riyadh 13713, Saudi Arabia
2. Department of Pharmacy, Tripura University (A Central University), Suryamaninagar 799022, Tripura (W), India; abumd97@gmail.com
3. Department of Pharmaceutical Science, Faculty of Pharmacy, Universitas Airlangga, Surabaya 60115, Indonesia; rr-retno-w@ff.unair.ac.id
4. Department of Pharmaceutical Science, Assam University (A Central University), Silchar 788011, Assam, India
5. Department of Pharmaceutical Sciences, Faculty of Science and Engineering, Dibrugarh University, Dibrugarh 786004, Assam, India; bedanta1994@gmail.com (B.B.); tirnapaul97@gmail.com (T.P.); bhargavdeka98@gmail.com (B.D.)
6. School of Pharmacy, Management and Science University, Seksyen 13, Shah Alam 40100, Selangor, Malaysia
7. Department of Pharmaceutics, Government College of Pharmacy, Bangalore 560027, Karnataka, India; vijay.joshi67@gmail.com
8. Basic Sciences Department, College of Applied Medical Sciences, King Saud bin Abdulaziz University for Health Sciences, Alahsa 31982, Saudi Arabia; Maam670@gmail.com
9. King Abdullah International Medical Research Center, Alahsa 31982, Saudi Arabia
10. Department of Clinical Pharmacy, Faculty of Pharmacy, Northern Border University, Rafha 91911, Saudi Arabia; kaliqs@gmail.com
11. Department of Pharmaceutical Chemistry, Faculty of Pharmacy, Northern Border University, Rafha 91911, Saudi Arabia; imran.pchem@gmail.com
12. Department of Pharmaceutical Sciences, College of Clinical Pharmacy, King Faisal University, Al-Hofuf, Al-Ahsa 31982, Saudi Arabia; sharsha@kfu.edu.sa (S.N.); anair@kfu.edu.sa (A.B.N.); mattimarad@kfu.edu.sa (M.A.); kvenugopala@kfu.edu.sa (K.N.V.)
13. Department of Pharmaceutics, Vidya Siri College of Pharmacy, Off Sarjapura Road, Bangalore 560035, India
14. Department of Biotechnology and Food Technology, Durban University of Technology, Durban 4001, South Africa

* Correspondence: ramkumar.sahu@aus.ac.in (R.K.S.); sasdag@mcst.edu.sa (S.M.B.A.); dr_santosh@msu.edu.my (S.F.)

Citation: Asdaq, S.M.B.; Ikbal, A.M.A.; Sahu, R.K.; Bhattacharjee, B.; Paul, T.; Deka, B.; Fattepur, S.; Widyowati, R.; Vijaya, J.; Al mohaini, M.; et al. Nanotechnology Integration for SARS-CoV-2 Diagnosis and Treatment: An Approach to Preventing Pandemic. *Nanomaterials* **2021**, *11*, 1841. https://doi.org/10.3390/nano11071841

Academic Editor: Angelo Ferraro

Received: 22 June 2021
Accepted: 14 July 2021
Published: 16 July 2021

Abstract: The SARS-CoV-2 outbreak is the COVID-19 disease, which has caused massive health devastation, prompting the World Health Organization to declare a worldwide health emergency. The corona virus infected millions of people worldwide, and many died as a result of a lack of particular medications. The current emergency necessitates extensive therapy in order to stop the spread of the coronavirus. There are various vaccinations available, but no validated COVID-19 treatments. Since its outbreak, many therapeutics have been tested, including the use of repurposed medications, nucleoside inhibitors, protease inhibitors, broad spectrum antivirals, convalescence plasma therapies, immune-modulators, and monoclonal antibodies. However, these approaches have not yielded any outcomes and are mostly used to alleviate symptoms associated with potentially fatal adverse drug reactions. Nanoparticles, on the other hand, may prove to be an effective treatment for COVID-19. They can be designed to boost the efficacy of currently available antiviral medications or to trigger a rapid immune response against COVID-19. In the last decade, there has been significant progress in nanotechnology. This review focuses on the virus's basic structure, pathogenesis, and current treatment options for COVID-19. This study addresses nanotechnology and its applications in diagnosis, prevention, treatment, and targeted vaccine delivery, laying the groundwork for a successful pandemic fight.

Keywords: COVID-19; nanotechnology; vaccines; sanitizers; biosensors

1. Introduction

The severe acute respiratory syndrome coronavirus 2 (SARS-CoV-2) was responsible for the recent coronavirus 2019 outbreak (COVID-19) [1]. This pandemic has resulted in tense situations around the country, resulting in regulations such as social distance, lockdowns, and other measures. According to a report published in November 2002 in Guangdong, China, the first disease confronted by the world in the twenty-first century was SARS [2]. Other four epidemics, such as SARS, Ebola, Swine Flu, and Middle East Respiratory Syndrome, do not appear to be able to match the fatality produced by COVID-19 [2]. COVID-19 has infected around 189 million individuals worldwide, including over 4 million deaths (https://www.worldometers.info/coronavirus, retrieved on 15 July 2021). This COVID-19 has quickly turned an epidemic into a pandemic. SARS-CoV-2 is a spherical encapsulated virus with a diameter of 60–140 nm that contains positive-sense single-stranded RNA [(+) ssRNA] [3]. COVID-19 is made up of several components, including an extracellular (E) protein, a nucleocapsid (N) protein, a spike (S) glycoprotein, and a matrix (M) protein [4]. Corona virus's globular protein has a size range of 70–90 mm [5]. S1 and S2 are two functional components of the S protein.

The [(+) ssRNA] interacts with N-protein to produce a helical nucleocapsid [6]. Fever, dry cough, loss of taste and smell, weariness, and difficulty breathing are some of the usual symptoms of COVID-19 infection. Ground-glass opacity in the lungs, pneumonia and other severe symptoms are examples. Some patients affected by COVID-19 are asymptomatic [7,8]. The reverse transcription-polymerase chain reaction (RT-PCR) technology is used to confirm the infection by detecting the COVID-19 virus's genetic material in blood and discharge samples. Rapid Antigen test (RAT), or rapid antigen test, is another method of diagnosis [9]. The reverse transcription-polymerase chain reaction (RT-PCR) approach is regarded the superior test for COVID-19 due to its excellent accuracy and specificity [9]. Chest-computed tomography (Chest-CT) scans may be used by some professionals for diagnosis [9]. When an infected person sneezes, talks, or coughs, tiny particles or droplets known as aerosols are released into the environment through their mouth or nose, allowing the virus to spread. Anyone within 6–8 feet of that person can breathe it into their lungs, and it is a direct transmission mode [10]. Indirect contact manifests itself by contacting a surface where a patient may have sneezed or coughed [11]. SARS-CoV-2 has yet to receive a precise treatment. Asymptomatic or mildly symptomatic people, on the other hand, should isolate themselves in a well-ventilated room. They should use a medical mask and discard it after around 8 h. Individuals should get enough rest and drink enough drinks to be hydrated. Daily temperature and oxygen levels should be checked. Take medical help if any deterioration of symptoms is noticed. Sanitization of the individual and the room should be done frequently. When fever, weakness is examined, paracetamol along with Azithromycin and multivitamins are given. Employment of nanotechnology can be seen in the extremity of viral diseases such as COVID-19 [12]. Food and Drug Administration (FDA) gives voice to nanotechnology systems that provide different biological and physicochemical properties, compared to micro and macro stuff [12]. For instance, nanocarriers encapsulated drugs can control the release rate to the targeted sites, which lessens the toxicity and enhances biocompatibility in healthy tissues [13,14]. Nanomaterials help avoid viral spoliation by air and may help sterilize purposes even in a hospital environment. One of the most attention-seeking nanomaterials is gold nano-particles (i.e., AuNPs). Their quantum mechanical properties can be traversed by various forms of biosensors, mainly those that depend on electrochemical, plasmonic, and colorimetric estimation [15,16]. Nanotechnology-based therapy techniques provide a potential way to address constraints in COVID-19 diagnosis, and treatment [17]. Nanomaterials can also be employed in diagnostic to produce simple, quick, and low-cost approaches for detecting SARS-CoV-2. Nanosystems are used to deliver small biomolecules and antiviral to the pulmonary system in a regulated manner, like, to suppress viral spreading [17]. We will discuss the virus's genetic morphological structure, etiopathogenesis, and current therapeutic intervention in this review. We also described and explored how nanotechnology could be used to fight this condition, as well as the most promising nanomaterials. We also expect that the nanotechnology-based strategies outlined in this study will aid researchers in developing novel COVID-19 diagnostic and treatment approaches.

2. Genetic Morphological Structure of SARS-CoV-2

SARS-CoV-2 is a crown-shaped, single-stranded RNA virus with active-sense single-stranded RNA. It belongs to the Coronaviridae family and has a diameter range of 600–1400 A° [18]. An electron microscope can be used to observe extracellular and free particles [19]. The spike (S), envelope (E), membrane (M), and nucleocapsid (N) components of the virus are responsible for pathogenesis (Figure 1) [20].

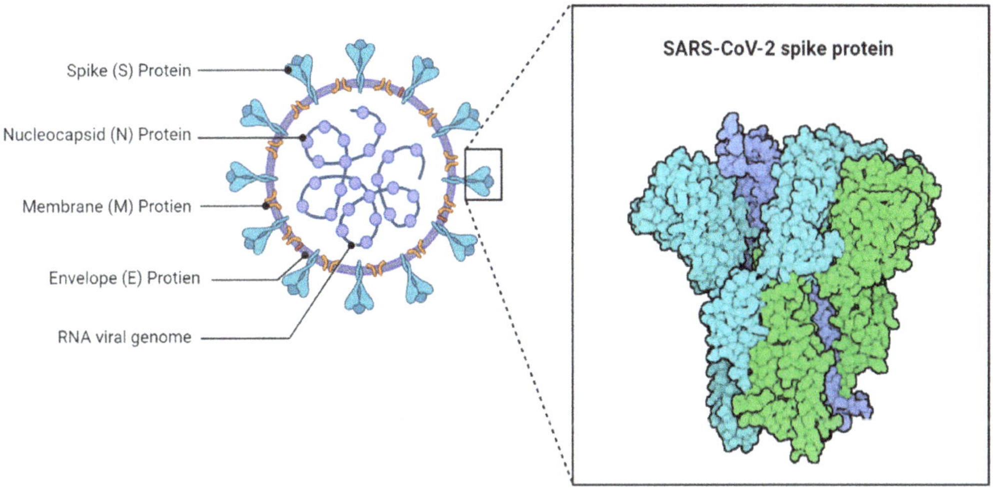

Figure 1. Structure of SARS-CoV-2.

By creating a nucleocapsid, the N protein aids in the replication process in the host cell. The three transmembrane segments of the M protein are crucial for the virus's shape, size, and replication. The E protein plays a vital part in the virus's pathogenic processes and the various stages of replication in infected cells [21]. In connection with this the S protein is made up of two subunits: S1 and S2. The subunits S1 play chief role for receptor binding and cell recognition, while the subunit S2 allows virus-cell fusion [22]. The structural analysis is required for the development of diagnostic procedures and vaccines [23].

3. Etiopathogenesis of SARS-CoV-2

The pathogenesis of the COVID-19 virus was disseminated to the host body by attachment, penetration, biosynthesis, maturity, and release of virus are all steps in the process [24]. The viral attachment with the interaction of viral spike protein and angiotensin-converting enzyme 2 (ACE2) receptor of the host cell. Direct membrane attachment or fusion occurs with the host receptor by undergoing proteolytic cleavage at the S1/S2 boundary, known to be penetration [25,26]. Then, in the host cell cytoplasm, the RNA is released, then RNA moves towards the nucleus and inside the nucleus; biosynthesis occurs by following transcription, translation, protein synthesis [27]. The synthesized protein moves through the endoplasmic reticulum, and the Golgi apparatus gets matured and liberated back to the body [28]. Mainly, lungs get infected by the COVID-19 virus with a dry cough, fever, body pain, dyspnea-like symptoms. The virus can infect a variety of vascular functions by penetrating the respiratory tract's epithelial cell layer and entering the circulation [29]. Consequently, the alveolar damage, hyaline membrane formation, desquamation of pneumocytes, and cellular fibromyxoid transudes can be observed in histopathology studies of an infected patient's lungs [30].

4. Current Therapeutic Strategies against SARS-CoV-2

COVID-19 has yet to receive a clinically recognized therapy. So far, the therapeutic options currently used have been aimed at minimizing and managing symptoms. A few antiviral medicines are now being tested in clinical trials. These medications may operate as immunomodulators or as antiviral agents directly. Since the SARS-CoV-2 outbreak, this antiviral drug research has continued. COVID-19 control treatments are primarily focused on the disease's pathogenesis. If targeting is done in essential viral proteins to upgrade virus internal structure and reproduction, medicines may be classified as virus targeting antivirals (VTA). If the target is a protein located inside the sick host, they may target antivirals (HTA) [31]. The pharmacological treatment for COVID-19 is currently through multiple phases of clinical studies to ensure its efficacy and safety. COVID-19 has yet to be targeted by specific antiviral medication, according to current knowledge. As a result, we must rely on our own safety rather than rely on treatment. Clinical trials should be conducted appropriately to avoid antiviral drug side effects and unpleasant responses. To avoid side effects, each patient's condition should be considered whether antivirals are given alone or in combination with other medications. However, because drug development is an expensive and time-consuming procedure, the current COVID-19 situation necessitates the use of freshly found and more effective medicinal replacements. As a result, nanotechnology might be seen as a powerful tool in the fight against COVID-19, as it offers up new diagnostic and therapeutic possibilities [32]. The availability of the current treatment options for SARS-CoV-2 are depicted in the Table 1.

Table 1. Summed up the current SARS-CoV-2 treatment options.

Drugs	Pharmacological Mechanism	Therapeutic Utility	Cons	Summary of Evidence	References
VIRUS TARGETING ANTIVIRAL (VTA)					
Remdesivir (Nucleotide analog)	Inhibit viral replication via interacting with RNA-dependent PNA Polymerase (RdRp)	SARS-CoV: Middle East respiratory syndrome-CoV (MERS-CoV): Influenza; and Ebola	There is no proof of safety or efficacy. Adverse effects like renal failure and increased liver enzymes	Remdesivir effectiveness was demonstrated in a study with 53 individuals infected with SARS-CoV-2, according to Gilead Sciences. The findings revealed a lower death rate in severe instances.	[33,34]
Ribavirin (Ribonucleic analog)	Interferes with DNA and RNA viral replication and suppresses natural guanosine synthesis by direct action on enzyme inosine monophosphate dehydrogenase.	SARS; MERS; Respiratory syncytial virus (RSV) infection; and Hepatitis C	Adverse effects like hematologic toxicity, high risk of anemia, and teratogenic effect in pregnant women.	Ribavirin is primarily used in combination with Ritonavir/Lopinavir and IFN-β.	[35]
Favipiravir (Viral RNA-dependent RNA polymerase inhibitor)	It prevents viral growth by inhibiting RNA polymerase activity.	Ebola; Influenza A	Efficacy is yet unknown. Adverse effects like hyperuricemia, increased liver enzymes, diarrhea, decreased neutrophil count, and teratogenicity.	According to a recent study, favipiravir suppressed COVID-19 development and favored viral clearance.	[34,36]
Ritonavir/ Lopinavir (Protease inhibitor)	Inhibition of the viral PL-pro or Mpro (3CL-pro)	HIV type 1	Adverse effects like hepatotoxicity, pancreatitis, cardiac conduction abnormalities, increase the risk of cardiac arrhythmia.	In a clinical study with 199 individuals, lopinavir did not show therapeutic improvement.	[37]
VIRUS-HOST INTERACTION					
Chloroquine and Hydroxychloroquine (Anti-malarial)	Both chloroquine and hydroxychloroquine block glycosylation of ACE2 receptor chains in SARS-CoV-2, reducing ligand binding and preventing the viral S protein from mediating cell entrance.	Chloroquine-Malaria; SARS-CoV; and extra-intestinal amoebiasis Hydroxychloroquine-Malaria; lupus erythematosus	Long-term use of chloroquine and hydroxychloroquine increasing the risk of retinal damage, hypoglycemia, and cardiac arrhythmias.	A recent investigation with 96,032 patients looked at the effectiveness of chloroquine or hydroxychloroquine, which is frequently administered in combination with a second-generation macrolide. The study found that chloroquine is linked to a greater incidence of ventricular arrhythmias.	[37,38]
Camostat Mesylate (Transmembrane protease, serine 2 inhibitors)	SARS-CoV-2 invasion into host cells is prevented by inhibiting TMPRSS2-mediated glycoprotein activation.	Chronic pancreatitis	Adverse effects like gastrointestinal intolerance, itching, skin rashes, and increased liver enzymes.	Camostat mesylate significantly inhibited the entrance of MERS-CoV, SARS-CoV, and SARS-CoV-2 into the lung cell line Calu-3, resulting in cytotoxic consequences. Furthermore, in cell line investigations, camostat mesylate therapy decreased SARS-CoV-2 infection.	[41]

Table 1. *Cont.*

Drugs	Pharmacological Mechanism	Therapeutic Utility	Cons	Summary of Evidence	References
Umifenovir (Fusion inhibitors)	Umifenovir prevents viral genes from entering the nucleus by impeding the interaction between ACE2/S protein and the fusion of the viral lipid envelope with the host cell membrane.	SARS-CoV-2; Influenza A and B	Common side effects like increased liver enzymes, gastrointestinal intolerance, and allergic reactions.	Umifenovir monotherapy was shown to be more effective than ritonavir/lopinavir in lowering viral load in COVID-19 patients.	[39,40]
HOST-TARGETING ANTI-VIRAL					
Convalescent plasma	Direct viral neutralization and immune system control molecules (cytokine storm, Th1/Th17 ratio, complement activation, etc.). Immunomodulation of a hypercoagulable condition.	SARS-CoV; MERS-CoV; H1N1; Spanish influenza A; and Sepsis	Expensive. The effects of passive vaccination are generally short-lived.	After receiving a convalescent plasma infusion, five patients with severe COVID-19 showed considerable clinical improvement. Using convalescent plasma therapies, 10 patients with COVID-19 were also cured in another clinical research.	[42]
Corticosteroids (Steroid hormones)	The interaction with glucocorticoid steroid (GR) receptors in the cell cytoplasm neutralizes pro-inflammatory cytokines, preventing inflammation.	Inflammation	Increase the risk of secondary infection	There is some benefit in the early stages of infection, and its effectiveness in the treatment of COVID-19 is still being debated by the World Health Organization.	[43]
Sarilumab (Recombinant humanized mAb, IL-6 antagonist)	Interaction with the IL-6 receptor inhibits IL-6 signaling.	Cytokine release syndrome; Rheumatoid arthritis	Still, there is not enough evidence to suggest sarilumab for COVID-19 therapy.	The use of sarilumab is under investigation.	[44]
Tocilizumab (Recombinant humanized mAb, IL-6 antagonist)	Interaction with the IL-6 receptor reduces inflammatory responses and protects against immunological dysregulation caused by inflammation.	Cytokine release syndrome; Rheumatoid arthritis	Expensive. Adverse effects like hepatotoxicity, gastrointestinal perforations, hypertension, and hypersensitivity reactions,	The research found that after a few doses of administration, C-reactive protein levels in 15 individuals with COVID-19 decreased, improving their medical state.	[45]

5. Ongoing Diagnostic Techniques for SARS-CoV-2

To date, various standard diagnostic methods have been used to detect SARS-CoV-2 infection. RT-PCR has long been considered the gold standard for detecting RNA viruses. This methodology was developed by the Centers for Disease Control and Prevention (CDC) in the United States. The RT-PCR test examined sputum, bronchoalveolar lavage, nasopharyngeal swab, pharyngeal swab, and saliva samples [46]. The electron microscope, cell culture, serological methods, (nucleic acid methods, and next-generation sequencing methods are the five types of RNA virus screening methodologies (Figure 2). The image of the virus captured under an electron microscope (EM) is being utilized to launch ground-breaking disease detection efforts. Immunoelectron microscopy (IEM), which detects a specific antibody-antigen combination, and solid-phase IEM (SPIEM), which identifies viral particles directly on the solid surface of a matrix, are both utilized in the diagnosis of RNA viruses [47]. EM, on the other hand, has some disadvantages, including high costs, the need for well-trained workers, and limited sensitivity [48]. Cell culture, a standard process, is employed as a validation standard for most viral diagnostics procedures that have been developed. On the other hand, cell culture is unsuitable for urgent situations due to its lack of specificity and lengthy incubation period [49]. Serology-based procedures and molecular approaches are the most often employed methodology for SARSCoV-2 diagnosis. The clinical samples are withdrawn from the respiratory secretions, body tissues, or blood for investigation of the genetic material of viruses. Further, advanced technology like gene sequencing has become a commonly utilized tool, especially in the epidemiology and characterization of viruses. This method is precise and trustworthy, but its higher cost make practical usefulness limited [50].

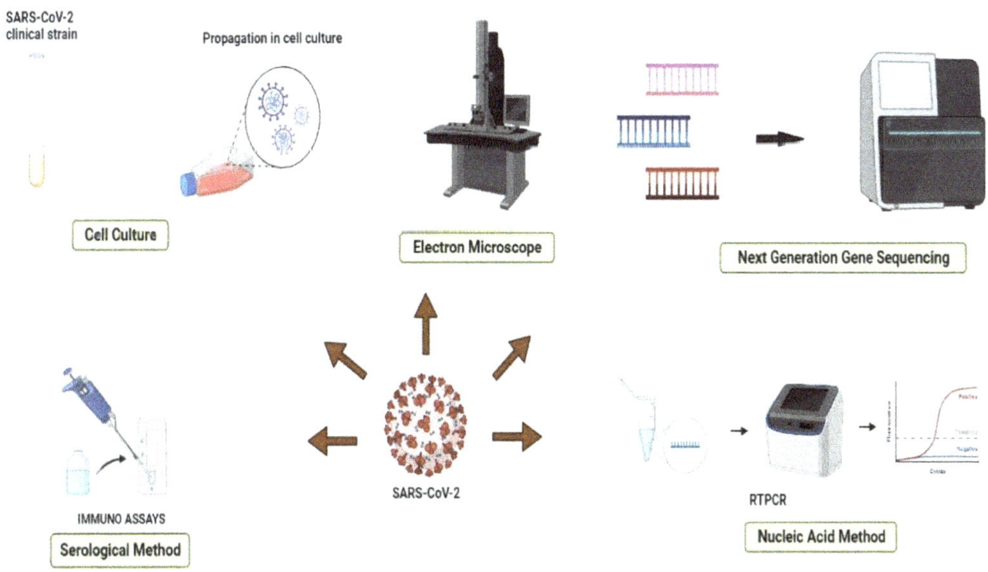

Figure 2. Methods for detecting RNA viruses.

5.1. Point-of-Care Diagnostics

In comparison to RT-PCR, point-of-care is a revolutionary diagnostic procedure that requires relatively little time. It assists physicians and medical workers develop suitable quarantine procedures for positive patients, allowing them to receive urgent medical attention and prevent the disease from spreading further. Point-of-Care techniques are classified as either serological antigen or molecular detection methods, with the former detecting

COVID-19 in blood via lateral flow immunoassay and colloidal gold immunochromatography, and the latter detecting COVID-19 in nasopharyngeal swab, throat swab, saliva, and nasal swab via PCR technique. WHO (World Health Organization) now recommends using these new point-of-care immunodiagnostic tests solely in research settings, based on current evidence. They should not be utilized in any other situation, including clinical decision-making, unless there is sufficient evidence to justify their use for specific reasons [46].

5.2. Lateral Flow Immunoassay

The FDA has approved the use of a lateral flow immunoassay-based point-of-care test to diagnose COVID-19 (EUA). One test, developed by Cellex Inc., is able to detect COVID-19 IgM/IgG antibodies to SARS-CoV-2. This method can be processed in very short time and used for the qualitative identification and discrimination of IgM and IgG antibodies of SARS-CoV-2 [51]. The other developed method is Chembio Diagnostic's DPP COVID-19 IgM/IgG, which provides results in about 15 min utilizing a simple finger–stick method [51].

5.3. Immunochromatography Assay by Colloidal Gold Method

This is a quick and straightforward detection approach that is commonly utilized for illness diagnosis [29,30]. With this technology, Aytu Bioscience established most reliable SARS-CoV-2 IgG/IgM Rapid Test, which can be confirmed within 10 min. Furthermore, it has been recorded higher sensitivity and specificity of IgM compared to other tests [51].

5.4. Signal Amplification Techniques

In a disease state, a nucleic acid-containing pathogen in samples is present in few numbers so amplification is needed to detect pathogens [52,53]. The amplification techniques are worked on the basis of following pattern:

(a) amplification of target nucleic acid,
(b) amplification of probes that interacts with the target nucleic acid,
(c) amplification of signals obtained from target nucleic acid [54].

The amplification techniques are designed into four different groups:

(a) branched DNA technique,
(b) tyramide signal amplification,
(c) hybrid capture,
(d) cleavage-based signal amplification [55].

The principle involved in the branched DNA technique is detecting a signal from target nucleic acid after immobilization and then amplification with the present multiple branched and labeled probes. In case of tyramide signal amplification, the hybridization of target nucleic acid and a biotinylated probe take place. It further promotes the complex of the formation of nucleic acid- biotinylated probe. Streptavidin containing hydrogen peroxidase is incorporated into the medium which binds to nucleic acid-biotinylated probe complex. The addition of inactivated tyramide as a substrate resulted in the formation of activated tyramide and precipitation takes place, precipitate amount indicates signal amplification [56]. The hybrid capture technique favors the hybridization of the complementary probe of signal–stranded target DNA. As a result, the transfer of the DNA: RNA hybrid initiated and transferred to the polyclonal anti-DNA: RNA hybrid antibody. The conjugated enzyme-labeled monoclonal antibody is used to label this complex. Finally, once the enzyme attaches to the chemiluminescence substrate, the signal is generated [57].

5.5. Molecular Methods

Molecular methods for detecting SARS-CoV-2 are mainly based on the proteomic and genomic composition of the virus. Two types of nucleic acid composition are currently available *viz* RT-PCR and RT-LAMP (reverse transcription loop-mediated isothermal amplification) [58]. Three main genomic targets are- (i) envelope protein gene (the E gene), (ii) RNA-dependent RNA polymerase gene (the RdRP gene), and (iii) nucleocapsid protein gene (the N gene) [52].

5.6. Loop-Based Isothermal Amplification

The loop-based isothermal amplification was created to obtain a test result that was quick, accurate, and cost-effective. Instead of the thermal cycle used in RT-PCR, amplification of the target site occurs without changing the temperature reaction [59]. The advantage of loop-mediated isothermal amplification is that the amount of DNA generated in RT-PCR is substantially higher, allowing for visualization without any additional processing. According to the study, when compared to RT-PCR, which is less sensitive, loop-based isothermal amplification has a sensitivity of more than 97% to the target open reading frame (ORF 1ab) gene. Another study found that loop-mediated isothermal amplification is precise, owing to the use of six to eight primers to determine the eight various sections of the target DNA [59,60].

5.7. Immunological Assays

The molecular methods are effective and sensitive for the diagnosis of COVID-19. However, it has some limitations like sampling failures, complex protocols, and expensive equipment, and these methods may sometimes give false negative results. Immunological assays are preferred to mitigate these limitations, while this technique requires comparatively more straightforward sampling with fewer technical experts and inexpensive equipment. Besides this, immunological assays are also used for antigen detection as an alternative for RT-PCR [61]. Immunological assays fall into various types, but the most common comprise antibody or antigens immobilized on a matrix that interacts with viral targets or antibodies in clinical samples. Then it is time to diagnose a virus-specific immunological response by adding more receptor protein to validate the antigen or antibody [61].

5.8. Antibody-Based Assays

After the virus infection has subsided, the body develops an immune response by manufacturing specific antibodies in reaction to the harmful organism. Antibody-based assays help diagnose this immune response, as well as detecting the earlier SARS-CoV-2 infection. Since the body takes time to elicit a reaction, antibody tests are rarely used to detect acute infections in their early stages because the body has not yet fully adapted to the associated antigen, which will require time to detect by establishing an immune response. As a result, antibody testing will yield false-negative findings at that time due to the lack of antibodies despite the presence of disease. Due to their short-lived nature, IgM antibodies were the first antibodies generated after a virus attack, and their identification necessitates a likely active or recent infection. Aside from that, IgG is the primary antibody that gives long-term protection against re-infection with the same virus [52] As a result, both IgG and IgM detection can provide information on the virus infection time course.

5.9. Antigen-Based Assay

Early and precise diagnosis of the disease and adequate quarantine conditions for both symptomatic and asymptomatic individuals are key factors in limiting the COVID-19 pandemic [62]. Since antibody-based detection assays have various limitations, such as when early detection is necessary, antibody detection will be impossible because antibodies will not form in the early stages of the disease. In order to detect viral RNA or viral antigens using antigen-based immunoassays, viral RNA or viral antigens are required. For simplicity of use, antibody-based assays should be accompanied by one of these techniques, and antigen tests that take less time are desired. Antibodies specific to viral antigens are added to the tests, which identify viral immunogen as a precursor to infection. When compared to other approaches, these are pretty quick and cost-effective. Antigen diagnosis tests may be employed for early detection of COVID-19 but not for prior exposure if the sensitivity and specificity are proven [62].

6. Approaches Based on Nanotechnology for Diagnosing SARS-CoV-2

At present, nanotechnology has gained much attention for detecting the novel coronavirus disease due to some known problems with the current diagnostic techniques available [63]. The current diagnostic techniques include RT-PCR (Reverse Transcriptase—Polymerase Chain Reaction), which is utilized to detect asymptomatic patients, but it was found that it also shows optimistic findings in the existence of SARS-CoV-2 to the samples. Another problem associated with this is the lack of PCR infrastructure for proving a sufficient number of kits for diagnosis in remote areas. Thus, the demand was not fulfilled [64,65]. WHO, in its report, mention the urgent need for diagnostic kits for detection of SARS-CoV-2. There are various nanotechnological methods available that can help to improve and meet the current high-level demand in diagnosis [9,66].

6.1. Nucleic Acid Testing

Instrumental RT-PCR diagnostic approaches are currently limited due to procedures that use isothermal conditions for nucleic acid amplification. Loop-mediated isothermal amplification is one of them, and it is thought to be a reasonably sensitive technique for detection [67,68].

Chen et al. investigated magnetic nanoparticles by using RT-PCR procedures to isolate SARS-CoV-2 viral RNA. For collecting RNA on its surface with carboxyl groups, these nanoparticles use poly(amino ester). For RT-PCR diagnosis, the carboxyl group is used in a complex of RNA magnetic nanoparticles. This procedure takes very little time and poses very little danger. Magnetic nanoparticles, rather than extraction, were utilized to isolate viral particles in another investigation. Because of the functionalized target receptors, the virus particles get sturdily linked to the nanoparticles, according to the findings. SPIONs (Superparamagnetic iron oxide nanoparticle) were used for this, as their external magnet aids in the isolation of virus particles. The other techniques such as quantitative Reverse transcription—polymerase chain reaction (qRT—PCR) assays, cell triggering assays, and immunochromatographic strip tests can be used to diagnose the virus [69]. Roh et al. combined quantum dots with fluorescent nanoparticles to develop a fast and sensitive diagnostic method. When the RNA aptamer is coupled with fluorescent quantum dots, it exhibits a difference in the optical signals of the quantum dots if the viral detection is positive. It is feasible to identify viral particle concentrations as low as 1 pmol μL^{-1} with this approach [70].

6.2. Point-of-Care Testing

Because it is a quick testing approach that eliminates the need to transfer samples to a laboratory, point-of-care testing is commonly employed in remote regions with limited facilities and infrastructure. The biosensors utilized are known as colorimetric biosensors because they detect color changes [71,72]. According to one study, paper-based DNA colorimetric sensors can detect virus samples relatively quickly. The cationic pyrrolidinyl peptide nucleic acid (PNA) probe employed in this procedure is more stable than DNA and RNA probes and is used to diagnose COVID-19 and MERS-CoV, both of which have lysine in their probe. Lysine, which is a cation, interacts with negatively charged silver nanoparticles and DNA. PNA-based nanoparticles aggregate with silver nanoparticles in the absence of viral DNA, but instead form complexes with viral particles in the presence of viral DNA. The result is based on the sample's color change (presence or absence situations) and is determined by a paper-based analytical device (PAD) (Figure 3). It is possible to detect up to 1.53 nM with point-of-care testing [73]. Teegam et al. used gold nanoparticles to diagnose COVID-19 and MERS-CoV virus infections. Thiol groups operate as probes in their research, interacting with the upstream E protein gene and capped gold nanoparticles. In the absence of virus, nanoparticles aggregate by changing color;

however, in the presence of virus, the virus forms a compound with viral DNA, blocking the aggregation process and limiting the change of optical characteristics of nanoparticles. Localized surface plasmon resonance shifts were used to diagnose the color variations. With a detection limit of 1 pmo/1.74, the method takes less time and is more cost-effective. The gold nanoparticles generate a solution that collects the virus, and the color change is used to detect it. It is a COVID-19 fast test. The main benefit of this detection method is that gold nanoparticles show specific colors by absorbing a specific wavelength. If the sample contains SARS-CoV-2, this results in a shift in the absorption peak, which causes color changes that can be seen with the naked eye when the virus concentration is high. When compared to conventional PCR procedures that need the extraction of RNA and amplification, these are a time-consuming process, and optical qualities are required for detection in point-of-care testing, hence gold nanoparticle-based techniques are chosen over other diagnostic approaches [74].

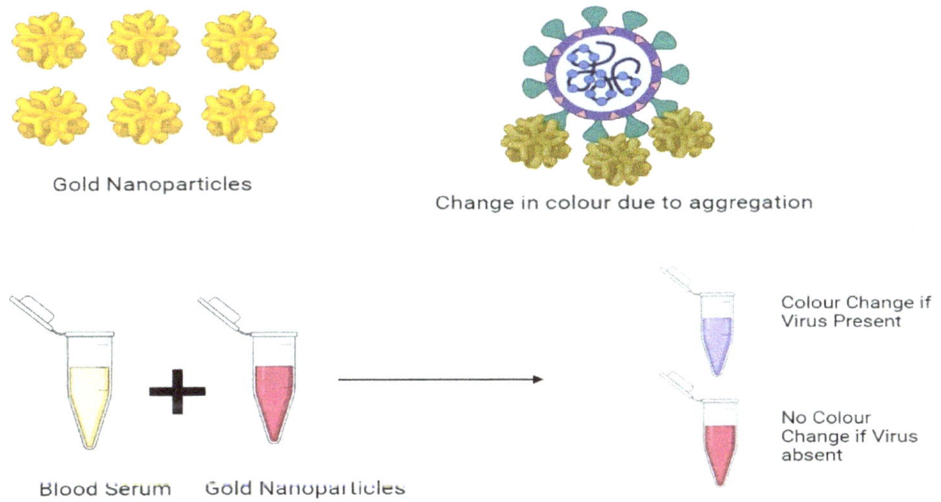

Figure 3. Nanoparticle-based virus detection using calorimetric assay.

6.3. Biosensors Based on Electrochemistry

Biosensors based on electrochemistry are thought to be one of the utmost effective tools for detecting the COVID-19 virus. A modified electrochemical sensing technique with gold nanoparticles maintaining the biomolecule-based functional moiety was used in this approach [75]. Layqah et al. developed an immunosensor using carbon electrodes covered with gold nanoparticles. In this technique COVID-19 spike protein interacts with the virus and the antibody present in a small proportion of the sample, and changes in the current are used to detect the virus. When there is no virus in the sample, the antibody attaches to the SI protein and creates the peak current; however, when the virus is present, the antibody binds to the virus and creates a stronger peak current. The present change is measured, and a conclusion is reached. The detection limit is between 0.4 and 1.0 pg/mL [76]. For the diagnosis of SARS and COVID-19 virus, Ishikawa et al. employed nanowires in the study rather than nanoparticles. Nucleocapsid proteins were detected using In_2O_3 nanowire sensors and fibronectin-based antibodies. In comparison to antibodies and aptamers, antibodies resemble proteins with a higher binding capacity. The virus is detected by changes in the electric current, which are noticed as a signal [77].

6.4. Chiral Biosensors

It has been reported that Chiral biosensors are utmost valuable technologies for diagnosing SARS and COVID-19 virus. According to a recent report, the COVID-19 virus was diagnosed using opened chiral zirconium quantum dots. In this concern the Zirconium quantum dots and magnetic nanoparticles interact with the virus in this manner, resulting in a magnetoplasmonic glow where the virus is present. The magneto plasmonic-fluorescent nanohybrids were then isolated using external magnets, and the virus was detected by measuring the fluorescence intensity. It has a 79.15 EID/50 L detection limit [78]. A chiral immunosensor with self-assembled layers of quantum dots and chiroplasmic gold nanoparticles was used in another study. COVID-19 may be detected in blood samples by mixing gold nanoparticles and quantum dots with the viral sample; chiral optical response changes are identified using the circular dichroism method, which has a low detection limit of 47.91 EID/50 L for coronavirus [79].

7. Nanotechnology-Based Approaches in the Treatment of SARS-CoV-2

For COVID-19 treatment, nanotechnology-based treatment has applications in membrane fusion, inhibition of virus-cell interaction, transcription, translation, viral replication, cell internalization, and activating such mechanism that causes irreversible damage to viruses. Nanotechnology-based treatment has drawn attention because of its properties. Some of the nanoparticles are discussed below. Figure 4 explores the future use of nanostructured materials to deliver recommended and repurposed antiviral medicines, which can improve treatment efficacy by reducing toxicity and allowing for controlled release.

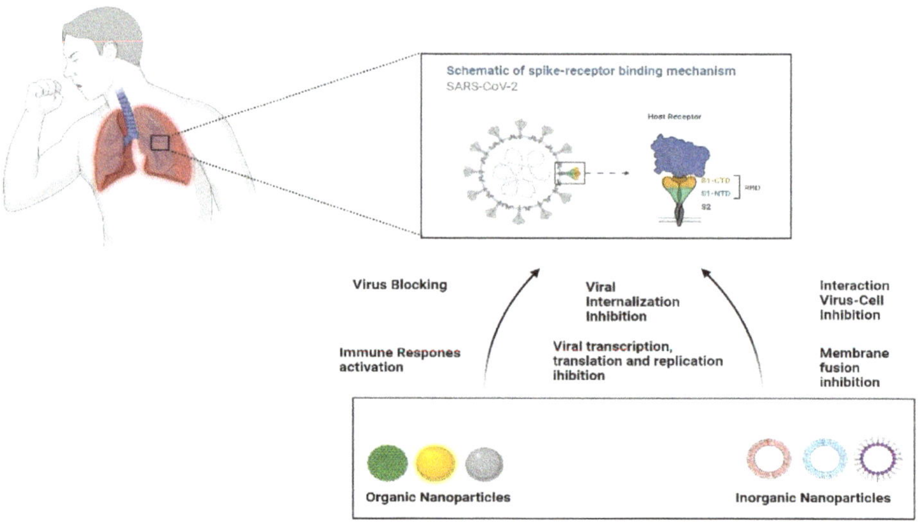

Figure 4. Types of nanoparticles for SARS-CoV-2 treatment.

7.1. Inorganic Nanoparticles

7.1.1. Silver Nanoparticles

The respiratory chain can be harmed by silver nanoparticles (AgNPs), and the electron chain enzymes particularly useful in combating several viruses [80]. AgNPs' greatest strength is that they can be used as antiviral medications with only little alterations. The researches documented that exposed AgNPs with a diameter of 55 nm have been found to have inhibitory effects for the Monkeypox Virus (MPV) and preventing virus internalization [81]. However, some restrictions of these bare AgNPs can be controlled by using polyvinylpyrrolidone (PVP), polyvinyl alcohol (PVA), etc. as coating materials.

7.1.2. Gold Nanoparticles

In in vivo and in vitro investigations, AuNPs with long linkages of sulfonate undecanesulfonic acid (MUS) and mercaptoethanesulfonic acid (MES) demonstrated irreversible deformation in various viruses, including respiratory syncytial virus. AuNPs combined with MUS, which is present as a multivalent binding with the virus, are exposed in a mechanistic interaction. It also causes the capsid's structure to collapse. As a result, this multivalent binding provides an engrossing course of action for COVID-19 therapy. AuNP was also employed against an RNA virus in one investigation. Another study discovered that the nanoparticle technology reduced membrane fusion caused by MERS-CoV. It also demonstrates the possibility of curing COVID-19 [82].

7.1.3. Nanoparticles of Mesoporous Silica

The Mesoporous silica nanoparticles (MsNPs) have different pore sizes that enable molecules to accommodate outside and inside for co-delivery. The dual nature provides an outstanding manifesto for treating COVID-19 infections. These nanoparticles interact with the various ligands to stop the viral uptake to the host cell and the target-specific liberation of those ligands, it also prevents replication of the virus. Lee et al. documented significant dual delivery characteristics of the MsNPs and showed improved antiviral activity [83].

7.1.4. Iron Oxide Nanoparticles

Antiviral properties of iron oxide nanoparticles (IONPs) are particularly depending on the alteration of the surface with sulfonates or other compounds. Bromberg et al. formulated core-shell silica magnetic nanoparticles containing poly (hexamethylene biguanide) functionalization (PHMBG) for improved antiviral activity. On different strains of the virus, NPs had a differential virucidal effect. It happened as a result of the genetic composition's amino alkylation. It also prevents gene replication, resulting in viral deactivation. Because many cases of COVID-19 have demonstrated iron dysregulation [84], IONPs need to be given more consideration. Recent research suggests that SARS-CoV-2 interacts with hemoglobin (and other receptors) in erythrocyte or red blood cell progenitors, causing hemoglobin denaturation and iron metabolism dysregulation. A key concern is an excessive iron burden in the tissue [85].

7.2. Organic Nanoparticles

Carbon Nanoparticles and Graphene Nanoparticles

Carbon-based nanoparticles (CNPs), such as graphenes, carbon nanotubes (CNTs), and fullerenes nanoparticles, have intriguing physicochemical features that can be used for innovative industrial and scientific applications [86]. When it comes to COVID-19 treatment, these CNTs have some toxicity limits. The activation of macrophages in the lungs leads to fibrosis and collagen accumulation in the lesions when exposed to the lower respiratory tract [87]. Intranasal administration of CNTs causes an amplification of influenza H1N1 virus infection in lung epithelial cells. When biomolecules (polymers and proteins) were functionalized on the surface of CNTs, it was found that the inflammatory process was significantly reduced, resulting in reduced cytotoxicity in the respiratory system [88]. CNTs conjugated with hyaluronic acid increase interactions with bronchus cells, and treatment reduces inflammation in the lungs. Furthermore, CNTs and graphene quantum dots functionalized with antiviral medicines (CHI360, CHI415, CHI499, and CDF119) inhibited reverse transcriptase enzyme activity in HIV-positive patients' cells [88]. These CNPs can be used to administer COVID 19 in the future.

7.3. Dendrimers

Dendrimers represent a significant advancement in nanotechnology since they can improve the efficacy of bioactive substances and medications. Dendrimers are synthetic nanoarchitects that are well-defined and highly branching [89]. It has unique physicochemical qualities such as efficient drug encapsulation, solubility, biodegradability, low polydispersity, and biocompatibility. The antiviral activity of dendrimers is amplified as a result of their strong contacts with viruses. HSV2 (Herpes simplex virus type 2) and HIV (human immunodeficiency virus) have both been treated with dendrimers, which prevent viral entrance fusion. Dendrimers made of polyamidoamine (PAMAM) are effective at preventing influenza in mice [90]. SPL7013 Gel is an example of a dendrimer available on the market (Viva Gel). It is a dendrimer with a divalent benzhydryl amine (BHA) core that works as a microbicide. The BHA core comprises four generations of lysine that bifurcate with an outermost branch, with 32 naphthalene di-sulfonic acid groups capped in the outermost branch, and it was designed to inhibit HSV and HIV infections. SPL7013 Gel's goal is to be utilized as a superficial vaginal microbicide that has been identified [91]. Dendrimers have not yet been utilized to combat SARS-CoV-2. It can; however, be employed for the treatment of the COVID-19 infection shortly.

7.4. Lipid-Based Nanoparticles

Because of their excellent biocompatibility and biodegradability, lipid-based nanoparticles have a large potential to be used in COVID-19 clinical situations [92]. Many lipid-based formulations are found in commercially available drugs. AmBisome, Doxil, and other similar products are examples. Because of their amphiphilic and hydrophobic character, several forms of lipid raw materials, such as fatty acids, waxes, oils, phospholipids, mono-, di-, and triglycerides, could be employed as nanocarriers. Since their encapsulation into various classes of antivirals mediates the carriage from the administering site towards the target site while modulating distinct biological retaliation, lipid-based nanoparticles have been explored for the treatment of herpes, hepatitis C (HCV) and B (HBV), HIV viruses, and other viruses [93,94]. Liposomes are ideal nanocarriers for both hydrophilic and lipophilic compounds since they have both aqueous compartments and hydrophobic layers. Polyunsaturated liposomes had more substantial virus inhibitory potential against three viral types: HCV, HBV, and HIV, according to a study. The findings indicated that virus-associated and cellular cholesterol levels were linked to a reduction in infection. Surface-modified liposomes can boost the potential of nanocarriers. LCNPs (liquid crystalline nanoparticles) are nonlamellar self-assembled systems made up of polar lipids such glyceryl monooleate (GMO) and phytantriol (PHYT) [95]. These nanocarriers have gotten a lot of attention since they can hold imaging agents as well as tiny and large molecules in their nano compartments [96]. Nanoemulsions (NE) and microemulsions (ME) are lipid-based nanodroplets that are identical in appearance but differ in physicochemical stability. Solid lipid NPs (SLNs) and nanostructured lipid carriers (NLCs) are made up of solid and liquid lipids or solid lipids, respectively. It has been discovered that alveolar macrophages can be produced utilizing SLNs, a type of cell-related with infectious lung diseases like COVID-19 [97].

7.5. Polymeric Nanoparticles

Polymeric nanoparticles are used to combat virus-borne illnesses, because they can be arranged to travel to specific intracellular or extracellular targets and impede virus attachment to the host cell receptor. These features are essential to stop viruses from causing diseases and to eradicate their negative consequences [98]. The use of nanoparticles to convey active compounds has been shown to be successful in preventing virus reproduction [99]. In in vivo and in vitro models, intravenous injection of poly(lactic-co- glycolic acid), PLGA (Poly lactic-co-glycolic acid) nanocarriers improved the anti-influenza applicability of dyphylline in H1N1 infection, resulting in sustained drug release in the lungs and efficient protection against noxious dosages of the virus. The optimized NPs PEGylation is a safe formulation as it prevents activation and clearance of classical macrophages from the lungs for at least one month [100,101]. It has been proposed that replacing mAbs (Monoclonal antibodies) and compounds can reduce virus uptake in host cells. As a result, NPs can be utilized to prevent infection in humans by blocking the interaction between COVID-19 and ACE2. As a result, NPs can inhibit COVID infections by preventing the virus's ACE2 binding. Compared to inorganic NPs, polymer-based nanoparticles or polymeric structures are subjected to pulmonary systemic absorptions and have lower cytotoxic effects in delicate cells [102]. Aerosol administration of corticosteroids-loaded polymeric NPs is utilized to treat asthma and chronic obstructive pulmonary disease with deep discernment in the lungs [102]. Because of its biocompatibility, polymeric NPs can be employed in inhalation therapy. The infected cells cause the activated immune cells in COVID-19 patients to produce a lot of inflammatory cytokines [103,104]. White blood cells engage in intense phagocytosis, resulting in the nebulization of PLGA nanoparticles overlaid with chitosan, which stops neutrophils and eosinophils from migrating to the lungs and prevents severe physiologic dysfunction [105]. Multifunctional polymeric nanocarriers have been certified as a new nano-sized stage for clinical solicitation, and they may be quickly updated for safe use in the recent SARS-CoV-2 epidemic [106].

8. Nano-Based Sanitizer and Disinfectants

Since the breakout of the new coronavirus, hand hygiene has been an essential tool in the fight against it. While the virus is transferred directly by coughing and sneezing, a study indicated that poor hand hygiene is responsible for 42% of infections. Hand hygiene is especially crucial in hospitals, where 40% of nosocomial infections are caused by poor hand hygiene [107]. As a result, especially in the current pandemic situation, an effective tool for maintaining adequate hand hygiene is essential. Hand sanitizers typically contain 70% alcohol, as well as other chemical or natural components. For many years, it has been the dominant force in this sector. Manufacturers are becoming more interested in nano-based sanitizers as nanoscience advances. As a result, silver nanoparticles have emerged as a viable solution for a wide range of COVID-19-related issues, including detection and disinfection [108]. Silver-based nanoparticles were found to have a wide range of bactericidal and antiviral activities in trials conducted worldwide. By inhibiting surface glycoproteins, it effectively inhibits viral RNA and reduces virulence [109]. Another benefit of nano-based hand sanitizers is that they can be made in a non-alcoholic formulation for customers who dislike alcohol-based hand sanitizers for various reasons [110]. Disinfecting the air, surfaces, and personal appliances is another strategy for neutralizing this virus [111]. While chemical disinfectants (such as chlorines, peroxides, quaternary amines, and alcohols) have been used for surface and personal equipment disinfection and sterilization, they are commonly associated with limitations such as high concentration requirements for viral inhibition and potential risk to public health and the environment [112,113]. As a result, metallic nanoparticles (silver and titanium dioxide) have been proposed as alternatives because of their broad range of applications, environmental and public safety, and efficacy at low concentrations [114]. Preliminary studies, for example, demonstrated that a coating of silver nanoclusters and silica composite on face masks had virus inhibition properties against SARS-CoV-2 [115]. In another instance, the Italian company NanoTechSurface developed a self-sterilizing solution for disinfecting surfaces using titanium dioxide and silver ions [116].

Similarly, a company named FN Nano Inc. in the United States has developed a titanium-based photo-catalytic coating that, when exposed to light, can destroy viruses by disrupting the viral membrane [116]. Nanomaterial has tremendous possibility as disinfectants owing to their unique properties. It includes intrinsic antiviral properties such as the production of reactive oxygen species to destroy viruses as well as photodynamic and photo-thermal properties. Biodegradable nanomaterials can also help to mitigate the detrimental impacts of metallic nanomaterials on human health and the environment.

9. Nano-Based Vaccine Development

Vaccines are biological substances that stimulate the body's immune system to produce antibodies and memory cells to resist foreign antigen particles. Immune cells generate a cascade that kills microorganisms with identical antigens, protecting the body from infection in the future. Nanoparticle-based vaccinations have diverse functionalities that can be tailored to specific challenges presented by microbes by boosting cellular immunity [17,82,117].

9.1. Sub-Unit Vaccines

Sub-unit vaccinations are composed of structural components of the COVID-19 virus combined with molecular adjuvants given to boost immunogenicity by stimulating the immune system [118]. Spike protein is essential for membrane fusion and receptor binding; hence a vaccination that targets spike protein is critical. Spike protein-targeted vaccines prevent infection by pre-blocking the interaction with the ACE2 receptor, limiting the virus's membrane fusion by the antibody [119–121]. Nanoparticles are created using the Novavax® unique recombinant nanoparticle vaccination technology with the spike protein and identical to immunogenic viruses [122]. In Brisbane, Australia, the University of Queensland is working on a novel subunit vaccination using "molecular clamp" technology. This vaccine works by pre-blocking viral protein binding [123]. The nanoparticle-based sub-unit virus is actively proceeding [124].

9.2. Nucleic Acid Vaccines

The antigen-antibody reaction occurs after the virus has triggered infection. After infection, the antigen encoded by the several unique nucleic acids is expressed in the host cell. The nucleic acid vaccine, a very effective vaccination technique, was developed based on the stated principle. To elicit an effective immune response, the vaccine employs chemically generated nucleic acid. Messenger ribonucleic acid (mRNA) vaccines have improved immunogenic characteristics because they can replicate the infectious process. The effects are maximized by mixing different mRNA in a single vaccination [125,126]. The use of nanotechnology in nucleic acid vaccines aids in vaccine delivery effectiveness. The introduction of nanotechnology, nanoparticles, such as cationic liposomes, polymeric nanoparticles, or dendrimer nanoparticles, can improve the stability and delivery efficacy of nucleic acid-based vaccines [127,128].

9.3. NP-Based Vaccines

Several investigations have shown that nanotechnology is being used to improve vaccine efficacy. In recent studies, virus-like particles are appropriate for vaccine and treatment procedure development. Nano-sized virus-like particles having viral character can be effectively delivered through lymph and capillaries [129,130]. By activating the B cell and boosting the immune system the nano-sized virus-like particles works in the host cell, which leads to the development of vaccines of nano-sized virus-like particles [131–133]. Nano-sized virus-like particles effectively prevent viral effects by boosting the immune system [134,135]. Using recombinant S, membrane, and envelope proteins, the nano-sized virus-like particles can be synthesized from the tested virus [136]. Nano-sized virus-like particles have an extensive range of applications and modulate vaccine safety and usefulness and have purpose-specific advantages [137].

10. Non-Invasive COVID-19 Detection

COVID-19 caused a viral pandemic by impacting the global economy, health system, transportation, and all elements of human life. The capacity to detect COVID-19 infections early via a non-invasive approach can function as an epidemic tool for control. Shan et al. developed a non-invasive intelligent nanomaterial-based hybrid sensor array with multiplexed capabilities for the detection and monitoring of COVID-19-specific volatile organic compounds (VOC) mixtures from exhaled breath. The potency of the device was investigated in the case control clinical research by the healthcare agencies. The sensor of the device contains various gold nanoparticles that are coupled to organic ligands. VOC exposure causes changes in the sensing layer, which might cause the organic ligand to expand or shrink. The inclusion of inorganic nanoparticles promotes electrical conductivity production. The changes in the electrical conductivity of the device indicate the presence of COVID-19 viral agents. The stated test could assist in rapid screening of large populations due to its simplicity and high sensitivity, and it could also be utilized as a valuable technique for COVID-19 screening [68]. The investigation of COVID-19 through the VOC mixtures from exhaled breath by using sensing device, it gives effective diagnosis of COVID-19 virions in less time. COVID-19 dissemination can be efficiently controlled if the device is integrated with the face mask. To achieve this aim, a concerted effort is required, but still the global community as a whole should strive for it [138]. Miripour et al. developed an efficient electrochemical sensor to assess the level of reactive oxygen species (ROS) in a sputum sample. Over 97% of positive patients were identified, and the sensors confirmed the diagnosis in under 30 s. During this pandemic, it could be utilized as a significant companion in rapidly detecting individuals who require additional medical evaluation, but it might be employed in the future to reduce the number of cases that require a CT-Scan for COVID-19 diagnosis. COVID-19 early-stage screening could benefit from these enhanced electronic biosensors [139].

11. Preventive Procedure from Virus

As the spreading of the COVID-19 virus is increasing globally, various guidelines are being issued by the World Health Organization (WHO) for disease control and prevention [140,141]. It is a personal feeling that virus prevention can be done by taking some measures and incorporating them into our daily life. In the localities where there is a widespread transmission of the COVID-19 virus, potential exposures are controlled by various preventive strategies. Temporary lockdowns are imposed to bring down the public gathering. Curfews are also being incorporated. A more significant number of COVID-19 positive cases are detected than these places marked as containment zones or hot spot regions, where traveling is wholly restricted to prevent further spread of the virus. Preventive measures are a new strategy to combat the virus. Early detection and proper diagnosis are necessary to prevent further spread. Preventive measures are chiefly focused on the isolation of COVID-19 positive patients. The primary strategy to combat the virus is to frequently wash hands, use portable hand sanitizer, and maintain social distance anywhere to avoid direct contact with face and mouth.

To create awareness among the general public posters and brochures prepared by many organizations related to preventing the COVID-19 virus should be widely used in all areas. The native should be asked to maintain social distancing and should stay at home as much as possible unless it is necessary to go out. In a public gathering maintaining 2 m (6 feet) distance from other people should be practiced. Such social distance is designed to bring down public interactions in a community. Individuals are suffering from infections and are yet to be identified and isolated [142,143]. By maintaining social distancing, chances of transmission can be controlled. Social distancing includes closure of offices, schools, colleges, canceling public gatherings, market suspension, encouraging online classes, office work from home, online shopping etc. Wearing of a mask by covering the nose and mouth in public is encouraged in many countries. Mask such as N95, surgical mask, and 3 ply masks prevents transmission of the virus. Wearing a double mask is encouraged for better protection. Respiratory hygiene is maintained by covering the face while coughing and sneezing. Frequent hand washing by soap or hand wash should be practiced, using a hand sanitizer containing not less than 70% of alcohol is an alternative to handwashing where water is not available, or handwashing is not possible if our hands are not filthy.

Proper handwashing should be done. We should remove the accessories such as wristwatches, finger rings before cleaning our hands. We are followed by wetting hands with water. Then, apply soap or hand washes and rub our hands thoroughly for at least

20 s and rinsing them well, followed by drying our hands with tissue paper or a clean towel and using the same tissue to turn off the faucet. Individuals are counseled to avoid touching their face, especially the nose, mouth, and eyes. We know that cleanliness is next to godliness. Here to prevent the infection, cleaning and disinfection play a vital role. High touching areas, including door handlers, window, bedside table, and switches, should be daily disinfected with surface sanitizers or household disinfectants containing dilute beach solution. The bathroom and toilet should be cleaned and disinfected regularly using a dilute bleach solution, having one part of bleach and nine parts of water. Our daily-wear clothes and hand and bathing towels should be cleaned using laundry soap or detergent. The wearing of disposable gloves while cleaning is also advised. All these measures should be taken to avoid the spread of the virus, such as wearing of mask in public, sanitizing of hands, maintaining social distancing etc. If any symptoms arise, immediately isolate yourself, consult a physician, check for temperature twice, and take proper care. Moreover, most importantly, "do not panic".

12. Future Perspectives for COVID-19 Treatment and Diagnosis

12.1. Theranostic Nanoparticles

The use of nanoparticles has proven to be transformative in the accurate detection and effective in treating diseases. Nanoparticles with reduced toxicity, chemical flexibility, electrical charge, and compact design serve to alleviate several issues with generic medication administration routes. COVID-19 can be treated by focusing on the entry and life cycle treatment. The virus's entry into the host cell can be inhibited by preventing membrane fusion, facilitated by the virus spike protein. We can use therapeutic nanoparticles to pre-block the entry of the COVID-19 virus by reducing the S protein's interaction to the host cell. Viruses such as HIV1 and HIV2 [144–146], hepatitis B virus [147–150], hepatitis C virus [151], influenza A and B viruses [152–154], Herpes simplex virus type 1 and 2 [155–157], EBOV [158], and human norovirus [159] can all be treated with nanotechnology, and it is already being utilized commercially. Dexamethasone is a medicine used to treat the COVID-19 virus developed utilizing nanotechnology and works as an anti-edema and anti-fibrotic [160].

12.2. Nanotechnology-Based Intranasal Delivery Therapy

Recent research has focused on creating innovative methods for safely and effectively delivering nanoparticles into the nasal cavity as a therapy for viral infection [161]. Because the SARS-CoV-2 virus infects the mucosal layer of the nasal cavity and eye, mucosal therapy is regarded as a viable technique in treating SARS-CoV-2. Nanoparticle delivery through the nasal canal is thought to be both simple and convenient. It is also non-invasive, and the nanoparticles are quickly absorbed due to the enormous surface area and the presence of a sizeable capillary plexus in the cavity [162]. When transporting nanoparticles into the nasal cavity, factors like form, size, and surface area play a crucial role in ensuring a safe and effective treatment [163]. Small animals are used in the studies, and nanoparticles are injected into their nasal cavities. To date, inorganic, organic, and virus-like nanoparticles have been developed. It assures that the delivery is safe and effective for therapeutic needs.

12.3. Treatment Using Virus-Like Nanoparticles

Capsids composed consisting of structural proteins and adjuvants derived from viruses are known as virus-like nanoparticles. The capacity of virus-like nanoparticles to generate an immunogenic motif increases immunogenicity. Virus-like nanoparticles can also act as adjuvants, and changing the adjuvants can result in a far more potent immune response than viruses [136,164,165]. Intranasal delivery of virus-like nanoparticles utilizing the influenza virus has been proven to operate as a vaccine by generating many T cells and antibodies that can trigger various types of immunological responses to increase immunity and reduce recurrence [166].

12.4. Treatment Using Cell-Derived Vesicles

Exosomes and cell membrane-derived nanovesicles have previously been shown to have the ability to bind and eliminate bacterial toxins [167–169]. On the other hand, biomimetic synthesis has recently been employed to synthesize cell membrane-derived nanovesicles containing proteins with the same structure and activity as native cells [170–172]. By expressing high quantities of ACE2 and a large number of cytokine receptors, cell membrane-generated nanovesicles can engage with host cells for virus and cytokine binding. In investigations, nanodecoy has been shown to successfully bind and neutralize inflammatory cytokines such GM-CSF and IL-6 and prevent SARS-CoV-2 proliferation and infection [173,174]. As a result, using cell membrane produced nanovesicles as a therapeutic alternative to SARS-CoV-2 and cytokine storms could be a realistic option.

12.5. Pulmonary Delivery Using Nanoparticles Inhalation Aerosols

One of the benefits of administering medicine through the nasal cavity is that it affects the nasal cavity's mucous membrane, where the infection is most prevalent. The lungs are the next target for COVID-19 infection treatment [175–177]. Inhaled aerosols can thus be utilized as a non-invasive and effective route of delivery. The targeted medication delivery system in the lungs also decreases the high drug concentration in the blood induced by oral and intravenous administration methods. A variety of nanotechnologies have been employed to create NPs that could be employed as lung inhalation aerosols. Drugs are given directly to the lungs via dry powder inhalers, colloidal dispersions sprays, and pressured metered-dose inhalers [174]. The delivery of high molecular weight drugs to the lungs is being studied [178,179]. Other factors to consider when delivering high molecular weight medications to the lungs include cytotoxicity and biocompatibility.

13. Conclusions

Many people have died due to the COVID-19 outbreak, with casualties coming from all corners of the world. The grievous nature of this pandemic has uncovered and necessitated new strategic approaches to curbing disease transmission. Because of the low efficacy and incidence of disastrous adverse drug reactions, novel COVID-19 therapies must be developed and implemented. Working on preventative and new technologies to combat infectious pandemics, which are equally as vital as the curative approach, is crucial when dealing with pandemics. The nano-based medicines are more effective compared to the conventional dosage forms. Despite these advantages, it has certain limitation such as stability, clinical study, primarily toxicity and large-scale production, all of which provide a challenge to manufacturers. Additionally, biocompatibility, nano-biointerface, safety, and regulatory issues are among the big obstacles for the manufacturers. As a result, research must be targeted in order to overcome these complications and produce better formulations. The different groups of scientists, particularly in nanomedicine, are working on their development using bioinformatics methods. Scientists have a lot of options when it comes to regulatory research. It has the potential to gain national and international exposure. The encouraging findings can also be used to patent the works in a major way.

The diagnosis of SARS-CoV-2 infection is frequently confused as either influenza or seasonal upper respiratory system viral infections. The following trustworthy, high-sensitivity, simple, rapid, and cost-effective diagnostic tools, namely RT-PCR, point-of-care, lateral flow immunoassay, immunochromatography assay, signal amplification, immunological assays, antibody-based assays, and antigen-based assay are used for the confirmation of COVID-19 infection. Furthermore, the study proved the applicability of nanotechnology-based treatment of SARS-CoV-2 employing diverse nano carriers such as inorganic nanoparticles, organic nanoparticles, dendrimers, lipid-based nanoparticles, and polymeric nanoparticles in nano formulations of antiviral drugs. When compared to traditional dosage forms, these formulations show superior therapeutic efficacy and better pharmacokinetic attributes. Similarly, nanoparticle-based vaccines, theranostic nanoparticles, nano-based intranasal delivery therapy, virus-like nanoparticles, cell-derived vesicles, and pulmonary delivery using nanoparticles inhalation aerosols all boost therapeutic efficacy significantly. While nanomedicine is a promising tool in the battle against pandemics, but it faces substantial challenges in clinical implementation, primarily toxicity and large-scale production. This review allows researchers to do study and develop better formulations for treating COVID-19 in society.

Author Contributions: The manuscript writing, editing, and original draft preparation were done by A.M.A.I., R.K.S., B.B., T.P., and B.D. The table and figure preparation was performed by S.M.B.A., S.F. and R.W.; conception and design were done by J.V., S.N., A.B.N., M.A. and K.N.V. The review and final editing of the manuscript was carried out by M.A.m., A.J.A. and M.I. All authors have read and agreed to the published version of the manuscript.

Funding: This research received no external funding.

Acknowledgments: The authors are thankful to AlMaarefa University, Riyadh for providing support to do this review article.

Conflicts of Interest: The authors declare no conflict of interest.

References

1. Lai, C.C.; Shih, T.P.; Ko, W.C.; Tang, H.J.; Hsueh, P.R. Severe acute respiratory syndrome coronavirus 2 (SARS-CoV-2) and coronavirus disease-2019 (COVID-19): The epidemic and the challenges. *Int. J. Antimicrob. Agents* **2020**, *55*, 105924. [CrossRef]
2. Block, P.; Hoffman, M.; Raabe, I.J.; Dowd, J.B.; Rahal, C.; Kashyap, R.; Mills, M.C. Social network-based distancing strategies to flatten the COVID-19 curve in a post-lockdown world. *Nat. Hum. Behav.* **2020**, *4*, 588–596. [CrossRef]
3. Ebrahim, S.H.; Memish, Z.A. COVID-19—The role of mass gatherings. *Travel Med. Infect. Dis.* **2020**, *34*, 101617. [CrossRef]
4. Cui, J.; Li, F.; Shi, Z.-L. Origin and evolution of pathogenic coronaviruses. *Nat. Rev. Microbiol.* **2019**, *17*, 181–192. [CrossRef] [PubMed]
5. Kim, J.-M.; Chung, Y.-S.; Jo, H.J.; Lee, N.-J.; Kim, M.S.; Woo, S.H.; Park, S.; Kim, J.W.; Kim, H.M.; Han, M.-G. Identification of Coronavirus Isolated from a Patient in Korea with COVID-19. *Osong Public Health Res. Perspect.* **2020**, *11*, 3–7. [CrossRef] [PubMed]
6. Swift, T.; People, M. ACE2: Entry Receptor for SARS-CoV-2. *Science* **2020**, *367*, 1444–1448.
7. Fehr, A.; Bickerton, E.; Britton, P. An Overview of Their Replication and Pathogenesis; Section 2 Genomic Organization. *Methods Mol. Biol.* **2015**, *1282*, 1–23.
8. Lan, J.; Song, Z.; Miao, X.; Li, H.; Li, Y.; Dong, L. Skin damage among healthcare workers managing coronavirus disease-2019. *J. Am. Acad. Dermatol.* **2020**, *82*, 1215–1216. [CrossRef] [PubMed]
9. Udugama, B.; Kadhiresan, P.; Kozlowski, H.N.; Malekjahani, A.; Osborne, M.; Li, V.Y.C.; Chen, H.; Mubareka, S.; Gubbay, J.B.; Chan, W.C.W. Diagnosing COVID-19: The Disease and Tools for Detection. *ACS Nano* **2020**, *14*, 3822–3835. [CrossRef]
10. Mackenzie, J.S.; Smith, D.W. COVID-19: A novel zoonotic disease caused by a coronavirus from China: What we know and what we don't. *Microbiol. Aust.* **2020**, *41*, 45–50. [CrossRef]
11. Jayaweera, M.; Perera, H.; Gunawardana, B.; Manatunge, J. Transmission of COVID-19 virus by droplets and aerosols: A critical review on the unresolved dichotomy. *Environ. Res.* **2020**, *188*, 109819. [CrossRef]
12. Lembo, D.; Donalisio, M.; Civra, A.; Argenziano, M.; Cavalli, R. Nanomedicine formulations for the delivery of antiviral drugs: A promising solution for the treatment of viral infections. *Expert Opin. Drug Deliv.* **2018**, *15*, 93–114. [CrossRef] [PubMed]
13. Singh, R.; Lillard, J.W., Jr. Nanoparticle-based targeted drug delivery. *Exp. Mol. Pathol.* **2009**, *86*, 215–223. [CrossRef]
14. Mora-Huertas, C.; Fessi, H.; Elaissari, A. Polymer-based nanocapsules for drug delivery. *Int. J. Pharm.* **2010**, *385*, 113–142. [CrossRef] [PubMed]
15. Quesada-González, D.; Stefani, C.; González, I.; de la Escosura-Muñiz, A.; Domingo, N.; Mutjé, P.; Merkoçi, A. Signal enhancement on gold nanoparticle-based lateral flow tests using cellulose nanofibers. *Biosens. Bioelectron.* **2019**, *141*, 111407. [CrossRef] [PubMed]
16. Huang, Y.; Xu, T.; Wang, W.; Wen, Y.; Li, K.; Qian, L.; Zhang, X.; Liu, G. Lateral flow biosensors based on the use of micro-and nanomaterials: A review on recent developments. *Microchim. Acta* **2020**, *187*, 1–25. [CrossRef]
17. Weiss, C.; Carriere, M.; Fusco, L.; Capua, I.; Regla-Nava, J.A.; Pasquali, M.; Scott, J.A.; Vitale, F.; Unal, M.A.; Mattevi, C.; et al. Toward Nanotechnology-Enabled Approaches against the COVID-19 Pandemic. *ACS Nano* **2020**, *14*, 6383–6406. [CrossRef]
18. Bhattacharjee, B.; Ikbal, A.M.A.; Sahu, R.K.; Ratra, M. Insight into Severe Acute Respiratory Syndrome Coronavirus-2 (SARS-CoV-2): Rationalized Review Special Reference to COVID-19. *Pharm. Biosci. J.* **2021**, *9*, 1–10. [CrossRef]
19. Meo, S.A.; Alhowikan, A.M.; Al-Khalawi, T.; Meo, I.M.; Halepoto, D.M.; Iqbal, M.; Usmani, A.M.; Hajjar, W.; Ahmed, N. Novel coronavirus 2019-nCoV: Prevalence, biological and clinical characteristics comparison with SARS-CoV and MERS-CoV. *Eur. Rev. Med. Pharmacol. Sci.* **2020**, *24*, 2012–2019.

20. Deka, B.; Dash, B.; Newar, A.; Ahmed, A.; Bharali, A.; Bordoloi, B.; Rohman, T.A.; Bingari, B.; Bhattacharjee, B.; Sadhukan, P. Coronavirus disease 2019: A comprehensive review on recent clinical updates and management tools for pharmacists. *Pharm. Biosci. J.* **2021**, *9*, 1–21.
21. Graham, R.L.; Donaldson, E.F.; Baric, R.S. A decade after SARS: Strategies for controlling emerging coronaviruses. *Nat. Rev. Microbiol.* **2013**, *11*, 836–848. [CrossRef] [PubMed]
22. Ziegler, C.G.K.; Allon, S.J.; Nyquist, S.K.; Mbano, I.M.; Miao, V.N.; Tzouanas, C.N.; Cao, Y.; Yousif, A.S.; Bals, J.; Hauser, B.M. SARS-CoV-2 receptor ACE2 is an interferon-stimulated gene in human airway epithelial cells and is detected in specific cell subsets across tissues. *Cell* **2020**, *181*, 1016–1035.e1019. [CrossRef] [PubMed]
23. Ahmed, S.F.; Quadeer, A.A.; McKay, M.R. Preliminary identification of potential vaccine targets for the COVID-19 coronavirus (SARS-CoV-2) based on SARS-CoV immunological studies. *Viruses* **2020**, *12*, 254. [CrossRef] [PubMed]
24. Kara, M.; Ekiz, T.; Ricci, V.; Kara, Ö.; Chang, K.-V.; Özçakar, L. 'Scientific Strabismus' or two related pandemics: Coronavirus disease and vitamin D deficiency. *Br. J. Nutr.* **2020**, *124*, 736–741. [CrossRef]
25. Tang, T.; Bidon, M.; Jaimes, J.A.; Whittaker, G.R.; Daniel, S. Coronavirus membrane fusion mechanism offers a potential target for antiviral development. *Antivir. Res.* **2020**, *178*, 104792. [CrossRef] [PubMed]
26. Astuti, I. Severe Acute Respiratory Syndrome Coronavirus 2 (SARS-CoV-2): An overview of viral structure and host response. Diabetes & Metabolic Syndrome. *Clin. Res. Rev.* **2020**, *14*, 407–412.
27. Iqbal, H.M.; Romero-Castillo, K.D.; Bilal, M.; Parra-Saldivar, R. The Emergence of Novel-Coronavirus and its Replication Cycle—An Overview. *J. Pure Appl. Microbiol.* **2020**, *14*, 13–16. [CrossRef]
28. Huang, M.; Tang, T.; Pang, P.; Li, M.; Ma, R.; Lu, J.; Shu, J.; You, Y.; Chen, B.; Liang, J.; et al. Treating COVID-19 with Chloroquine. *J. Mol. Cell Biol.* **2020**, *12*, 322–325. [CrossRef]
29. Hussain, A.; Kaler, J.; Tabrez, E.; Tabrez, S.; Tabrez, S.S. Novel COVID-19: A Comprehensive Review of Transmission, Manifestation, and Pathogenesis. *Cureus* **2020**, *12*. [CrossRef]
30. Xu, Z.; Shi, L.; Wang, Y.; Zhang, J.; Huang, L.; Zhang, C.; Liu, S.; Zhao, P.; Liu, H.; Zhu, L.; et al. Pathological findings of COVID-19 associated with acute respiratory distress syndrome. *Lancet Respir. Med.* **2020**, *8*, 420–422. [CrossRef]
31. Conteduca, V.; Sansonno, D.E.; Russi, S.; Pavone, F.; Dammacco, F. Therapy of chronic hepatitis C virus infection in the era of direct-acting and host-targeting antiviral agents. *J. Infect.* **2014**, *68*, 1–20. [CrossRef]
32. Catanzaro, M.; Fagiani, F.; Racchi, M.; Corsini, E.; Govoni, S.; Lanni, C. Immune response in COVID-19: Addressing a pharmacological challenge by targeting pathways triggered by SARS-CoV-2. *Signal Transduct. Target. Ther.* **2020**, *5*, 1–10. [CrossRef]
33. Guan, W.-J.; Ni, Z.-Y.; Hu, Y.; Liang, W.-H.; Ou, C.-Q.; He, J.-X.; Liu, L.; Shan, H.; Lei, C.-L.; Hui, D.S.C.; et al. Clinical Characteristics of Coronavirus Disease 2019 in China. *N. Engl. J. Med.* **2020**, *382*, 1708–1720. [CrossRef] [PubMed]
34. Jorgensen, S.C.; Kebriaei, R.; Dresser, L.D. Remdesivir: Review of pharmacology, pre-clinical data, and emerging clinical ex-perience for COVID-19. Pharmacotherapy. *J. Hum. Pharmacol. Drug Ther.* **2020**, *40*, 659–671. [CrossRef] [PubMed]
35. Khalili, J.S.; Zhu, H.; Mak, N.S.A.; Yan, Y.; Zhu, Y. Novel coronavirus treatment with ribavirin: Groundwork for an evaluation concerning COVID-19. *J. Med. Virol.* **2020**, *92*, 740–746. [CrossRef] [PubMed]
36. Cai, Q.; Yang, M.; Liu, D.; Chen, J.; Shu, D.; Xia, J.; Liao, X.; Gu, Y.; Cai, Q.; Yang, Y.; et al. Experimental Treatment with Favipiravir for COVID-19: An Open-Label Control Study. *Engineering* **2020**. [CrossRef] [PubMed]
37. Nitulescu, G.M.; Paunescu, H.; Moschos, S.A.; Petrakis, D.; Nitulescu, G.; Ion, G.N.D.; Spandidos, D.A.; Nikolouzakis, T.K.; Drakoulis, N.; Tsatsakis, A. Comprehensive analysis of drugs to treat SARS-CoV-2 infection: Mechanistic insights into current COVID-19 therapies. *Int. J. Mol.* **2020**, *46*, 467–488. [CrossRef]
38. Mehta, N.; Mazer-Amirshahi, M.; Schulman, C.; O'Connell, F.; Pourmand, A. Sex-based pharmacotherapy in acute care set-ting, a narrative review for emergency providers. *Am. J. Emerg. Med.* **2020**, *38*, 1253–1256. [CrossRef] [PubMed]
39. Jin, Z.; Liu, J.-Y.; Feng, R.; Ji, L.; Jin, Z.-L.; Li, H.-B. Drug treatment of coronavirus disease 2019 (COVID-19) in China. *Eur. J. Pharmacol.* **2020**, *883*, 173326. [CrossRef]
40. Dong, L.; Hu, S.; Gao, J. Discovering drugs to treat coronavirus disease 2019 (COVID-19). *Drug Discov. Ther.* **2020**, *14*, 58–60. [CrossRef]
41. Özdemir, A.T.; Özdemir, R.B.O.; Kırmaz, C.; Sarıboyacı, A.E.; Halbutoğlları, Z.S.U.; Özel, C.; Karaöz, E. The paracrine immunomodulatory interactions between the human dental pulp derived mesenchymal stem cells and CD4 T cell subsets. *Cell. Immunol.* **2016**, *310*, 108–115. [CrossRef]
42. Rojas, M.; Rodríguez, Y.; Monsalve, D.M.; Acosta-Ampudia, Y.; Camacho, B.; Gallo, J.E.; Rojas-Villarraga, A.; Ramírez-Santana, C.; Díaz-Coronado, J.C.; Manrique, R.; et al. Convalescent plasma in COVID-19: Possible mechanisms of action. *Autoimmun. Rev.* **2020**, *19*, 102554. [CrossRef]
43. Russell, B.; Moss, C.; Rigg, A.; Van Hemelrijck, M. COVID-19 and treatment with NSAIDs and corticosteroids: Should we be limiting their use in the clinical setting? *Ecancermedicalscience* **2020**, *14*, 1023. [CrossRef]
44. Lee, E.B. A review of sarilumab for the treatment of rheumatoid arthritis. *Immunotherapy* **2018**, *10*, 57–65. [CrossRef]
45. Luo, P.; Liu, Y.; Qiu, L.; Liu, X.; Liu, D.; Li, J. Tocilizumab treatment in COVID-19: A single center experience. *J. Med. Virol.* **2020**, *92*, 814–818. [CrossRef]
46. Wang, X.Y.; Gao, R.; Lu, R.; Han, K.; Wu, G.; Tan, W. Detection of SARS-CoV-2 in different types of clinical specimens. *JAMA* **2020**, *323*, 1843–1844. [CrossRef] [PubMed]

47. Lewis, D.C.; Lightfoot, N.F.; Pether, J.V. Solid-phase immune electron microscopy with human immunoglobulin M for serotyping of Norwalk-like viruses. *J. Clin. Microbiol.* **1988**, *26*, 938–942. [CrossRef]
48. Beniac, D.R.; Siemens, C.G.; Wright, C.J.; Booth, T.F. A Filtration Based Technique for Simultaneous SEM and TEM Sample Preparation for the Rapid Detection of Pathogens. *Viruses* **2014**, *6*, 3458–3471. [CrossRef] [PubMed]
49. Cella, L.N.; Blackstock, D.; Yates, M.A.; Mulchandani, A.; Chen, W. Detection of RNA Viruses: Current Technologies and Future Perspectives. *Crit. Rev. Eukaryot. Gene Expr.* **2013**, *23*, 125–137. [CrossRef] [PubMed]
50. Malik, Y.S.; Verma, A.K.; Kumar, N.; Touil, N.; Karthik, K.; Tiwari, R.; Bora, D.P.; Dhama, K.; Ghosh, S.; Hemida, M.; et al. Advances in Diagnostic Approaches for Viral Etiologies of Diarrhea: From the Lab to the Field. *Front. Microbiol.* **2019**, *10*, 1957. [CrossRef]
51. Wang, Y.; Lee, Y.; Yang, T.; Sun, J.; Shen, C.; Cheng, C. Current diagnostic tools for coronaviruses–From laboratory diagnosis to POC diagnosis for COVID-19. *Bioeng. Transl. Med.* **2020**, *5*, e10177. [CrossRef]
52. Alpdagtas, S.; Ilhan, E.; Uysal, E.; Sengor, M.; Ustundag, C.B.; Gunduz, O. Evaluation of current diagnostic methods for COVID-19. *APL Bioeng.* **2020**, *4*, 041506. [CrossRef]
53. Gullett, J.C.; Nolte, F.S. Quantitative Nucleic Acid Amplification Methods for Viral Infections. *Clin. Chem.* **2015**, *61*, 72–78. [CrossRef]
54. Nolte, F.S.; Wittwer, C.T. Nucleic Acid Amplification Methods Overview. *Mol. Microbiol.* **2016**, 3–18. [CrossRef]
55. Niesters, H.G.M.; Van Leeuwen, W.B. *Quantitative Isothermal Molecular Amplification Techniques*; Springer Science and Business Media LLC: Berlin/Heidelberg, Germany, 2019; pp. 321–337.
56. Griffiths, K.; Laboratories, A.G.A.; Partis, L.; Croan, D.; Wang, N.; Emslie, K. *Review of Technologies for Detecting Genetically Modified Materials in Commodities and Food*; Department of Agriculture, Fisheries and Forestry-Australia: Queensland, Australia, 2002.
57. Wang, Y.F. Signal Amplification Techniques: bDNA, Hybrid Capture. In *Advanced Techniques in Diagnostic Microbiology*; Springer Science and Business Media LLC: Berlin/Heidelberg, Germany, 2007; pp. 228–242.
58. Park, G.S.; Ku, K.; Baek, S.H.; Kim, S.J.; Kim, S.I.; Kim, B.T.; Maeng, J.S. Development of reverse transcription loop-mediated isothermal amplification assays targeting severe acute respiratory syndrome coronavirus 2 (SARS-CoV-2). *J. Mol. Diagn.* **2020**, *22*, 729–735. [CrossRef] [PubMed]
59. Boehme, C.C.; Nabeta, P.; Henostroza, G.; Raqib, R.; Rahim, Z.; Gerhardt, M.; Sanga, E.; Hoelscher, M.; Notomi, T.; Hase, T.; et al. Operational Feasibility of Using Loop-Mediated Isothermal Amplification for Diagnosis of Pulmonary Tuberculosis in Microscopy Centers of Developing Countries. *J. Clin. Microbiol.* **2007**, *45*, 1936–1940. [CrossRef]
60. González, E.G.; Mayorga, I.M.L.; Sánchez, I.P.R.; Zhang, Y.S.; Chapa, S.O.M.; Santiago, G.T.; Alvarez, M.M. Colorimetric loop-mediated isothermal amplification (LAMP) for cost-effective and quantitative detection of SARS-CoV-2: The change in color in LAMP-based assays quantitatively correlates with viral copy number. *Anal. Methods* **2021**, *13*, 169–178. [CrossRef] [PubMed]
61. Pourali, F.; Afshari, M.; Alizadeh-Navaei, R.; Javidnia, J.; Moosazadeh, M.; Hessami, A.; Pourali, F.; Afshari, M.; Alizadeh-Navaei, R.; Javidnia, J.; et al. Relationship between blood group and risk of infection and death in COVID-19: A live meta-analysis. *N. Microbes N. Infect.* **2020**, *37*, 100743. [CrossRef] [PubMed]
62. Guo, L.; Ren, L.; Yang, S.; Xiao, M.; Chang, D.; Yang, F.; Cruz, C.S.D.; Wang, Y.; Wu, C.; Xiao, Y.; et al. Profiling Early Humoral Response to Diagnose Novel Coronavirus Disease (COVID-19). *Clin. Infect. Dis.* **2020**, *71*, 778–785. [CrossRef] [PubMed]
63. Belhouideg, S. Impact of 3D printed medical equipment on the management of the COVID 19 pandemic. *Int. J. Health Plan. Manag.* **2020**, *35*, 1014–1022. [CrossRef]
64. Purohit, B.; Vernekar, P.R.; Shetti, N.P.; Chandra, P. Biosensor nanoengineering: Design, operation, and implementation for biomolecular analysis. *Sensors Int.* **2020**, *1*, 100040. [CrossRef]
65. Pokhrel, P.; Hu, C.; Mao, H. Detecting the Coronavirus (COVID-19). *ACS Sens.* **2020**, *5*, 2283–2296. [CrossRef] [PubMed]
66. Kumar, R.; Nagpal, S.; Kaushik, S.; Mendiratta, S. COVID-19 diagnostic approaches: Different roads to the same destination. *VirusDisease* **2020**, *31*, 97–105. [CrossRef]
67. Mujawar, M.A.; Gohel, H.; Bhardwaj, S.K.; Srinivasan, S.; Hickman, N.; Kaushik, A. Aspects of nano-enabling biosensing systems for intelligent healthcare; towards COVID-19 management. *Mater. Today Chem.* **2020**, *17*, 100306. [CrossRef] [PubMed]
68. Shan, B.; Broza, Y.Y.; Li, W.; Wang, Y.; Wu, S.; Liu, Z.; Wang, J.; Gui, S.; Wang, L.; Zhang, Z.; et al. Multiplexed Nanomaterial-Based Sensor Array for Detection of COVID-19 in Exhaled Breath. *ACS Nano* **2020**, *14*, 12125–12132. [CrossRef] [PubMed]
69. Chen, H.-W.; Fang, Z.-S.; Chen, Y.-T.; Chen, Y.-I.; Yao, B.-Y.; Cheng, J.-Y.; Chien, C.-Y.; Chang, Y.-C.; Hu, C.-M.J. Targeting and Enrichment of Viral Pathogen by Cell Membrane Cloaked Magnetic Nanoparticles for Enhanced Detection. *ACS Appl. Mater. Interfaces* **2017**, *9*, 39953–39961. [CrossRef]
70. Roh, C.; Jo, S.K. Quantitative and sensitive detection of SARS coronavirus nucleocapsid protein using quantum dots-conjugated RNA aptamer on chip. *J. Chem. Technol. Biotechnol.* **2011**, *86*, 1475–1479. [CrossRef] [PubMed]
71. Konwar, A.N.; Borse, V. Current status of point-of-care diagnostic devices in the Indian healthcare system with an update on COVID-19 pandemic. *Sensors Int.* **2020**, *1*, 100015. [CrossRef]
72. Asif, M.; Ajmal, M.; Ashraf, G.; Muhammad, N.; Aziz, A.; Iftikhar, T.; Wang, J.; Liu, H. The role of biosensors in COVID-19 outbreak. *Curr. Opin. Electrochem.* **2020**, *23*, 174–184. [CrossRef]
73. Kim, H.; Park, M.; Hwang, J.; Kim, J.H.; Chung, D.R.; Lee, K.S.; Kang, M. Development of label-free colorimetric assay for MERS-CoV using gold nanoparticles. *ACS Sens.* **2019**, *4*, 1306–1312. [CrossRef]

74. Teengam, P.; Siangproh, W.; Tuantranont, A.; Vilaivan, T.; Chailapakul, O.; Henry, C.S. Multiplex paper-based colorimetric DNA sensor using pyrrolidinyl peptide nucleic acid-induced AgNPs aggregation for detecting MERS-CoV, MTB, and HPV oligonucleotides. *Anal. Chem.* **2017**, *89*, 5428–5435. [CrossRef]
75. Abdul, W.; Muhammad, A.; Ullah, K.A.; Asmat, A.; Abdul, B. Role of nanotechnology in diagnosing and treating COVID-19 during the Pandemi. *Int. J. Clin. Virol.* **2020**, *4*, 065–070. [CrossRef]
76. Layqah, L.A.; Eissa, S. An electrochemical immunosensor for the corona virus associated with the Middle East respiratory syndrome using an array of gold nanoparticle-modified carbon electrodes. *Microchim. Acta* **2019**, *186*, 224. [CrossRef]
77. Ishikawa, F.N.; Chang, H.K.; Curreli, M.; Liao, H.; Olson, A.; Chen, P.C.; Zhang, R.; Roberts, R.W.; Sun, R.; Cote, R.J. Label-free, electrical detection of the SARS virus N-Protein with nanowire biosensors utilizing antibody mimics as capture probes. *ACS Nano* **2009**, *3*, 1219–1224. [CrossRef] [PubMed]
78. Katz, R.; Graeden, E.; Abe, K.; Attal-Juncqua, A.; Boyce, M.; Eaneff, S. Mapping stakeholders and policies in response to deliberate biological events. *Heliyon* **2018**, *4*, e01091. [CrossRef] [PubMed]
79. Ahmed, S.R.; Nagy, É.; Neethirajan, S. Self-assembled star-shaped chiroplasmonic gold nanoparticles for an ultrasensitive chiro-immunosensor for viruses. *RSC Adv.* **2017**, *7*, 40849–40857. [CrossRef]
80. Rudramurthy, G.R.; Swamy, M.K.; Sinniah, U.R.; Ghasemzadeh, A. Nanoparticles: Alternatives against drug-resistant path-ogenic microbes. *Molecules* **2016**, *21*, 836. [CrossRef]
81. Rogers, J.V.; Parkinson, C.V.; Choi, Y.W.; Speshock, J.L.; Hussain, S.M. A preliminary assessment of silver nanoparticle inhibition of monkeypox virus plaque formation. *Nanoscale Res. Lett.* **2008**, *3*, 129–133. [CrossRef]
82. Huang, C.; Wang, Y.; Li, X.; Ren, L.; Zhao, J.; Hu, Y.; Zhang, L.; Fan, G.; Xu, J.; Gu, X. Clinical features of patients infected with 2019 novel coronavirus in Wuhan, China. *Lancet* **2020**, *395*, 497–506. [CrossRef]
83. Lee, E.C.; Nguyen, C.T.; Strounina, E.; Davis-Poynter, N.; Ross, B.P. Structure–activity relationships of GAG mimet-ic-functionalized mesoporous silica nanoparticles and evaluation of acyclovir-loaded antiviral nanoparticles with dual mechanisms of action. *ACS Omega* **2018**, *3*, 1689–1699. [CrossRef]
84. Bromberg, L.; Bromberg, D.; Hatton, T.A.; Bandín, I.; Concheiro, A.; Alvarez-Lorenzo, C. Antiviral Properties of Polymeric Aziridine- and Biguanide-Modified Core–Shell Magnetic Nanoparticles. *Langmuir* **2012**, *28*, 4548–4558. [CrossRef] [PubMed]
85. Cavezzi, A.; Troiani, E.; Corrao, S. COVID-19: Hemoglobin, Iron, and Hypoxia beyond Inflammation. A Narrative Review. *Clin. Pract.* **2020**, *10*, 24–30. [CrossRef] [PubMed]
86. Comparetti, E.J.; Pedrosa, V.; Kaneno, R. Carbon Nanotube as a Tool for Fighting Cancer. *Bioconjugate Chem.* **2018**, *29*, 709–718. [CrossRef] [PubMed]
87. Mangum, J.B.; Turpin, E.A.; Antao-Menezes, A.; Cesta, M.F.; Bermudez, E.; Bonner, J.C. Single-Walled Carbon Nanotube (SWCNT)-induced interstitial fibrosis in the lungs of rats is associated with increased levels of PDGF mRNA and the formation of unique intercellular carbon structures that bridge alveolar macrophages In Situ. *Part. Fibre Toxicol.* **2006**, *3*, 15. [CrossRef] [PubMed]
88. Hussain, S.; Ji, Z.; Taylor, A.J.; Graff, L.M.D.; George, M.; Tucker, C.J.; Chang, C.H.; Li, R.; Bonner, J.C.; Garantziotis, S. Multiwalled carbon nanotube functionalization with high molecular weight hyaluronan significantly reduces pulmonary injury. *ACS Nano* **2016**, *10*, 7675–7688. [CrossRef]
89. Telwatte, S.; Moore, K.; Johnson, A.; Tyssen, D.; Sterjovski, J.; Aldunate, M.; Gorry, P.R.; Ramsland, P.A.; Lewis, G.R.; Paull, J.R.; et al. Virucidal activity of the dendrimer microbicide SPL7013 against HIV-1. *Antivir. Res.* **2011**, *90*, 195–199. [CrossRef]
90. Landers, J.J.; Cao, Z.; Lee, I.; Piehler, L.T.; Myc, P.P.; Myc, A.; Hamouda, T.; Galecki, A.T.; Baker, J.R., Jr. Prevention of Influenza Pneumonitis by Sialic Acid–Conjugated Dendritic Polymers. *J. Infect. Dis.* **2002**, *186*, 1222–1230. [CrossRef]
91. Kim, Y.; Park, E.J.; Na, D.H. Recent progress in dendrimer-based nanomedicine development. *Arch. Pharmacal Res.* **2018**, *41*, 571–582. [CrossRef]
92. Puri, A.; Loomis, K.; Smith, B.; Lee, J.H.; Yavlovich, A.; Heldman, E.; Blumenthal, R. Lipid-based nanoparticles as pharmaceutical drug carriers: From concepts to clinic. *Crit. Rev. Ther. Drug Carr. Syst.* **2009**, *26*, 523–580. [CrossRef]
93. Pollock, S.; Nichita, N.B.; Böhmer, A.; Radulescu, C.; Dwek, R.A.; Zitzmann, N. Polyunsaturated liposomes are antiviral against hepatitis B and C viruses and HIV by decreasing cholesterol levels in infected cells. *Proc. Natl. Acad. Sci. USA* **2010**, *107*, 17176–17181. [CrossRef]
94. Shah, M.R.; Imran, M.; Ullah, S. *Lipid-Based Nanocarriers for Drug Delivery and Diagnosis*; William Andrew: Norwich, NY, USA, 2017.
95. Azmi, I.D.M.; Moghimi, S.M.; Yaghmur, A. Cubosomes and hexosomes as versatile platforms for drug delivery. *Ther. Deliv.* **2015**, *6*, 1347–1364. [CrossRef]
96. Zhai, J.; Fong, C.; Tran, N.; Drummond, C.J. Non-Lamellar Lyotropic Liquid Crystalline Lipid Nanoparticles for the Next Generation of Nanomedicine. *ACS Nano* **2019**, *13*, 6178–6206. [CrossRef]
97. Chuan, J.; Li, Y.; Yang, L.; Sun, X.; Zhang, Q.; Gong, T.; Zhang, Z. Enhanced rifampicin delivery to alveolar macrophages by solid lipid nanoparticles. *J. Nanopart. Res.* **2013**, *15*, 1–9. [CrossRef]
98. Singh, L.; Kruger, H.G.; Maguire, G.E.; Govender, T.; Parboosing, R. The role of nanotechnology in the treatment of viral infections. *Ther. Adv. Infect. Dis.* **2017**, *4*, 105–131. [CrossRef]

99. Dai, M.; Sui, B.; Xue, Y.; Liu, X.; Sun, J. Cartilage repair in degenerative osteoarthritis mediated by squid type II collagen via immunomodulating activation of M2 macrophages, inhibiting apoptosis and hypertrophy of chondrocytes. *Biomaterials* **2018**, *180*, 91–103. [CrossRef]
100. Hu, C.-M.J.; Chen, Y.-T.; Fang, Z.-S.; Chang, W.-S.; Chen, H.-W. Antiviral efficacy of nanoparticulate vacuolar ATPase inhibitors against influenza virus infection. *Int. J. Nanomed.* **2018**, *13*, 8579–8593. [CrossRef] [PubMed]
101. Shen, T.W.; Fromen, C.A.; Kai, M.P.; Luft, J.C.; Rahhal, T.B.; Robbins, G.R.; DeSimone, J.M. Distribution and Cellular Uptake of PEGylated Polymeric Particles in the Lung Towards Cell-Specific Targeted Delivery. *Pharm. Res.* **2015**, *32*, 3248–3260. [CrossRef]
102. Ajdary, M.; Moosavi, M.A.; Rahmati, M.; Falahati, M.; Mahboubi, M.; Mandegary, A.; Jangjoo, S.; Mohammadinejad, R.; Varma, R.S. Health Concerns of Various Nanoparticles: A Review of Their in Vitro and in Vivo Toxicity. *Nanomaterials* **2018**, *8*, 634. [CrossRef] [PubMed]
103. Singh, A.P.; Biswas, A.; Shukla, A.W.; Maiti, P. Targeted therapy in chronic diseases using nanomaterial-based drug delivery vehicles. *Signal. Transduct. Target. Ther.* **2019**, *4*, 1–21. [CrossRef] [PubMed]
104. Hirano, T.; Murakami, M. COVID-19: A New Virus, but a Familiar Receptor and Cytokine Release Syndrome. *Immunity* **2020**, *52*, 731–733. [CrossRef] [PubMed]
105. Santiago, L.A.; Hillaireau, H.; Grabowski, N.; Mura, S.; Nascimento, T.L.; Dufort, S.; Coll, J.-L.; Tsapis, N.; Fattal, E. Compared in vivo toxicity in mice of lung delivered biodegradable and non-biodegradable nanoparticles. *Nanotoxicology* **2016**, *10*, 292–302. [CrossRef] [PubMed]
106. Rosa, S.G.V.; Santos, W.C. Clinical trials on drug repositioning for COVID-19 treatment. *Rev. Panam. Salud Pública* **2020**, *44*, e40. [CrossRef] [PubMed]
107. Rundle, C.W.; Presley, C.L.; Militello, M.; Barber, C.; Powell, D.L.; Jacob, S.E.; Atwater, A.R.; Watsky, K.L.; Yu, J.; Dunnick, C.A. Hand hygiene during COVID-19: Recommendations from the American Contact Dermatitis Society. *J. Am. Acad. Dermatol.* **2020**, *83*, 1730–1737. [CrossRef]
108. Liu, J.-F.; Yu, S.-J.; Yin, Y.-G.; Chao, J.-B. Methods for separation, identification, characterization and quantification of silver nanoparticles. *TrAC Trends Anal. Chem.* **2012**, *33*, 95–106. [CrossRef]
109. Jeremiah, S.S.; Miyakawa, K.; Morita, T.; Yamaoka, Y.; Ryo, A. Potent antiviral effect of silver nanoparticles on SARS-CoV-2. *Biochem. Biophys. Res. Commun.* **2020**, *533*, 195–200. [CrossRef]
110. Oyeniyi, Y.; Mumuni, A. Formulation development of an herbal hand sanitizer containing Moringa olifera silver nanoparticles. *Braz. J. Technol.* **2021**, *4*, 36–49.
111. Peng, X.; Xu, X.; Li, Y.; Cheng, L.; Zhou, X.; Ren, B. Transmission routes of 2019-nCoV and controls in dental practice. *Int. J. Oral Sci.* **2020**, *12*, 9. [CrossRef]
112. Fathizadeh, H.; Maroufi, P.; Heravi, M.; Dao, S.; Ganbarov, K.; Pagliano, P.; Esposito, S.; Kafil, H.S. Protection and disinfection policies against SARS-CoV-2 (COVID-19). *Infez. Med.* **2020**, *28*, 185–191. [PubMed]
113. Du, W.; Zhang, Q.; Shang, Y.; Wang, W.; Li, Q.; Yue, Q.; Gao, B.; Xu, X. Sulfate saturated biosorbent-derived Co-S@ NC nanoarchitecture as an efficient catalyst for peroxymonosulfate activation. *Appl. Catal. B Environ.* **2020**, *262*, 118302. [CrossRef]
114. Shahzadi, S.; Zafar, N.; Sharif, R. Antibacterial Activity of Metallic Nanoparticles. *Bact. Pathog. Antibact. Control.* **2018**, *51*. [CrossRef]
115. Balagna, C.; Perero, S.; Percivalle, E.; Nepita, E.V.; Ferraris, M. Virucidal effect against coronavirus SARS-CoV-2 of a silver nanocluster/silica composite sputtered coating. *Open Ceram.* **2020**, *1*, 100006. [CrossRef]
116. Nanotechnology in Battle Against Coronavirus. Available online: https://statnano.com/nanotechnology-in-battle-against-coronavirus (accessed on 26 May 2021).
117. Palestino, G.; García-Silva, I.; González-Ortega, O.; Rosales-Mendoza, S. Can nanotechnology help in the fight against COVID-19? *Expert Rev. Anti-Infect. Ther.* **2020**, *18*, 849–864. [CrossRef] [PubMed]
118. Zhang, J.Y.; Zeng, H.; Gu, J.; Li, H.B.; Zheng, L.X.; Zou, Q.M. Progress and prospects on vaccine development against SARS-CoV-2. *Vaccines* **2020**, *8*, 153. [CrossRef]
119. Jiang, S.B.; Bottazzi, M.E.; Du, L.Y.; Lustigman, S.; Tseng, C.T.K.; Curti, E.; Jones, K.; Zhan, B.; Hotez, P.J. Roadmap to developing a recombinant coronavirus S protein receptor-binding domain vaccine for severe acute respiratory syndrome. *Expert Rev. Vaccines* **2012**, *11*, 1405–1413. [CrossRef]
120. Lu, R.; Zhao, X.; Li, J.; Niu, P.; Yang, B.; Wu, H.; Wang, W.; Song, H.; Huang, B.; Zhu, N.; et al. Genomic characterisation and epidemiology of 2019 novel coronavirus: Implications for virus origins and receptor binding. *Lancet* **2020**, *395*, 565–574. [CrossRef]
121. Zhou, P.; Yang, X.-L.; Wang, X.-G.; Hu, B.; Zhang, L.; Zhang, W.; Si, H.-R.; Zhu, Y.; Li, B.; Huang, C.-L.; et al. A pneumonia outbreak associated with a new coronavirus of probable bat origin. *Nature* **2020**, *579*, 270–273. [CrossRef] [PubMed]
122. Coleman, C.; Liu, Y.V.; Mu, H.; Taylor, J.K.; Massare, M.; Flyer, D.C.; Glenn, G.M.; Smith, G.E.; Frieman, M.B. Purified coronavirus spike protein nanoparticles induce coronavirus neutralizing antibodies in mice. *Vaccine* **2014**, *32*, 3169–3174. [CrossRef]
123. Takashima, Y.; Osaki, M.; Ishimaru, Y.; Yamaguchi, H.; Harada, A. Artificial Molecular Clamp: A Novel Device for Synthetic Polymerases. *Angew Chem. Int. Ed.* **2011**, *50*, 7524–7528. [CrossRef] [PubMed]
124. Tai, W.; He, L.; Zhang, X.; Pu, J.; Voronin, D.; Jiang, S.; Zhou, Y.; Du, L. Characterization of the receptor-binding domain (RBD) of 2019 novel coronavirus: Implication for development of RBD protein as a viral attachment inhibitor and vaccine. *Cell. Mol. Immunol.* **2020**, *17*, 613–620. [CrossRef]

125. Liu, C.; Zhou, Q.; Li, Y.; Garner, L.V.; Watkins, S.; Carter, L.J.; Smoot, J.; Gregg, A.C.; Daniels, A.D.; Jervey, S.; et al. Research and Development on Therapeutic Agents and Vaccines for COVID-19 and Related Human Coronavirus Diseases. *ACS Cent. Sci.* **2020**, *6*, 315–331. [CrossRef]
126. Chen, W.H.; Strych, U.; Hotez, P.J.; Bottazzi, M.E. The SARS-CoV-2 vaccine pipeline: An overview. *Curr. Trop. Med. Rep.* **2020**, *7*, 61–64. [CrossRef]
127. Pardi, N.; Hogan, M.; Porter, F.W.; Weissman, D. mRNA vaccines—A new era in vaccinology. *Nat. Rev. Drug Discov.* **2018**, *17*, 261–279. [CrossRef]
128. Zeng, C.; Hou, X.; Yan, J.; Zhang, C.; Li, W.; Zhao, W.; Du, S.; Dong, Y. Leveraging mRNAs sequences to express SARS-CoV-2 antigens in vivo. *bioRxiv* **2020**. [CrossRef]
129. Reddy, S.; Andr, E.; Van Der Vlies, J.; Simeoni, E.; Angeli, V.; Randolph, G.J.; O'Neil, C.P.; Lee, L.K. Exploiting lymphatic transport and complement activation in nanoparticle vaccines. *Nat. Biotechnol.* **2007**, *25*, 1159–1164. [CrossRef]
130. Bachmann, M.F.; Jennings, G.T. Vaccine delivery: A matter of size, geometry, kinetics and molecular patterns. *Nat. Rev. Immunol.* **2010**, *10*, 787–796. [CrossRef] [PubMed]
131. Heesters, B.A.; Myers, R.C.; Carroll, M.C. Follicular dendritic cells: Dynamic antigen libraries. *Nat. Rev. Immunol.* **2014**, *14*, 495–504. [CrossRef]
132. Amanna, I.J.; Raué, H.-P.; Slifka, M.K. Development of a new hydrogen peroxide–based vaccine platform. *Nat. Med.* **2012**, *18*, 974–979. [CrossRef]
133. Hanson, M.C.; Crespo, M.P.; Abraham, W.; Moynihan, K.D.; Szeto, G.L.; Chen, S.H.; Melo, M.B.; Mueller, S.; Irvine, D.J. Nanoparticulate STING agonists are potent lymph node–targeted vaccine adjuvants. *J. Clin. Investig.* **2015**, *125*, 2532–2546. [CrossRef] [PubMed]
134. Ma, C.; Wang, L.; Tao, X.; Zhang, N.; Yang, Y.; Tseng, C.-T.K.; Li, F.; Zhou, Y.; Jiang, S.; Du, L. Searching for an ideal vaccine candidate among different MERS coronavirus receptor-binding fragments—The importance of immunofocusing in subunit vaccine design. *Vaccine* **2014**, *32*, 6170–6176. [CrossRef]
135. Wang, L.S.; Shi, W.; Joyce, M.G.; Modjarrad, K.; Zhang, Y.; Leung, K.; Lees, C.R.; Zhou, T.; Yassine, H.M.; Kanekiyo, M. Evaluation of candidate vaccine approaches for MERS-CoV. *Nat Commun.* **2015**, *6*, 1–11. [CrossRef] [PubMed]
136. Kato, T.; Takami, Y.; Deo, V.K.; Park, E.Y. Preparation of virus-like particle mimetic nanovesicles displaying the S protein of Middle East respiratory syndrome coronavirus using insect cells. *J. Biotechnol.* **2019**, *306*, 177–184. [CrossRef]
137. Darnell, M.E.R.; Subbarao, K.; Feinstone, S.M.; Taylor, D.R. Inactivation of the coronavirus that induces severe acute respir-atory syndrome, SARS-CoV. *J. Virol. Methods* **2004**, *121*, 85–91. [CrossRef]
138. Giovannini, G.; Haick, H.; Garoli, D. Detecting COVID-19 from Breath: A Game Changer for a Big Challenge. *ACS Sens.* **2021**, *6*, 1408–1417. [CrossRef] [PubMed]
139. Miripour, Z.S.; Sarrami-Forooshani, R.; Sanati, H.; Makarem, J.; Taheri, M.S.; Shojaeian, F.; Eskafi, A.H.; Abbasvandi, F.; Namdar, N.; Ghafari, H. Real-time diagnosis of reactive oxygen species (ROS) in fresh sputum by electrochemical tracing; correlation between COVID-19 and viral-induced ROS in lung/respiratory epithelium during this pandemic. *Biosens. Bioelectron.* **2020**, *165*, 112435. [CrossRef] [PubMed]
140. Lahiri, A.; Jha, S.S.; Bhattacharya, S.; Ray, S.; Chakraborty, A. Effectiveness of preventive measures against COVID-19: A systematic review of In Silico modeling studies in indian context. *Indian J. Public Health* **2020**, *64*, 156–S167. [CrossRef]
141. Dadras, O.; Alinaghi, S.A.S.; Karimi, A.; MohsseniPour, M.; Barzegary, A.; Vahedi, F.; Pashaei, Z.; Mirzapour, P.; Fakhfouri, A.; Zargari, G.; et al. Effects of COVID-19 prevention procedures on other common infections: A systematic review. *Eur. J. Med. Res.* **2021**, *26*, 1–13. [CrossRef]
142. Wilder-Smith, A.; Freedman, D.O. Isolation, quarantine, social distancing and community containment: Pivotal role for old-style public health measures in the novel coronavirus (2019-nCoV) outbreak. *J. Travel Med.* **2020**, *27*, 1–4. [CrossRef] [PubMed]
143. Gross, J.V.; Judith Mohren, J.; Erren, T.C. COVID-19 and healthcare workers: A rapid systematic review into risks and preventive measures. *BMJ Open* **2021**, *11*, e042270. [CrossRef]
144. Pereira de Oliveira, M.; Garcion, E.; Venisse, N.; BenoIt, J.P.; Couet, W.; Olivier, J.C. Tissue distribution of indinavir adminis-tered as solid lipid nanocapsule formulation in mdr1a (+/+) and mdr1a (-/-) CF-1 mice. *Pharm Res.* **2005**, *22*, 1898–1905. [CrossRef] [PubMed]
145. Rodriguez, B.; Asmuth, D.M.; Matining, R.M.; Spritzler, J.; Jacobson, J.; Mailliard, R.; Li, X.D.; Martinez, A.I.; Tenorio, A.; Lori, F. Safety, tolerability, and immunogenicity of repeated doses of DermaVir, a candidate therapeutic HIV vaccine, in HIV-infected patients receiving combination antiretroviral therapy: Results of the ACTG 5176 trial. *J. Acquir Immune Defic.* **2013**, *64*, 351–359. [CrossRef]
146. Orkin, C.; Squires, K.E.; Molina, J.-M.; Sax, P.E.; Wong, W.-W.; Sussmann, O.; Kaplan, R.; Lupinacci, L.; Rodgers, A.; Xu, X.; et al. Doravirine/Lamivudine/Tenofovir Disoproxil Fumarate is Non-inferior to Efavirenz/Emtricitabine/Tenofovir Disoproxil Fumarate in Treatment-naive Adults with Human Immunodeficiency Virus–1 Infection: Week 48 Results of the DRIVE-AHEAD Trial. *Clin. Infect. Dis.* **2019**, *68*, 535–544. [CrossRef]
147. Wieland, S.F.; Vega, R.G.; Muller, R.; Evans, C.F.; Hilbush, B.; Guidotti, L.G.; Sutcliffe, J.G.; Schultz, P.G.; Chisari, F.V. Searching for interferon-induced genes that inhibit hepatitis B virus replication in transgenic mouse hepatocytes. *J. Virol.* **2003**, *77*, 1227–1236. [CrossRef] [PubMed]

148. Liang, T.J.; Block, T.M.; McMahon, B.J.; Ghany, M.G.; Urban, S.; Guo, J.-T.; Locarnini, S.; Zoulim, F.; Chang, K.-M.; Lok, A.S. Present and future therapies of hepatitis B: From discovery to cure. *Hepatology* **2015**, *62*, 1893–1908. [CrossRef] [PubMed]
149. Lingala, S.; Lau, D.T.-Y.; Koh, C.; Auh, S.; Ghany, M.G.; Hoofnagle, J.H. Long-term lamivudine therapy in chronic hepatitis B. *Aliment. Pharmacol. Ther.* **2016**, *44*, 380–389. [CrossRef] [PubMed]
150. Marcellin, P.; Chang, T.T.; Lim, S.G.; Tong, M.J.; Sievert, W.; Shiffman, M.I.; Jeffers, L.; Goodman, Z.; Wulfsohn, M.S.; Xiong, S. Adefovir dipivoxil for the treatment of hepatitis B e antigen-positive chronic hepatitis B. *N. Engl. J. Med.* **2003**, *348*, 808–816. [CrossRef]
151. Szunerits, S.; Barras, A.; Khanal, M.; Pagneux, Q.; Boukherroub, R. Nanostructures for the Inhibition of Viral Infections. *Molecules* **2015**, *20*, 14051–14081. [CrossRef]
152. Barik, S. New treatments for influenza. *BMC Med.* **2012**, *10*, 104. [CrossRef]
153. Levina, A.S.; Repkova, M.N.; Mazurkova, N.A.; Zarytova, V.F. Nanoparticle-Mediated Nonviral DNA Delivery for Effective Inhibition of Influenza a Viruses in Cells. *IEEE Trans. Nanotechnol.* **2016**, *15*, 248–254. [CrossRef]
154. Hendricks, G.L.; Weirich, K.L.; Viswanathan, K.; Li, J.; Shriver, Z.H.; Ashour, J.; Ploegh, H.L.; Jones, E.A.K.; Fygenson, D.K.; Finberg, R.W. SialylneolactoN-tetraose c (LSTc)-bearing liposomal decoys capture influenza A virus. *J. Biol. Chem.* **2013**, *288*, 8061–8073. [CrossRef]
155. Cavalli, R.; Donalisio, M.; Bisazza, A.; Civra, A.; Ranucci, E.; Ferruti, P.; Lembo, D. Enhanced Antiviral Activity of Acyclovir Loaded into Nanoparticles. *Methods Enzymol.* **2012**, *509*, 1–19. [CrossRef]
156. Lembo, D.; Swaminathan, S.; Donalisio, M.; Civra, A.; Pastero, L.; Aquilano, D.; Vavia, P.; Trotta, F.; Cavalli, R. Encapsulation of Acyclovir in new carboxylated cyclodextrin-based nanosponges improves the agent's antiviral efficacy. *Int. J. Pharm.* **2013**, *443*, 262–272. [CrossRef]
157. Hu, R.; Li, S.; Kong, F.; Hou, R.; Guan, X.; Guo, F. Inhibition effect of silver nanoparticles on herpes simplex virus 2. *Genet. Mol. Res.* **2014**, *13*, 7022–7028. [CrossRef]
158. Dunning, J.; Sahr, F.; Rojek, A.; Gannon, F.; Carson, G.; Idriss, B.; Massaquoi, T.; Gandi, R.; Joseph, S.; Osman, H.K.; et al. Experimental Treatment of Ebola Virus Disease with TKM-130803: A Single-Arm Phase 2 Clinical Trial. *PLoS Med.* **2016**, *13*, e1001997. [CrossRef] [PubMed]
159. Broglie, J.J.; Alston, B.; Yang, C.; Ma, L.; Adcock, A.F.; Chen, W.; Yang, L. Antiviral Activity of Gold/Copper Sulfide Core/Shell Nanoparticles against Human Norovirus Virus-Like Particles. *PLoS ONE* **2015**, *10*, e0141050. [CrossRef]
160. Lammers, T.; Sofias, A.M.; Van Der Meel, R.; Schiffelers, R.; Storm, G.; Tacke, F.; Koschmieder, S.; Brümmendorf, T.H.; Kiessling, F.; Metselaar, J.M. Dexamethasone nanomedicines for COVID-19. *Nat. Nanotechnol.* **2020**, *15*, 622–624. [CrossRef] [PubMed]
161. Alshweiat, R.A.A.I.C.A.; Ambrus, R.; Csóka, I. Intranasal Nanoparticulate Systems as Alternative Route of Drug Delivery. *Curr. Med. Chem.* **2019**, *26*, 6459–6492. [CrossRef] [PubMed]
162. Costantino, H.R.; Illum, L.; Brandt, G.; Johnson, P.H.; Quay, S.C. Intranasal delivery: Physicochemical and therapeutic aspects. *Int. J. Pharm.* **2007**, *337*, 1–24. [CrossRef]
163. Marasini, N.; Kaminskas, L. Subunit-based mucosal vaccine delivery systems for pulmonary delivery—Are they feasible? *Drug Dev. Ind. Pharm.* **2019**, *45*, 882–894. [CrossRef] [PubMed]
164. Murata, K.; Lechmann, M.; Qiao, M.; Gunji, T.; Alter, H.J.; Liang, T.J. Immunization with hepatitis C virus-like particles protects mice from recombinant hepatitis C virus-vaccinia infection. *Proc. Natl. Acad. Sci. USA* **2003**, *100*, 6753–6758. [CrossRef]
165. Quan, F.-S.; Compans, R.W.; Nguyen, H.H.; Kang, S.-M. Induction of Heterosubtypic Immunity to Influenza Virus by Intranasal Immunization. *J. Virol.* **2008**, *82*, 1350–1359. [CrossRef] [PubMed]
166. Lee, Y.T.; Ko, E.J.; Lee, Y.; Kim, K.H.; Kim, M.C.; Lee, Y.N.; Kang, S.M. Intranasal vaccination with M2e5x virus-like particles induces humoral and cellular immune responses conferring cross-protection against heterosubtypic influenza viruses. *PLoS ONE* **2018**, *13*, e0190868. [CrossRef]
167. Henry, B.D.; Neill, D.; Becker, K.A.; Gore, S.; Bricio-Moreno, L.; Ziobro, R.; Edwards, M.J.; Mühlemann, K.; Steinmann, J.; Kleuser, B.; et al. Engineered liposomes sequester bacterial exotoxins and protect from severe invasive infections in mice. *Nat. Biotechnol.* **2015**, *33*, 81–88. [CrossRef] [PubMed]
168. Hu, C.-M.J.; Fang, R.H.; Copp, J.; Luk, B.T.; Zhang, L. A biomimetic nanosponge that absorbs pore-forming toxins. *Nat. Nanotechnol.* **2013**, *8*, 336–340. [CrossRef]
169. Keller, M.D.; Ching, K.; Liang, F.-X.; Dhabaria, A.; Tam, K.; Ueberheide, B.; Unutmaz, D.; Torres, V.J.; Cadwell, K. Decoy exosomes provide protection against bacterial toxins. *Nat. Cell Biol.* **2020**, *579*, 260–264. [CrossRef]
170. Zhang, P.; Chen, Y.; Zeng, Y.; Shen, C.; Li, R.; Guo, Z.; Li, S.; Zheng, Q.; Chu, C.; Wang, Z.; et al. Virus-mimetic nanovesicles as a versatile antigen-delivery system. *Proc. Natl. Acad. Sci. USA* **2015**, *112*, E6129–E6138. [CrossRef]
171. Wang, J.; Li, P.; Yu, Y.; Fu, Y.; Jiang, H.; Lu, M.; Sun, Z.; Jiang, S.; Lu, L.; Wu, M.X. Pulmonary surfactant-biomimetic nano-particles potentiate heterosubtypic influenza immunity. *Science* **2020**, *367*, eaau0810. [CrossRef]
172. Zhang, P.F.; Zhang, L.; Qin, Z.E.; Hua, S.; Guo, H.; Chu, C.; Lin, H.; Zhang, Y.; Li, W.; Zhang, X. Genetically engineered liposome-like nanovesicles as active targeted transport platform. *Adv. Mater.* **2018**, *30*, 242. [CrossRef] [PubMed]
173. Rao, L.; Xia, S.; Xu, W.; Tian, R.; Yu, G.; Gu, C.; Pan, P.; Meng, Q.F.; Cai, X.; Qu, D. Decoy nanoparticles protect against COVID-19 by concurrently adsorbing viruses and inflammatory cytokines. *Proc. Natl. Acad. Sci. USA* **2020**, *117*, 27141–27147. [CrossRef]
174. Zhang, Q.; Honko, A.; Zhou, J.; Gong, H.; Downs, S.N.; Vasquez, J.H.; Fang, R.H.; Gao, W.; Griffiths, A.; Zhang, L. Cellular nanosponges inhibit SARS-CoV-2 infectivity. *Nano Lett.* **2020**, *20*, 5570–5574. [CrossRef] [PubMed]

175. Yang, W.; Peters, J.I.; Williams, R.O., 3rd. Inhaled nanoparticles–A current review. *Int. J. Pharm.* **2008**, *356*, 239–247. [CrossRef]
176. Patton, J.S.; Byron, P.R. Inhaling medicines: Delivering drugs to the body through the lungs. *Nat. Rev. Drug Discov.* **2007**, *6*, 67–74. [CrossRef] [PubMed]
177. Cavalcanti, I.D.L.; de Fatima Ramos Dos Santos Medeiros, S.M.; Dos Santos Macedo, D.C.; Ferro Cavalcanti, I.M.; de Britto Lira Nogueira, M.C. Nanocarriers in the delivery of hydroxychloroquine to the respiratory system: An alternative to COVID-19? *Curr. Drug Deliv.* **2020**, *17*. [CrossRef] [PubMed]
178. Bai, S.; Thomas, C.; Ahsan, F. Dendrimers as a Carrier for Pulmonary Delivery of Enoxaparin, a Low-Molecular Weight Heparin. *J. Pharm. Sci.* **2007**, *96*, 2090–2106. [CrossRef] [PubMed]
179. Rudolph, C.; Lausier, J.; Naundorf, S.; Rosenecker, J. In vivo gene delivery to the lung using polyethylenimine and fractured polyamidoamine dendrimers. *J. Gene Med.* **2000**, *2*, 269–278. [CrossRef]

nanomaterials MDPI

Review

Synthesis of Cerium Oxide Nanoparticles Using Various Methods: Implications for Biomedical Applications

Mpumelelo Nyoka, Yahya E. Choonara, Pradeep Kumar, Pierre P. D. Kondiah and Viness Pillay *

Wits Advanced Drug Delivery Platform Research Unit, Department of Pharmacy and Pharmacology, School of Therapeutics Sciences, Faculty of Health Sciences, University of the Witwatersrand, 7 York Road, Parktown, Johannesburg 2193, South Africa; mpumelelo.nyoka1@students.wits.ac.za (M.N.); yahya.choonara@wits.ac.za (Y.E.C.); pradeep.kumar@wits.ac.za (P.K.); pierre.kondiah@wits.ac.za (P.P.D.K.)

* Correspondence: viness.pillay@wits.ac.za; Tel.: +27-11-717-2274

Received: 29 November 2019; Accepted: 14 January 2020; Published: 29 January 2020

Abstract: Cerium oxide nanoparticles have been used in a number of non-medical products over the years. The therapeutic application of these nanoparticles has mainly been due to their oxidative stress ameliorating abilities. Their enzyme-mimetic catalytic ability to change between the Ce^{3+} and Ce^{4+} species makes them ideal for a role as free-radical scavengers for systemic diseases as well as neurodegenerative diseases. In this review, we look at various methods of synthesis (including the use of stabilizing/capping agents and precursors), and how the synthesis method affects the physicochemical properties, their behavior in biological environments, their catalytic abilities as well as their reported toxicity.

Keywords: neurodegenerative disease; cerium oxide nanoparticles; Parkinson's disease; oxidative stress; physicochemical properties; blood-brain barrier; synthesis methods

1. Introduction

During the latter years of the 1960s, scientists dedicated to miniaturized delivery systems introduced nanoparticle-based drug delivery systems and vaccines [1]. Since that application of nanoparticulates into medicine, the use of nanomaterials has yielded great advances in the diagnosis and treatment of numerous pathologies [2,3]. Cerium belongs to a class of lanthanide metals in the periodic table [4]. In oxide form, cerium has a fluorite structure. The nanoscale form, cerium oxide nanoparticles retain the fluorite structure with oxygen deficiencies. This yields cerium oxide nanoparticles with (CeO^{2-x}) vacancies which provide sites for reduction-oxidation reactions. The arrangement of the surfaces of the fluorite structure determines the catalytic performance of the nanoparticles. The (100), (110), and (111) are the best possible surfaces on a cerium oxide nanocrystal (see Figure 1 below) [5]. These represent the lattice arrangement of ions (Ce^{3+} and O^{2-}) in the fluorite structure. The (111) and the (100) possess the o-terminal endings, while the (110) arrangement exposes the Ce center and the O ions [5,6]. These properties enable these nanoparticles to be very useful in industrial applications such as the removal of carbon monoxide, hydrocarbons, and nitric oxide species from the exhaust gas. This property is facilitated by the ability of cerium to occur in trivalent (Ce^{+3}) and tetravalent (Ce^{+4}) states [7,8]. Cerium oxide nanoparticles have been shown to mitigate oxidative stress damage, which has been linked to the development of neurodegenerative diseases such as Parkinson's and Alzheimer's disease. The ability of cerium oxide nanoparticles to switch between valency states enables them to mimic specific enzyme functions such as superoxide dismutase (SOD), Catalase (CAT) and phosphatase, oxidase peroxidase, and phosphotriesterase [9,10].

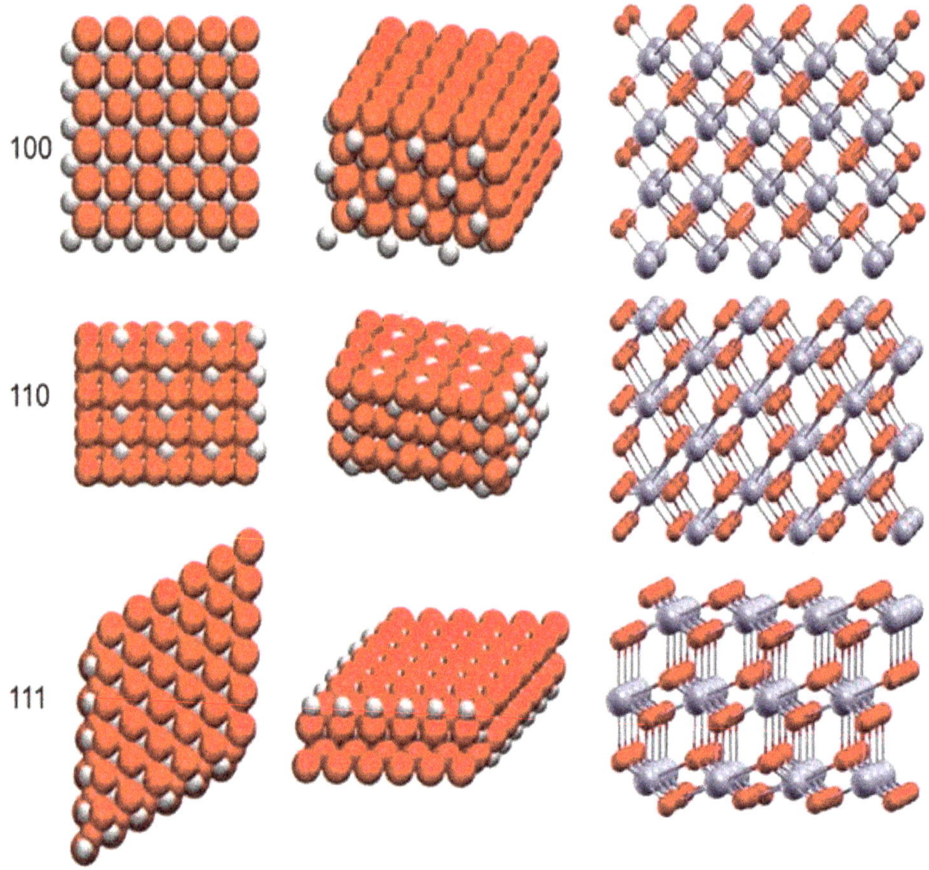

Figure 1. Crystal lattice and (100), (110), and (111) surfaces of cerium oxide nanoparticles [11], used with no changes under the terms of the Creative Commons Attribution 3.0 International License (Creative Commons Attribution 3.0 License).

Cerium oxide nanoparticles are cost-effective and maintain their catalytic properties under harsh environments (Figure 2). Although they are not naturally occurring enzymes, they possess powerful SOD-like activities. Their ability to scavenge reactive oxygen species in an enzyme-like fashion renders them a suitable biocompatible alternative for natural bio-scavengers such as superoxide dismutase [12]. The surface properties of nanoparticles impact on their enzyme mimetic activities. The formation of nanostructures of cerium oxide alters the oxygen non-stoichiometry ratio of Ce^{3+}/Ce^{4+} on the surface of nanostructures. The activity of cerium oxide nanoparticles correlates with the oxygen non-stoichiometry of nanoparticles. The SOD-like enzyme-mimetic activities of cerium oxide nanoparticles have been reported to be dependent on the Ce^{3+} fraction [10,13]. Another report evaluates the surface properties of cerium oxide nanoparticles using x-ray photoelectron spectroscopy (XPS) and UV-vis. The study illustrates that treatment of the nanoparticles with hydrogen peroxide decreases the Ce^{3+}/Ce^{4+} ratio which correlates directly with the loss of SOD-mimetic activity [13,14].

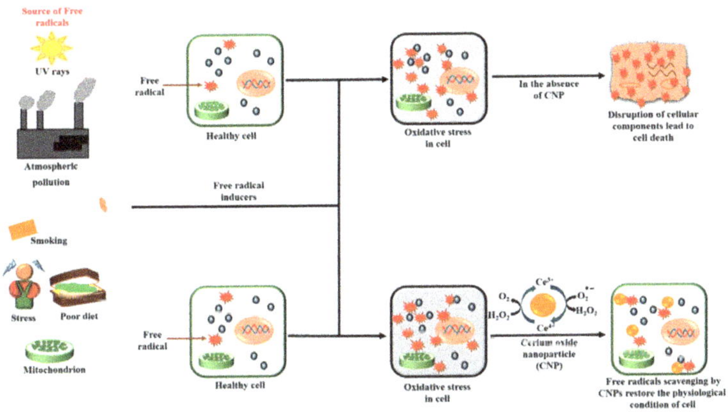

Figure 2. A depiction of some of the causes of oxidative stress and the redox action of cerium oxide nanoparticles (CNPs) [15], used with no changes under the terms of the Creative Commons Attribution 4.0 International License (http://creativecommons.org/publicdomain/zero/1.0/).

Cerium oxide nanoparticles have also shown catalase-mimicking activities. A recent study exploring the mimetic activities of these nanoparticles employed the pulsed electron evaporation as the method of synthesis. The catalase-mimicking activities were measured by the addition of hydrogen peroxide to the cerium oxide nanoparticle (CNP) suspension and was measured at the 380 nm absorption. Vazirov and coworkers suggested that surface doping of metals onto the CNPs increases the Ce^{3+}/Ce^{4+} ratio and the number of vacancies on the nanoparticle surface [16]. This consequently increases the points of interaction and reaction on the surfaces of the nanoparticles. The catalase-like activities are dependent on the Ce^{4+} fraction. A study assessed the catalase-like activities of CNPs using the Amplex Red assay to detect the physiological levels/concentration of peroxide. The test demonstrated catalase-like activity in the suspension with high levels of Ce^{4+} compared to a suspension with high Ce^{3+} [14,17].

In general, phosphorylation and dephosphorylation mechanisms are very valuable in the regulation of critical physiological conditions. The process involves the addition and removal of phosphate groups for energy maintenance. This, in turn, is the fundamental process in the synthesis of adenosine triphosphate (ATP) molecule. Adenosine diphosphate (ADP) is the hydrolyzed form of ATP. Upon hydrolysis, energy (H^+) is released and the phosphate is used in multiple cellular applications such as cellular physiological and metabolism. Cerium oxide nanoparticles and HMT-CNPs (Hexamethylenetetramine-cerium oxide nanoparticles) were tested for phosphate-like activity. HMT-CNPs decrease the levels of ATPs, thus exhibiting substantial ATPase (phosphatase) activity. This causes a decrease in cell viability which leads to an increase ADP, resulting in toxicity [16].

Moreover, pro and antioxidant properties are dependent on the method of synthesis, the ratio of +3/+4 and their ability to switch between the two species, which most studies have found to be the increase of this ratio that provides (+3) antioxidant properties after internalization of the nanoceria [9,18].

2. Synthesis of Cerium Oxide Nanoparticles and Their Properties

Studies have demonstrated different methods for the synthesis of cerium oxide nanoparticles for different applications. The differences in the conditions of synthesis influence the end product. This then implies that the resulting nanostructures will possess different physical/morphological as well as chemical properties, thus affecting the behavior of each. It is very important to consider the intended application of the nanostructures in order to determine the method of synthesis. The final features of the nanostructures are very important, especially in the medical application, as their properties affect the interaction at the biological interface [19].

While numerous studies have documented the therapeutic and beneficial effects of cerium oxide nanoparticles, some studies have suggested that cerium oxide nanoparticles may exhibit toxic and harmful effects on cells [20,21]. However, there are studies that show that the resultant properties such as pro/antioxidant and toxicity of cerium oxide are mainly dependent on synthesis conditions such as pH, temperature, and method of synthesis which confer behavior-altering physicochemical properties such as size/agglomeration, morphology, surface chemistry, and zeta potential [22–25]. These properties then dictate the behavior and therapeutic effectiveness upon application.

2.1. *Size of Nanoparticles*

In general, the size of nanoparticles enables the delivery of therapeutics across biological membranes without compromising the integrity of the membrane. Changes in nanoparticle sizes directly affect their characteristics and behavior. These changes influence biological parameters such as the biological half-life, diffusivity, and immunogenicity [26]. Decreases in the sizes of the particles increases the surface area to volume ratio. The large surface to volume ratio influences the catalytic properties of cerium oxide nanoparticles. The large surface area to volume ratio of minute nanoparticles enables particles to entrap free radicals on the surface [27]. Studies suggest that as the size of the cerium oxide nanoparticles decreases it promotes the reducibility and high oxygen storage capacity [28]. The increase in the oxygen vacancies enhances the movement of oxygen through the crystal lattice, thus assisting in the ability of ceria nanoparticles to oxidize or reduce molecules. It has been reported that a decrease in the size of cerium oxide nanoparticles correlates with an expansion of the lattice [15]. This arrangement increases the oxygen vacancies on the surfaces of the nanoparticles [29].

A study by Morones and coworkers demonstrated that the smaller the size of the nanoparticles the more potent the bacterial properties of silver nanoparticles [30]. Eriksson and coworkers demonstrated that cerium oxide nanoparticles between 3–5 nm offer exceptional antioxidative properties when doped into gadolinium for MRI imaging [26]. Studies have also identified size as one of the causes of nanoparticle toxicity. For these reasons, the size of nanoparticles is important in engineering the desired effects of nanotherapeutics. Zhang et al. were able to control the sizes of cerium oxide nanoparticles by adjusting a couple of parameters. In order to obtain particle sizes of between 3–12 nm, they adjusted the reaction times. However, in order to obtain larger nanoparticles, they adjusted temperature was adjusted between 400–800 °C for 30 min [29]. Dhall and Self suggested that careful consideration of the interaction of cerium oxide nanoparticles with the biological milieu is needed when synthesizing nanoparticles for therapeutic intervention [31]. It has also been suggested that the size of the nanoparticles contributes in the vascular clearance of cerium oxide nanoparticles by the immune system. "Bare" or un-coated cerium oxide nanoparticles tend to aggregate, which increases the size. However, applying a stabilizer prevents aggregation and increases the retention time in the body. This increase in size due to aggregation causes rapid clearance by the reticuloendothelial system [32]. Xu and Qu suggested that the body does not possess mechanisms for clearance of cerium oxide as it is not naturally produced in the human body [33]. However, when cerium oxide nanoparticles are ingested, they are not quickly absorbed and they are excreted via feces [32]. Negatively charged cerium oxide nanoparticles under the size of 6 nm have been also found to be excreted via the renal system [34].

2.2. *Aggregation and Agglomeration of the Particle System*

The use and meaning of the words 'aggregation' and 'agglomeration' are discipline-specific. Here we define 'aggregation' as strongly bonded or fused collection of particles while 'agglomeration' involves loosely clumped nanoparticles (or aggregates) held by weaker van der Waals forces. The biological environment and nanoparticle physicochemical characteristics (e.g., surface properties, particle size) are believed to influence the agglomeration and aggregation of the nanoparticles [35,36].

The Brownian motion is a major factor in this physical behavior as it causes continuous collisions between the particles. As a result of this agglomeration occurs as the energy of attraction exceeds the energy of repulsion [37]. The forces involved in the collisions include Born repulsion, diffuse double layer potential, and van der Waals attraction. It is also known that aggregate size in the solution depends on properties such as initial particle size and concentration [38]. Also, the aggregate size may vary among different particle types, zinc oxide nanoparticles (NPs) dispersed in aqueous solutions aggregate in a wide range of sizes [39–41] whereas TiO2 NPs showed a uniform distribution and agglomeration [42].

2.3. Particle Morphology

The particle morphology is one of the vital physical properties that are to be considered for efficacious nano-therapeutics in biological systems (Figure 3) [19]. Interaction between nanoparticles and biological components can be affected by the morphology of the nanoparticles. Studies have suggested that the method of synthesis may affect particle morphology. Nanoparticles exist in various shapes including spherical, polygonal, cube, or rod shapes. Nanoparticles with sharp edges are less biocompatible as they may inflict mechanical damage on cell membranes and organelles. Particle shape is known to play an important role in the fate and behavior of manufactured NPs into their environment. This could be as a result of the differences in diffusion rates of the material change with the aspect ratio of the material (e.g., higher drag on a tubular structure compared to a perfect sphere) or because of steric hindrance in the collisions as the morphology limits inter-particular interactions [43]. Several reports have addressed the role of shape and size on cellular internalization [44–46].

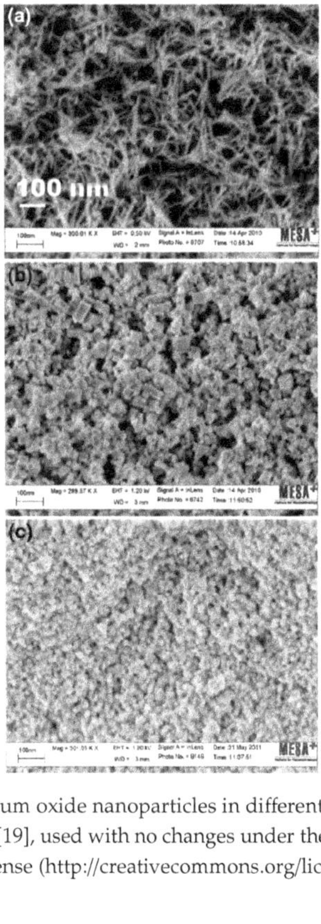

Figure 3. HRSEM images of cerium oxide nanoparticles in different morphologies; (**a**) nanorods, (**b**) nanocubes, and (**c**) nanospheres [19], used with no changes under the terms of the Creative Commons Attribution 4.0 International License (http://creativecommons.org/licenses/by/4.0/).

2.4. Chemical Composition

The chemical composition of the medium is known to have a crucial impact on the electrostatic surface charge of the nanoparticles, thereby affecting the rate at which these nanoparticles agglomerate/aggregate and thus affecting nanoparticle stability. Most of the manufactured nanoparticles nowadays are coated with surfactants to increase the stability of the suspension. The presence of a surface coating on manufactured NPs may significantly modify their surface chemistry, compared with the uncoated equivalents [18,47]. Similar kinds of changes may happen when pristine particles are treated with complex media such as humic acids. For example, adsorption of humic acids on the surface of small aggregates of silver NPs is known to result in the disaggregation of the NPs [48]. Similarly, using stabilizers on the surfaces of cerium oxide nanoparticle limits inter-particle interactions which may result in aggregation [21]. The authors also suggest that the application of stabilizers prevents agglomeration-induced toxicity. Andreescu and colleagues suggested that the environment to which cerium oxide nanoparticles are exposed to may affect their surface chemistry. The interaction of the hydrogen ions in the solution as well as the oxygen atoms in the facets of the lattice alters the surface chemistry of the nanoparticles and consequently alters their catalytic abilities (see Figure 1). The authors also suggest that the nanoparticle surface chemistry is highly sensitized to components of an environment such as pH, proteins, oxidizing, and/reducing agents [49]. Szymanski and company showed that in a biological environment, cerium oxide nanoparticles showed different oxidation states in different parts of the cells or organelles. Nanoparticles in the intracellular environment were shown to have a high Ce^{3+}/Ce^{4+} ratio compared to those in the extracellular environment. This suggests that a noticeable reduction of the cerium oxide nanoparticles occurs in the cells. The authors postulate that differences in pH of the endocytic vesicles (pH = 7.4) versus lysosome (pH = 4.5) may also be the reason for these differences in surface chemistry [50]. Qi and company synthesized PPEG (Phosphonated-Polyethylene glycol oligomers) surface modified nano-powder for re-dispersion in aqueous solutions. The authors showed that the coated cerium oxide nanoparticles were successfully re-dispersed in various organic solvents such as acetone, chloroform, and ethanol. The complexing of the oligomer increased the stability of these nanoparticles when compared to their uncoated/bare counterparts [51]. These examples suggest that the chemical composition of the environment to which cerium oxide nanoparticles are exposed, impacts the surface chemistry and ultimately the behavior of the nanoparticles.

2.5. Surface Chemistry and Physical Properties

Redox state: Redox reactions (oxidation and reduction processes) occurring on the surface of nanoparticles result in an altered crystalline nature. For example, cerium oxide NPs. Cerium occurs in both trivalent (III) as well as tetravalent (IV) states and has the unique ability to switch readily between these two states [18,52]. This low energy change confers unique catalytic properties on the cerium oxide nanoparticles [53]. However, it is known that the oxidation state is spatially variable within an individual particle and is dependent on size [54], so understanding of the redox mechanism on the particle surface is crucial.

Zeta potential: 'zeta potential' is the potential difference between the dispersion medium and the layer of fluid attached to the dispersed particle and is often used as an analog for colloidal stability, although this is only relevant where NPs are charge stabilized.

Solubility/Dissolution: Dissolution is a dynamic process in which the contents of the dissolving substance migrate from the surface to the bulk solution through a diffusion layer [55]. The thermodynamic parameter that controls this process is described as solubility [56]. Metal-based NPs such as zinc oxide are known to dissolve quickly and release ions that are themselves known to be toxic. Thus, the extent of dissolution and the relative toxicities of both the nanoparticulate and dissolved forms need to be considered to better understand the potential NP effects on organisms over time [57].

Surface functionalization of nanostructures has been used to synthesize non-toxic, stable, and biocompatible nano-therapeutics for applications in various conditions. Polymer chemistry in nano-therapeutics is a means of improving bioavailability, biocompatibility, solubility, as well as bio-retention of water-soluble and insoluble therapeutics for biomedical applications [58,59]. A number of these biodegradable polymers have been applied in the synthesis of multiple stable cerium oxide nanoparticle-systems [60,61]. Cerium oxide nanoparticles can be covered by hydroxyl groups. Therefore, biocompatible and biodegradable polymers that intrinsically possess hydroxyl moieties are capable of stabilizing the CNPs.

Applying the polymers as capping/stabilizing agents can assist in logically controlling the diameter of NPs [62]. The major challenge of suspending nanoparticles in solution is the ability to maintain their physicochemical properties. The surface charge of the nanoparticles, the low volume to surface area ratio of the nanoparticles causes them agglomerate, thus increasing the size of the particles from nano-range to micro-range [63,64]. This then negatively impacts on their therapeutic properties. In order to mitigate this, the surface coating is often employed to prevent the particles from agglomerating. This application of biocompatible coating enhances the solubility of the nanoparticles. Applying these coatings onto the surface of particles increases the stability and mitigates the toxicity of nanoparticles for efficient therapy.

2.6. Precipitation Method

Numerous studies have looked at various precursors to synthesize cerium oxide nanoparticles using the precipitation method, Table 1. Changes in any of the factors such as precursor, pH, and temperature alter the physicochemical properties of the cerium oxide nanoparticles. Researchers have investigated and reported on the precipitation method for the synthesis of cerium oxide nanoparticles using various chemical ingredients. Some of the reagents that are used in the precipitation method include precursors such as cerium nitrate hexahydrate and sodium hydroxide [26,65,66].

Table 1. Table showing the synthesis of cerium oxide nanoparticles using the precipitation method.

Capping Agent	Precursors	Particle Size (nm)	Morphology	Ref.
-	Cerium nitrate hexahydrate	9–18	Cubic Hexagonal	[67]
PVP	Cerium nitrate hexahydrate	27	Spherical	[67]
MEEETES Samarium	Cerium nitrate hexahydrate	10	Polyhedral	[68]
Ethylene glycol Samarium	Cerium nitrate hexahydrate	5–10	Square	[69]
Dextran	Cerium nitrate hexahydrate	3–5	Spherical	[70]
Sarium	Cerium nitrare hexahydrate	10–13	Spherical	[22,71,72]

Metal oxide powders are acquired by the addition of a ligand such as ammonia into solution of metal ions. In order to obtain a precipitate from this solution or reaction, the point of solubility must be surpassed. In a study by Chen and Chang, the effect of the atmosphere, as well as the reaction temperature in the synthesis of cerium oxide nanoparticles, were studied. The precursor aqueous cerium nitrate and ammonia water were used. The temperature, stirring rate, the atmosphere in the reaction chamber, and the pH of 9.5 was set. The results suggest that increases in temperature change the morphology of the nanoparticles from cubic to hexagonal. Increases in oxygen content or/and decrease in temperature of the reaction atmosphere decreases the size of the nanoparticles [68].

The application of surfactants, capping, or surface doping agents has also been demonstrated in controlling the physicochemical and even catalytic properties of nanoparticles. In a study by Ramachandran and coworkers sought to synthesize cerium oxide nanoparticles with a preferred morphology. It is to be noted that controlling the resulting morphology of nanostructures can be a challenging task. In order to synthesize the nanoparticles, the pH was varied between one and 12 in order to investigate its effect on nanoparticle properties. The synthesis was carried out using cerium nitrate, polyvinyl/pyrrilidone as a surfactant. The synthesis was conducted at different pH ranges of between nine and 12. The XPS results suggest that at the pH of 12, the Ce^{4+} ion was abundant on the surfaces of the particles in the sample. The results also showed that the size of nanoparticles with spherical morphology decreases as the pH increases towards the pH 12 [69].

Cerium oxide nanoparticles possess catalytic, optical, and luminescence properties that have been assessed in imaging for therapeutics. A number of studies have proven that doping metals onto the surfaces of nanoparticles may enhance the properties of the nanoparticles. In order to improve these imaging qualities while maintaining the oxygen vacancies that provide them with their catalytic properties, Nanda sought to surface modify cerium oxide nanoparticles. In their study, they used ammonia-induced ethylene glycol-assisted precipitation to synthesize Samarium-doped cerium oxide nanoparticles (Sm-doped cerium oxide nanoparticles). The dynamic light scattering results from this study revealed that the nanoparticles had a hydrodynamic size of 236 nm for the Sm-doped nanoparticles. This was larger compared to that of the (6-{2-[2-(2-methoxy-ethoxy)-ethoxy]-ethoxy}-hexyl) triethoxysilane (MEEETES)-Sm-doped nanoparticles (MEEETES-Sm-doped nanoparticles), which has a hydrodynamic size of 116.4 nm [70]. Along with the doped cerium oxide nanoparticles, a hydrophilic MEEETES was used as a functional moiety on the surface. The application of biocompatible surface modifying moiety, MEEETES inhibits aggregation and enables dispersion in media. Their results comparing the zeta potential of the nanoparticles at day 0 and at day 180 showed that the zeta potential changed from between −11 and −12 mV at day 0 to between −14 and −13 mV on day 180 [71]. These insignificant changes over a period of 180 days suggests that the zeta potential of nanoparticle can be stabilized over time by the application of stabilizing/doping agents. These studies also suggest that using hydrophilic moieties such as MEEETES on the surface of the nanoparticles may influence not only the hydrodynamic size but also the stability of the zeta potential.

The pH is a vital factor to consider when synthesizing cerium oxide nanoparticles for therapeutic applications. The pH of environment to which the nanoparticles are applied can affect the catalytic properties of the nanoparticles. Therapeutic properties of dextran-coated nanoparticles were investigated in an in vitro study using osteosarcoma cells. The synthesis was carried out using cerium nitrate solution, 0.1 M dextran and ammonium hydroxide. The resulting product consisted of spherical particles of sizes between 3–5 nm. The investigation was carried out in an acidic pH of 3, 5, and 6 to replicate the acidic cancerous conditions, neutral pH of 7, and in an alkaline pH of 9. Characterization of the dextran-coated nanoparticles using XPS showed that the nanoparticles with the Ce^{3+} oxidation states on there were synthesized using this method. However, the effect of the pH was observed when the surfaces of the nanoparticles had Ce^{4+} in abundance at pH 6 and pH 7. The pH was also shown to have an effect on the zeta potential. The recorded zeta potential at pH 7 was 0.58 mV compared to that of 16.68 mV at the pH of 6 and 6.32 mV at pH 9. A study by Yan and company suggests that altering the synthesis pH can alter the morphology and increases the surface area of the nanoparticles. This shows that applying a capping agent such as dextran and controlling the synthesis conditions such as pH may yield cerium oxide nanoparticles with predetermined properties and may enhance the therapeutic properties [72–74].

The abovementioned studies of cerium oxide nanoparticle synthesis using the precipitation method show that a number of studies use cerium nitrate hexahydrate as a precursor for synthesis. The morphology was predominantly spherical. Employing the precipitation method produced nanoparticles with sizes 3–27 nm. The use of capping agents/solvents/doping agents narrowed the polydispersity of the nanoparticles. The studies also suggest that the synthesis conditions also play a vital role in the resulting nanoparticles. Therefore, careful control of the synthesis conditions such as temperature, concentration of the reactants, pH, atmosphere, as well as stabilizing agents influence the properties of nanoparticles when employing the simple and economic precipitation method.

2.7. Microemulsification Method

Micro-emulsion is a method that utilizes a polar aqueous medium (water) and a non-polar aqueous medium (oil) in the presence of a surfactant. This method produces nanoparticles of controlled size and structure. The application of this method has yielded a number of insoluble products such as nanoparticles, Table 2, quantum dots of metal oxides, metals, silica, as well as lanthanide fluorides [75–78].

Table 2. Table showing the synthesis of cerium oxide nanoparticles using the microemulsion method.

Capping Agent	Precursors	Particle Size (nm)	Morphology	Ref.
AOT DDAB DTAB Brij35	Cerium nitrate hexahydrate/ Cerous chloride	6–13 (surfactant) 21 (No surfactant)	Cubic	[75]
OP-10	Cerium nitrate hexahydrate	2–6	-	[77]
Hexamethyl tetraamine	Cerium nitrate hexahydrate	7–10	Spherical	[79]
Ethylene glycol Aβ	Cerium nitrate hexahydrate	3–5	Spherical	[80]

Properties of nanoparticles are crucial in their function. Some of those properties are surface area and oxygen storage capacity. Nanoparticles with high surface area and high oxygen storage capacity enhance the catalytic properties of cerium oxide nanoparticles. However, increased temperature treatment of between 600–800 °C increases particle size thus decreases the surface area. Bumajdad and coworkers sought to synthesize cerium oxide nanoparticles with exposed surfaces and high thermal stability. The microemulsion method included sodium bis(2-ethylhexyl) sulfosuccinate (AOT), di-n-didodecyldimethylammonium bromide (DDAB), and DTAB+Brij 35 as surfactants. The results suggested that the application of surfactants in the synthesis process yielded cerium oxide nanoparticles with high surface area compared to the surfactant-free synthesis. Moreover, application of surfactant such as DDAB increased the surface area and produced cerium oxide nanoparticles which are stable at high temperatures. The use of cerium nitrate hexahydrate and cerium chloride heptahydrate as precursors in the synthesis of nanoparticles, solvents, and surfactants were used at the temperature of 50 °C. After calcination at 400 °C, the XPS results suggest that the surface of the nanoparticles consists of both Ce^{4+} and Ce^{3+}. Upon calcination, the Ce^{4+}/Ce^{3+} ratio increased [75].

In a study by Kockrick and coworkers, the team employed the microemulsion method to synthesize cerium oxide nanoparticles. The nanoparticles were calcined at a temperature of between 100–600 °C and the results suggest a size-dependent catalytic activity. The increase in temperature also aligns with an increase in size of the nanoparticles and a decrease in catalytic activity. The size was narrowly controlled using surfactants to between 6–16 nm [76]. In another study, the pH treatment of the nanoparticles was conducted and the zeta potential was measured. The zeta potential of the nanoparticles synthesized using this method was negative compared to a positive zeta potential of nanoparticles obtained from hydrothermal synthesis. The use of NH_4OH resulted in the zeta potential of −16.26 mV. The study also found that positive zeta potential assists in protein adsorption as opposed to a negative zeta potential [78].

Arya and coworkers used cerium oxide nanoparticles in primary cortical cells as a means of protection against apoptosis via mitochondrial membrane depolarization. The authors also looked at the NADH/NAD+ and ATP concentration post cerium oxide nanoparticle treatment. Precursors such as cerium nitrate hexahydrate and hexamethyltetraamine were used in the microemulsion method of synthesis. The results reveal that spherical nanostructure with a diameter of between 7–10 nm was synthesized. The cortical primary cultures were treated with cerium oxide nanoparticles and the results suggest that cerium oxide nanoparticles prevented apoptosis by stabilizing the mitochondrial membrane potential [79].

Masui and coworkers prepared ultrafine cerium oxide nanoparticles with narrow size distribution and increased oxidative activity using the microemulsion method. The narrow size distribution was achieved by controlling the concentration of starting reagents. The ultra-fine nanoparticles synthesized by this microemulsion were not found to have the quantum-size effect. This was seen when the particle size had no direct and indirect optical energy gaps [80].

Cimini and coworkers investigated the neuroregenerative properties of anti-Aβ-conjugated polyethylene glycol (PEG)-coated cerium oxide nanoparticles in a model of oxidative stress-induced Alzheimer's disease. The XPS results showed that the amine-functionalized nanoparticles had mixed oxidation state (Ce^{4+}/Ce^{3+}) on the surface. The zeta potential of amine-functionalized was +16 mV for bare cerium oxide nanoparticles and −37 mV for PEG-functionalized nanoparticles [81].

These studies suggest that the temperature, capping agents, and pH in the synthesis of cerium oxide nanoparticles using the microemulsion method of synthesis can control and narrow the particle size, morphology, and catalytic activity of cerium oxide nanoparticles. The various studies mentioned above show that the particle size was limited to between 2–21 nm. The predominant shape of the nanoparticles is seen to be spherical.

2.8. Hydrothermal Method

Particle sizes of the cube-shaped nanoparticles increases as the concentration of the solvent increased from 10–15 M. The hydrothermal method used produced rod-shaped nanoparticles at temperatures between 70–100 degrees Celsius as shown in Table 3. At a constant temperature of 100 degrees Celsius, the size of the nanosized particles decreased. Upon characterization of the nanoparticles using the temperature program reduction (TPR), the results showed an increase in the catalytic activity of nanorods versus nanocubes. This was due to the high concentration of surface oxygen on the surfaces of the nanoparticles. The high Ce^{3+} on the surfaces of the nanoparticles was confirmed by XPS analysis [82].

Table 3. Table showing the synthesis of cerium oxide nanoparticles using the hydrothermal method.

Capping Agent	Precursors	Particle Size (nm)	Morphology	Ref.
-	Cerium nitrate hexahydrate	8–16	Cubes; rods	[79]
-	Cerium hydroxide/Cerium acetate	5–54	Cubes; amorphous	[81]
Citric acid	Cerium chloride	<5	Spherical	[82]
Trisodium phosphate dodecahydrate	Cerium nitrate hexahydrate	5–60	Rods; Cubes; octahedral	[83]
Dithio-polydopamine	Cerium nitrate hexahydrate	L = 60 d = 5.8	Rods	[80]
-	Cerium nitrare hexahydrate	3–5	Spherical; Cubes; Rods	[84]

In a study that applied the hydrothermal method, Tok and coworkers sought to investigate the use of cerium hydroxide and/or cerium acetate as precursors. Thermal degradation and the duration of synthesis on the shape and crystallinity were investigated. The authors explored the effects of acidic (pH 4) and basic (pH 10) pH on the synthesis using either precursor prior to hydrothermal treatment. The results suggest that the crystalline morphology of the cerium oxide nanoparticles synthesized from both precursors yielded cubic structured nanostructures across different time-points (0 h, 6 h, 12 h, 18 h, and 24 h). At the 24 h time-point, the results suggest that the cerium acetate-derived/synthesized nanostructures were larger compared to the cerium hydroxide-synthesized nanostructures. Thermal treatments of cerium hydroxide-synthesized nanostructures at 500 °C and 1000 °C for 2 h grew the crystalline size to 8.8 and 47.4 nm, respectively from 5–6 nm. Thermal treatment of the cerium acetate-synthesized grew the crystalline size to 17.7 and 53.6 nm, respectively from 10–15 nm. The result also showed that in both precursor-derived nanostructures, agglomeration was observed [83].

Masui and coworkers investigated the application of citric acid as a protectant against particle growth. The surface of the particles was homogeneously covered with citric acid. Large particles formed prior to hydrothermal treatment when using Ce (III) salts were smaller in comparison with synthesis involving citric acid. This is because of the adsorption of citric acid onto the surface of the nanoparticles, thus controlling the nucleation process. The particle sizes of 5 nm were obtained from this method, with less agglomeration observed [80].

Cerium oxide nanoparticles with different morphologies were synthesized using hydrothermal synthesis. Different synthesis parameters such as temperature and concentrations were applied in order to manipulate the morphology of the resulting nanoparticles. The resulting nanostructures were nanocubes, nanorods, nano-octahedral, and submicronic cerium oxide nanoparticles. Thus, suggesting that altering synthesis parameters does alter the morphology and size of the nanoparticles. An analysis of the nanoparticles using XPS showed an oxidation state of Ce^{3+} on the surfaces of less than 1% [84].

In a different study, Zhang and coworkers synthesized degradable and stimuli-responsive polymer-coated cerium oxide nanorods for cancer therapy. The authors used dithio-polydopamine for the coating of the nanorods. The hydrothermal method of synthesis using cerium nitrate in mili-Q water (18.2 MΩ). In order to target cancerous cells, a lactose derivative moiety was attached to the surface of the nanoparticles. The resulting nanoparticles (spherical, cube, and rods) had hydrodynamic radii of 370, 194, and 192 nm and a set potential of −9.98, −10.86, and −5.9 mV [85].

Singh and coworkers evaluated the morphology and size-dependent cellular internalization at the nano-bio interface. The synthesis method used was hydrothermal synthesis. Varying concentrations of cerium nitrate hexahydrate (0.45–0.6 M) in deionized water were used. The results suggest that the resulting nanoparticles had a diameter of 3–5 nm. The results also suggest that the concentration altered the shape of the nanoparticles from spherical to nanorod to nanocube [86].

A myriad of morphologies can result from employing the hydrothermal method. Depending on the type of solvents, stabilizing agents, and synthesis conditions, resultant nanoparticles possess different properties.

2.9. Green Synthesis Routes

The green synthesis method is an environmentally innocuous technique that employs naturally occurring substances such as plant extracts, sugars, biodegradable polymers, and microorganisms in the synthesis of nanoparticles, Table 4. These naturally occurring components function as reducing as well as capping agents [87,88].

Table 4. Table showing the synthesis of cerium oxide nanoparticles using the green synthesis method.

Capping Agent	Precursors	Particle Size (nm)	Morphology	Ref.
Acalypha indica	Cerium chloride heptahydrate	8–54	Eliptical spherical	[87]
Fructose; Glucose; Lactose	Ammonnium Cerium nitrate	2–6	Spherical/Agglomerate	[86]
Hibiscus Sabdariffa	Cerium nitrate hexahydrate	3.9	Amorphous	[89]
Egg White	Cerium nitrate hexahydrate	25	Spherical	[90]

In a study by Thovhogi and coworkers, physicochemical properties of cerium oxide nanoparticles synthesis from Hibiscus Sabdariffa flower extract. The green synthesis method yielded nano-spheres with a diameter of 3.9 nm. An XPS investigation of the oxidation states on the surface of nanoparticles post-functionalization was undertaken. The results revealed a low 3+/4+ ratio. This shows the ability of environmentally friendly-green synthesis to alter the physical properties of nanoparticles [89].

Cerium oxide nanopowder with spherical particle sizes of 25 nm was synthesized using fresh egg white as a capping agent. The fresh egg white was used to control the particle distribution of the nanopowder. The eco-friendly capping and stabilizing agent is bio-degradable and possesses numerous amino acids. The ovalbumin and lysosome are the two major proteins that are available on the egg white. The interaction of the egg white with water and its ability to associate well with metal ions make it suitable for application as a shape-controlling stabilizing agent. The cell-viability of periodontal fibroblast cells treated with a range of CNP concentration of between 12.5–800 μg/mL was done. The results showed no toxicity of CNP concentrations below 800 μg/mL [90].

2.10. Solvothermal Method

Solvothermal synthesis employs the use of organic solvents in a chamber under high pressure and temperature to produce nanomaterials of varying sizes, Table 5. Teflon-lined autoclaves are used as they can withstand high temperature and pressure. Properties such as size were regulated by manipulating the temperature, reaction time, the concentration of the reagents, as well as the type of solvent. The 1,4-butanediol prohibited particle growth because of its viscous nature compared to water. This suggests that the viscous nature of a capping agent/solvent influences particle growth upon nucleation [91,92].

Table 5. Table showing the synthesis of cerium oxide nanoparticles using the solvothermal method.

Capping Agent	Precursors	Particle Size (nm)	Morphology	Ref.
1,4-butanediol	Ceric ammonium nitrate	5–10	-	[92]
Ethylenediamine	Cerium nitrate hexahydrate	2.5–8	-	[91]
Ethylene glycol	Cerium nitrate hexahydrate	-	Plate; Spherical	[93]

Kar and coworkers sought to synthesize ultra-small, water-soluble cerium oxide nanoparticles with the application of ethylenediamine as a surface capping/stabilizing agent at room temperature. The relatively monodispersed nanoparticles of 2.5 nm diameter showed mixed valence on the nanoparticle surface. The ultra-small particle exhibited a mixed oxidation state/valence on the surface of the nanoparticles [94]. The use of a solvent controls the physical properties of the cerium oxide nanoparticles such as morphology, particle size, and crystal growth. In this study, solvents/media such as water, ethanol, and ethylene glycol at different compositions were used. These were used to control the morphology, oxygen storage capacity, and electrical properties of cerium oxide nanoparticles. Water produced plate-shaped nanoparticles, spherical-shaped cerium oxide nanoparticles with a decreased size of the nanoparticles were produced with a 70:30 (water:ethanol) and the 70:30 (water:ethylene glycol) which resulted in porous nanoparticles [93].

2.11. Sol-Gel Method

Precursors such as cerium nitrate hexahydrate $Ce(NO)_3.6H2O$ undergo express hydrolysis, resulting in the production of a cerium hydroxide. The metal hydroxide then undergoes condensation thus forming a gel, Table 6. The gel is then exposed to a drying process to yield the final product [95].

Table 6. Table showing the synthesis of cerium oxide nanoparticles using the solvothermal method.

Capping Agent	Precursors	Particle Size (nm)	Morphology	Ref.
Diphenyl ether/oleylamine	Cerium nitrate hexahydrate	1.2–35	Spherical; tadpole; wire	[95]
Methanol	Cerium chloride heptahydrate	8–30	Spherical; Sheet-like	[96]
Lu seeds	Cerium nitrate hexahydrate	21–32	-	[97]
Pullulan	Cerium nitrate hexahydrate	-	-	[98]
Carrageenan	Cerium nitrate hexahydrate	18–60–	Spherical	[99]

Yu and coworkers used cerium (III) nitrate in a reaction with diphenyl ether as well as surfactants such as oleylamine to produce nanoparticles with spherical, wire, and tadpole morphology. The synthesis of nanostructures with different morphologies was achieved by using cerium nitrate and diphenyl ether in the presence of other surfactants. To obtain spherical morphologies oleylamine was used at the temperature of 320 °C. Nanowires were obtained from the application of the co-surfactants, oleylamine, and oleic acid at the temperature of 320 °C. Longer nanowires were synthesized by the addition of more oleic acid. Tadpole-like nanostructures were synthesized by increasing the amount of aleylamine in the co-surfactant mixture. The morphology and size of these nanostructures were synthesized/controlled by altering experimental parameters such as surfactant molar ratio as well as the duration of the reaction [95]. Another study investigated the effect of methanol as a solvent and the temperature of 400 °C on the size of nanoparticle morphology and optical properties. The results showed that the high ratio of UV to visible emission suggested good nanocrystal quality, i.e., good surface density. Nanoparticles with uniform spherical morphology with a size of 8 nm were synthesized [96].

A solution of cerium nitrate hexahydrate in a mixture of distilled water and Lu seeds and calcined at between 400–600 °C. The biosynthesis of these nanoparticles using the Lu seeds provided nanoparticles with altered physicochemical properties. Elahi and coworkers attached the 99Tc radio-labeller onto the nanoparticles in order to track the biological mobility/bioavailability of the nanoparticles in vivo. The results show on agglomerated morphology of the nanoparticles with an average size of 21–32 nm aligning with the calcination temperature of between 400–600 °C. These suggest that calcination, as well as the calcination temperature, has an effect on particle size. The smallest size showed no significant cytotoxicity on A549 cells. A bio-distribution to the kidneys and poor uptake by the stomach and thyroid were observed on an in vivo study using Wistar rats [97].

Another in vivo study on Wistar rats investigated the toxic effects of sol-gel prepared cerium oxide nanoparticles. The oxidative stress status, biochemical as well as hematological parameters were investigated. Pullulan was used as a capping agent to the cerium oxide nanoparticles. The resulting cerium oxide nanoparticles had a diameter of 12 nm. The results showed that pullulan-mediated cerium oxide nanoparticles possessed no cytotoxic effects upon acute administration on Neuro2A cells. There were also no toxic effects observed in the liver, spleen, and other organs. There was no damaged reported on the relevant antioxidant enzymes. The cerium oxide nanoparticles showed oxidative stress-protective properties and no hemotoxic effects or tissue damage upon histological analysis [98].

The green synthesis of cerium oxide nanoparticle-embedded hydrogel was conducted. The hydrogel was synthesized from carrageen as a stabilizing agent for the sol-gel synthesis of cerium oxide nanoparticles. Nourmohammadi and coworkers investigated the effectivity of carrageen hydrogel as capping agents on cerium oxide nanoparticles. They also investigated the cytotoxicity of these hydrogel-embedded nanoparticles on WEHI 164 cancer cells. The synthesis was conducted using cerium nitrate hexahydrate in double-distilled water and carrageen powder. The resulting nanoparticles possessed a spherical morphology with an average size of 34 nm. The results also showed no significant signs of toxicity in vitro even at high concentrations. However, the cellular metabolic activity was decreased with an increase in dose [99].

In a different study, starch was used as a capping agent on CNPs. The 6 nm spherical CNPs obtained were shown to have no cytotoxic effects in an in vitro study using the Neuro2A cell line [96]. The van der Waals forces between the particles increase the particle size during storage and upon contact with media. This causes the particles to be toxic to cells. Therefore, the application of starch as a stabilizing agent limited the inter-particular interactions which often result in aggregation and an increase in particle size. The cell-viability studies demonstrated no toxic effects on CNPs at a concentration of 5 μg/mL.

2.12. *Ball Milling*

A laboratory aluminar-zirconia ceramic mill and zirconia balls (ZrO2wt % 95%, diameter 0.5–0.3 mm, Mohs hardness 9) are used to create nanosized cerium oxide powder of 50 nm from cerium oxide powder with average particle sizes of 7.4 μm [100].

In a study by Yadav and coworkers, nanoparticles were produced from micron-sized cerium powder following 30 h of high energy ball milling, Table 7. The process yielded spherical nanoparticles with sizes between 8–12 nm with no defects. The conditions that yielded optimum results were noted to be 30 h of high energy ball milling at ice-cold temperatures [100]. Another study applied poly acrylic acid (PAA) to use as a dispersant of the cerium oxide nanoparticles. Normally, ball milling has been shown to produce nanoparticles with a diameter of 210 nm [101]. Li and coworkers used hydrated cerium carbonate and sodium hydroxide during high energy ball milling. In a different study, an organic base was used to synthesize cerium oxide nanoparticles with a nearly-spherical shape as opposed to an inorganic base during ball milling [102]. In this study, by He and coworkers, the use of PAA reduced the size of the nanoparticles from 128 nm to 50 nm and provided more stable cerium oxide nanoparticle slurry [103].

Table 7. Table showing the synthesis of cerium oxide nanoparticles using the various synthesis methods.

Synthesis Method	Capping Agent	Precursors	Particle Size (nm)	Morphology	Ref.
Ball Milling	-	Cerium powder (5μm)	10	Spherical	[104]
Ball Milling	Tetrabutyl ammonium hydroxide	Cerium ammonium nitrate	5–56	Spherical	[100]
Flame spray pyrolysis	Folic acid	Cerium acetate hydrate	7	-	[102]
Reverse phase evaporation	Lipid anionic mixture	Cerium nitrate hexahydrate	12–230	Spherical	[103]

2.13. *Flame Spray Pyrolysis*

In an interesting study, Vassie and coworkers synthesized folic acid (FA)-coated cerium oxide nanoparticles by flame spray pyrolysis, Table 7. The aim was to investigate whether FA-coated cerium oxide nanoparticles can modulate intracellular activities and control ROS levels when compared to bare/uncoated cerium oxide nanoparticles. The liquid precursor, cerium 2-ethylhexanoate in xylene was used for the experiment. Using a syringe, the mixture was fed through the flame at a determined rate. Folic acid was also conjugated onto the nanoparticles to form FA-cerium oxide nanoparticles. The results show cytoplasmic localization of both the FA-cerium oxide nanoparticles and bare nanoparticles. There was also an increase in the uptake of the FA-coated nanoparticles in ovarian cancer cells that led to the induction of cell death via an increased concentration of ROS [105].

2.14. Reverse-Phase Evaporation

Liposome-entrapped cerium oxide nanoparticles were synthesized to enhance their therapeutic properties, Table 7. Cerium oxide nanoparticles with a size of 12 nm were loaded into liposomes. The sizes of nanoparticle-loaded liposomes were 230 nm. The liposomes enhanced the lipophilicity of the cerium oxide nanoparticles. The in vitro toxicity test was conducted on normal human dermal fibroblast (NHDF) cells. The study findings showed that the cerium oxide nanoparticle-loaded liposomes displayed colloidal stability and retained their antioxidant properties [106].

2.15. Reverse Micelle

An in vivo model of cardiotoxicity using rats was created to elucidate the ameliorative effects of cerium oxide nanoparticles, Table 7. The rat model was created by subcutaneous injection of isoproterenol hydrochloride (30 mg/kg) in saline. This study compared the ameliorative effects of Captopril and cerium oxide nanoparticles in oxidative-stress-induced cardiotoxicity. The cerium oxide loaded reverse micelles with sizes of 5 nm were synthesized. The results showed that cerium oxide-loaded micelles showed better prophylactic and ameliorative effects of cardio cytotoxicity compared with Captopril [107].

3. Toxicity and Cytotoxicity of Cerium Oxide Nanoparticles

The toxicity and cytotoxicity of cerium oxide has long been a point of contention among researchers exploring the application of cerium oxide nanoparticles in biomedicine. The lanthanide-derived nanoparticles have been applied immensely in industrial or non-medical applications. These applications include environmental chemistry, cosmetics, gas-emissions, fuel cells, oxygen sensors, and corrosion protectors [31,33].

Studies have suggested that the successful application of these metal oxides in the abovementioned fields may not infer positive application in biomedical systems. This is may be due to the complex nature of physiological systems. The properties of these nanoparticles and the manner in which interact with the biological milieu needs to be considered in order to synthesize biocompatible nanoparticles for biomedical application. This is because the properties and the manner in which these nanoparticles interact with the biological system may have toxic effects [108,109]. A study by Asati and coworkers investigated the surface charge-dependent internalization of polymer-coated cerium oxide nanoparticles into normal versus malignant cells. Transformed lung cells A549 and MCF-7 breast cells as well as untransformed HEK293 and H9c2 were chosen. Poly-(acrylic acid), aminated poly (acrylic acid), and/or dextran were applied as coating on the nanoparticles. These polymers furnished the nanoparticles with different charge characteristics such as negative, positive and neutral, respectively.

The internalization of the nanoparticles was observed in every cell type except the MCF-7 cell line. The negatively charged nanoparticles were internalized in the A549 cell line only and were not taken up by the MCF-7, thus the toxicity to the A549 lung carcinoma cells. Neutral nanoparticles were localized in the cytoplasm of all cell types in this experiment and were non-toxic. The study thus found that factors such as internalization, localization, and surface charge of the cerium oxide nanoparticles play an important role in the toxicity of cerium oxide nanoparticles [110]. In another study, alkaline-precipitation as well as the inverse micro-emulsion methods were employed to synthesize ultrafine cerium oxide nanoparticles. Li and coworkers synthesized water-insoluble (CeO2-P), water-soluble, dextran coated, PAA coated, and EDA coated nanoparticles were synthesized. The polymers provided different functional groups on nanoparticle surfaces such as –OH, –COOH, and -NH_2. The results suggest a concentration-dependent cytotoxicity. The results also suggest that the "bare" cerium oxide nanoparticles (without functional groups) induced acute cytotoxicity on BGC-803 cell lines. The dextran, PAA, and EDA coated nanoparticles showed some viability as high concentrations (20 mg/mL) [14].

The toxicity of cerium oxide nanoparticles has also been attributed to their size. The smaller the size the more toxic they are. This has been suggested to be due to the large surface area to volume ratio. This means they possess more catalytic activity on the surface, thus making them very reactive and consequently toxic. Their size may also result in increased internalization into the cells which may result in toxicity. The tendency of the cerium oxide nanoparticles to agglomerate has also been suggested to be one of the causes of their toxicity [111]. The interaction of these metal oxide nanoparticles with their biological milieu may induce toxic effects. The synthesis method and the storage conditions have also been implicated in their toxicity [112]. These studies suggest that the physico-chemical properties of cerium oxide nanoparticles as well as the manner they interact with the biological environment may cause them to be toxic.

4. Conclusions and Future Perspectives

These studies clearly demonstrate the therapeutic effects and the behavior of cerium oxide nanoparticles in a bio-relevant environment. They also demonstrate the effects of synthesis and surface modification on the therapeutic effects of cerium oxide nanoparticles in the biological interface. This highlights the importance of the synthesis method or post-synthesis modifications of cerium oxide nanoparticles. The application of biocompatible and biodegradable capping and stabilizing agents clearly prove necessary in the synthesis of nanoparticles that are able to exert therapeutic effects. These capping agents have to be considered when synthesizing biologically-relevant nanoparticles in order to avoid unwanted interaction between the nanoparticles as well as between the nanoparticles and their environment. More studies looking into the therapeutic effects of cerium oxide nanoparticles in systemic conditions caused inter alia by oxidative stress, inflammation, and bacteria. Therapeutic effects of these nanoparticles in diseases that require tissue regeneration (scaffolds) need to be further explored.

Author Contributions: M.N., Y.E.C., P.K., P.P.D.K., V.P. conceptualized the review; M.N. wrote the first draft; Y.E.C., P.K., P.P.D.K., V.P. reviewed and revised the manuscript draft; Y.E.C., P.K., P.P.D.K., V.P. supervised the project. All authors have read and agreed to the published version of the manuscript.

Funding: This work was funded by the National Research Foundation (NRF) of South Africa and the South African Medical Research Council (SAMRC).

Conflicts of Interest: The authors declare no conflicts of interest.

References

1. Kreuter, J. Nanoparticles—A Hist. *Perspect. Int J. Pharm* **2007**, *331*, 1–10. [CrossRef]
2. Attia, M.S.; Al-Radadi, N.S. Nano optical sensor binuclear Pt-2-pyrazinecarboxylic acid -bipyridine for enhancement of the efficiency of 3-nitrotyrosine biomarker for early diagnosis of liver cirrhosis with minimal hepatic encephalopathy. *Biosens. Bioelectron.* **2016**, *86*, 406–412. [CrossRef]
3. Panja, S. Biological application of green silver nanoparticle synthesized from leaf extract of Rauvolfia serpentina Benth. *Asian Pac. J. Trop. Dis.* **2016**, *6*, 549–556. [CrossRef]
4. Weeks, M.E. The discovery of the elements. *Xvi Rare Earth Elem. J. Chem. Educ.* **1751**, *1932*, 9. [CrossRef]
5. Yao, S.Y.; Xu, W.Q.; Johnston-Peck, A.C.; Zhao, F.Z.; Liu, Z.Y.; Luo, S.; Senanayake, S.D.; Martínez-Arias, A.; Liu, W.J.; Rodriguez, J.A. Morphological effects of nanostructured ceria supports on the activity and stability of CuO/ CeO2 catalysts for the water-gas shift reaction. *Phys. Chem. Chem. Phys.* **2014**, *16*, 17183–17195. [CrossRef] [PubMed]
6. Huang, W.; Gao, Y. Morphology-dependent surface chemistry and catalysis of CeO_2 nanocrystals. *Catal. Sci. Technol.* **2014**, *4*, 3772–3784. [CrossRef]
7. Medalia, A. Byrne Spectrophotometric, B. of Cerium (IV). *Anal. Chem.* **1951**, *23*, 453–456. [CrossRef]
8. Dowding, J.M.; Das, S.; Kumar, A.; Dosani, T.; McCormack, R.; Gupta, A.; Sayle, T.X.; Sayle, D.C.; von Kalm, L.; Seal, S.; et al. Cellular interaction and toxicity depend on physicochemical properties and surface modification of redox-active nanomaterials. *Acs Nano* **2013**, *7*, 4855–4868. [CrossRef]
9. Yang, Y.; Mao, Z.; Huang, W.; Liu, L.; Li, J.; Li, J.; Wu, Q. Redox enzyme-mimicking activities of CeO_2 nanostructures: Intrinsic influence of exposed facets. *Sci. Rep.* **2016**, *6*, 35344. [CrossRef]

10. Baldim, V.; Bedioui, F.; Mignet, N.; Margaill, I.; Berret, J.F. The enzyme-like catalytic activity of cerium oxide nanoparticles and its dependency on Ce3+ surface area concentration. *Nanoscale* **2018**, *10*, 6971–6980. [CrossRef]
11. Naaz, F.; Farooq, U.; Ahmad, T. Ceria as an Efficient Nanocatalyst for Organic Transformations. In *Nanocatalysts*; IntechOpen: London, UK, 2019.
12. Charbgoo, F.; Ahmad, M.B.; Darroudi, M. Cerium oxide nanoparticles: Green synthesis and biological applications. *Int. J. Nanomed.* **2017**, *12*, 1401–1413. [CrossRef] [PubMed]
13. Heckert, E.G.; Karakoti, A.S.; Seal, S.; Self, W.T. The role of cerium redox state in the SOD mimetic activity of nanoceria. *Biomaterials* **2008**, *29*, 2705–2709. [CrossRef] [PubMed]
14. Li, C.; Shi, X.; Shen, Q.; Guo, C.; Hou, Z.; Zhang, J. Hot Topics and Challenges of Regenerative Nanoceria in Application of Antioxidant Therapy. *J. Nanomater.* **2018**, *2018*, 1–12. [CrossRef]
15. Thakur, N.; Manna, P.; Das, J. Synthesis and biomedical applications of nanoceria, a redox active nanoparticle. *J. Nanobiotechnol.* **2019**, *17*, 84. [CrossRef] [PubMed]
16. Vazirov, R.A. Physicochemical characterization and antioxidant properties of cerium oxide nanoparticles. In *Journal of Physics Conference Series*; IOP Publishing: Bristol, UK, 2018.
17. Pirmohamed, T.; Dowding, J.M.; Singh, S.; Wasserman, B.; Heckert, E.; Karakoti, A.S.; King, J.E.; Seal, S.; Self, W.T. Nanoceria exhibit redox state-dependent catalase mimetic activity. *Chem. Commun. (Camb. Engl.)* **2010**, *46*, 2736–2738. [CrossRef]
18. Pulido-Reyes, G.; Rodea-Palomares, I.; Das, S.; Sakthivel, T.S.; Leganes, F.; Rosal, R.; Seal, S.; Fernández-Piñas, F. Untangling the biological effects of cerium oxide nanoparticles: The role of surface valence states. *Sci. Rep.* **2015**, *5*, 15613. [CrossRef]
19. Kovacevic, M.; Mojet, B.L.; van Ommen, J.G.; Lefferts, L. Effects of Morphology of Cerium Oxide Catalysts for Reverse Water Gas Shift Reaction. *Catal. Lett.* **2016**, *146*, 770–777. [CrossRef]
20. Kumari, M.; Singh, S.P.; Chinde, S.; Rahman, M.F.; Mahboob, M.; Grover, P. Toxicity study of cerium oxide nanoparticles in human neuroblastoma cells. *Int. J. Toxicol.* **2014**, *33*, 86–97. [CrossRef]
21. Berret, J.-F. In vitro toxicity of nanoceria: Effect of coating and stability in biofluids. *Nanotoxicology* **2014**, *8*, 799–811.
22. Alpaslan, E.; Geilich Benjamin, M.; Yazici Hilal Webster Thomas, J. pH-Controlled Cerium Oxide Nanoparticle Inhibition of Both Gram-Positive and Gram-Negative Bacteria Growth. *Sci. Rep.* **2017**, *7*, 45859. [CrossRef]
23. Kalyanaraman, V.; Naveen, S.V.; Mohana, N.; Balaje, R.M.; Navaneethakrishnan, K.R.; Brabu, B.; Murugan, S.S.; Kumaravel, T.S. Biocompatibility studies on cerium oxide nanoparticles-combined study for local effects, systemic toxicity and genotoxicity via implantation route. *Toxicol. Res. (Camb.)* **2019**, *8*, 25–37. [CrossRef] [PubMed]
24. Karakoti, A.S.; Singh, S.; Kumar, A.; Malinska, M.; Kuchibhatla, S.V.; Wozniak, K.; Self, W.T.; Seal, S. PEGylated nanoceria as radical scavenger with tunable redox chemistry. *J. Am. Chem. Soc.* **2009**, *131*, 14144–14145. [CrossRef] [PubMed]
25. Yang, X.; Pan, H.; Wang, P.; Zhao, F.J. Particle-specific toxicity and bioavailability of cerium oxide (CeO_2) nanoparticles to Arabidopsis thaliana. *J. Hazard. Mater.* **2017**, *322*, 292–300. [CrossRef] [PubMed]
26. Eriksson, P.; Tal, A.A.; Skallberg, A.; Brommesson, C.; Hu, Z.; Boyd, R.D.; Olovsson, W.; Fairley, N.; Abrikosov, I.A.; Zhang, X.; et al. Cerium oxide nanoparticles with antioxidant capabilities and gadolinium integration for MRI contrast enhancement. *Sci. Rep.* **2018**, *8*, 6999. [CrossRef]
27. Das, S.; Dowding, J.M.; Klump, K.E.; McGinnis, J.F.; Self, W.; Seal, S. Cerium oxide nanoparticles: Applications and prospects in nanomedicine. *Nanomedicine* **2013**, *8*, 1483–1508. [CrossRef]
28. Arul, N. Sabari, Strong quantum confinement effect in nanocrystalline cerium oxide. *Mater. Lett.* **2011**, *65*, 2635–2638. [CrossRef]
29. Zhang, F.; Chan, S.W.; Spanier, J.E.; Apak, E.; Jin, Q.; Robinson, R.D.; Herman, I.P. Cerium Oxide Nanoparticles: Size-Selective Formation and Structure Analysis. *Appl. Phys. Lett.* **2002**, *80*, 127–129. [CrossRef]
30. Morones, J.R.; Elechiguerra, J.L.; Camacho, A.; Holt, K.; Kouri, J.B.; Ramirez, J.T.; Yacaman, M.J. The bactericidal effect of silver nanoparticles. *Ssrn Electron. J.* **2005**, *16*, 2346–2353. [CrossRef]
31. Dhall, A.; Self, W. Cerium Oxide Nanoparticles: A Brief Review of Their Synthesis Methods and Biomedical Applications. *Antioxidants* **2018**, *7*, 97. [CrossRef]

32. Molina, R.M.; Konduru, N.V.; Jimenez, R.J.; Pyrgiotakis, G.; Demokritou, P.; Wohlleben, W.; Brain, J.D. Bioavailability, distribution and clearance of tracheally instilled, gavaged or injected cerium dioxide nanoparticles and ionic cerium. *Environ. Sci. Nano* **2014**, *1*, 561–573. [CrossRef]
33. Xu, C.; Qu, X. Cerium oxide nanoparticle: A remarkably versatile rare earth nanomaterial for biological applications. *Npg Asia Mater.* **2014**, *6*, e90. [CrossRef]
34. Walkey, C.; Das, S.; Seal, S.; Erlichman, J.; Heckman, K.; Ghibelli, L.; Traversa, E.; McGinnis, J.F.; Self, W.T. Catalytic Properties and Biomedical Applications of Cerium Oxide Nanoparticles. *Environ. Sci. Nano* **2015**, *2*, 33–53. [CrossRef] [PubMed]
35. Rasmussen, K.; Rauscher, H.; Mech, A.; Riego Sintes, J.; Gilliland, D.; Gonzalez, M.; Kearns, P.; Moss, K.; Visser, M.; Groenewold, M.; et al. Physico-chemical properties of manufactured nanomaterials-Characterisation and relevant methods. An outlook based on the OECD Testing Programme. *Regul. Toxicol. Pharm.* **2018**, *92*, 8–28. [CrossRef] [PubMed]
36. Farré, M.; Sanchís, J.; Barceló, D. Analysis and assessment of the occurrence the fate and the behavior of nanomaterials in the environment. *Trac. Trends Anal. Chem.* **2011**, *30*, 517–527. [CrossRef]
37. Azimi, S.S.; Kalbasi, M. A molecular dynamics simulation of Brownian motion of a nanoparticle in a nanofluid. *Nanoscale Microscale Thermophys. Eng.* **2017**, *21*, 263–277. [CrossRef]
38. Phenrat, T.; Saleh, N.; Sirk, K.; Tilton, R.D.; Lowry, G.V. Aggregation and Sedimentation of Aqueous Nanoscale Zerovalent Iron Dispersions. *Environ. Sci. Technol.* **2007**, *41*, 284–290. [CrossRef]
39. Pipan-Tkalec, Ž. Zinc bioaccumulation in a terrestrial invertebrate fed a diet treated with particulate ZnO or ZnCl2 solution. *Toxicology* **2010**, *269*, 198–203. [CrossRef]
40. Wang, H.; Wick, R.L.; Xing, B. Toxicity of nanoparticulate and bulk ZnO, Al_2O_3 and TiO_2 to the nematode Caenorhabditis elegans. *Env. Pollut.* **2009**, *157*, 1171–1177. [CrossRef]
41. Lin, D.; Xing, B. Root uptake and phytotoxicity of ZnO nanoparticles. *Env. Sci. Technol.* **2008**, *42*, 5580–5585. [CrossRef]
42. Jemec, A.; Drobne, D.; Remškar, M.; Sepčić, K.; Tišler, T. Effects of ingested nano-sized titanium dioxide on terrestrial isopods (Porcellio scaber). *Environ. Toxicol. Chem. Setac* **2008**, *27*, 1904–1914. [CrossRef]
43. Handy, R.D.; Von der Kammer, F.; Lead, J.R.; Hassellöv, M.; Owen, R.; Crane, M. The ecotoxicology and chemistry of manufactured nanoparticles. *Ecotoxicology* **2008**, *17*, 287–314. [CrossRef]
44. Barreneche, C. Influence of nanoparticle morphology and its dispersion ability regarding thermal properties of water used as phase change material. *Appl. Therm. Eng.* **2018**, *128*, 121–126. [CrossRef]
45. Geng, Y.A.N.; Dalhaimer, P.; Cai, S.; Tsai, R.; Tewari, M.; Minko, T.; Discher, D.E. Shape effects of filaments versus spherical particles in flow and drug delivery. *Nat. Nanotechnol.* **2007**, *2*, 249–255. [CrossRef] [PubMed]
46. Champion, J.A.; Mitragotri, S. Role of target geometry in phagocytosis. *Proc. Natl. Acad. Sci. USA* **2006**, *103*, 4930–4934. [CrossRef] [PubMed]
47. Badawy, A.M.E.; Luxton, T.P.; Silva, R.G.; Scheckel, K.G.; Suidan, M.T.; Tolaymat, T.M. Impact of environmental conditions (pH, ionic strength, and electrolyte type) on the surface charge and aggregation of silver nanoparticles suspensions. *Env. Sci. Technol.* **2010**, *44*, 1260–1266. [CrossRef] [PubMed]
48. Fabrega, J.; Fawcett, S.R.; Renshaw, J.C.; Lead, J.R. Silver Nanoparticle Impact on Bacterial Growth: Effect of pH, Concentration, and Organic Matter. *Environ. Sci. Technol.* **2009**, *43*, 7285–7290. [CrossRef]
49. Andreescu, D.; Bulbul, G.; Özel, R.E.; Hayat, A.; Sardesai, N.; Andreescu, S. Applications and implications of nanoceria reactivity: Measurement tools and environmental impact. *Environ. Sci. Nano* **2014**, *1*, 445–458. [CrossRef]
50. Szymanski, C.J.; Munusamy, P.; Mihai, C.; Xie, Y.; Hu, D.; Gilles, M.K.; Tyliszczak, T.; Thevuthasan, S.; Baer, D.R.; Orr, G. Shifts in oxidation states of cerium oxide nanoparticles detected inside intact hydrated cells and organelles. *Biomaterials* **2015**, *62*, 147–154. [CrossRef]
51. Qi, L.; Sehgal, A.; Castaing, J.C.; Chapel, J.P.; Fresnais, J.; Berret, J.F.; Cousin, F. Redispersible Hybrid Nanopowders: Cerium Oxide Nanoparticle Complexes with Phosphonated-PEG Oligomers. *ACS Nano* **2008**, *2*, 879–888. [CrossRef]
52. Suzuki, T.; Kosacki, I.; Anderson, H.U.; Colomban, P. Electrical Conductivity and Lattice Defects in Nanocrystalline CeO_2 Thin Films. *J. Am. Ceram. Soc.* **2004**, *84*, 2007–2014. [CrossRef]
53. Turner, S.; Lazar, S.; Freitag, B.; Egoavil, R.; Verbeeck, J.; Put, S.; Strauven, Y.; Van Tendeloo, G. High resolution mapping of surface reduction in ceria nanoparticles. *Nanoscale* **2011**, *3*, 3385–3390. [CrossRef] [PubMed]

54. Paier, J.; Penschke, C.; Sauer, J. Oxygen defects and surface chemistry of ceria: Quantum chemical studies compared to experiment. *Chem Rev.* **2013**, *113*, 3949–3985. [CrossRef]
55. Borm, P.; Klaessig, F.C.; Landry, T.D.; Moudgil, B.; Pauluhn, J.; Thomas, K.; Trottier, R.; Wood, S. Research strategies for safety evaluation of nanomaterials, part V: Role of dissolution in biological fate and effects of nanoscale particles. *Toxicol. Sci.* **2006**, *90*, 23–32. [CrossRef] [PubMed]
56. Misra, S.K.; Dybowska, A.; Berhanu, D.; Luoma, S.N.; Valsami-Jones, E. The complexity of nanoparticle dissolution and its importance in nanotoxicological studies. *Sci. Total Env.* **2012**, *438*, 225–232. [CrossRef] [PubMed]
57. Avramescu, M.L.; Rasmussen, P.E.; Chenier, M.; Gardner, H.D. Influence of pH, particle size and crystal form on dissolution behaviour of engineered nanomaterials. *Env. Sci. Pollut. Res. Int.* **2017**, *24*, 1553–1564. [CrossRef] [PubMed]
58. Li, Y.; Liao, C.; Tjong, S.C. Synthetic Biodegradable Aliphatic Polyester Nanocomposites Reinforced with Nanohydroxyapatite and/or Graphene Oxide for Bone Tissue Engineering Applications. *Nanomaterials* **2019**, *9*, 590. [CrossRef] [PubMed]
59. Chen, C.-L.; Zuckermann, R.N.; DeYoreo, J.J. Surface-Directed Assembly of Sequence-Defined Synthetic Polymers into Networks of Hexagonally Patterned Nanoribbons with Controlled Functionalities. *ACS Nano* **2016**, *10*, 5314–5320. [CrossRef]
60. Soppimath, K.S.; Aminabhavi, T.M.; Kulkarni, A.R.; Rudzinski, W.E. Biodegradable polymeric nanoparticles as drug delivery devices. *J. Control. Release* **2001**, *70*, 1–20. [CrossRef]
61. Fathi, M.; Barar, J. Perspective highlights on biodegradable polymeric nanosystems for targeted therapy of solid tumors. *Bioimpacts* **2017**, *7*, 49–57. [CrossRef]
62. Wang, J.; Xia, H.; Zhang, Y.; Lu, H.; Kamat, R.; Dobrynin, A.V.; Cheng, J.; Lin, Y. Nucleation-Controlled Polymerization of Nanoparticles into Supramolecular Structures. *J. Am. Chem. Soc.* **2013**, *135*, 11417–11420. [CrossRef]
63. Khan, S.; Ansari, A.A.; Rolfo, C.; Coelho, A.; Abdulla, M.; Al-Khayal, K.; Ahmad, R. Evaluation of in vitro cytotoxicity, biocompatibility, and changes in the expression of apoptosis regulatory proteins induced by cerium oxide nanocrystals. *Sci. Technol. Adv Mater.* **2017**, *18*, 364–373. [CrossRef] [PubMed]
64. Gatoo, M.A.; Naseem, S.; Arfat, M.Y.; Mahmood Dar, A.; Qasim, K.; Zubair, S. Physicochemical properties of nanomaterials: Implication in associated toxic manifestations. *Biomed. Res. Int.* **2014**, *2014*, 498420. [CrossRef] [PubMed]
65. Pujar, M.S.; Hunagund, S.M.; Desai, V.R.; Patil, S.; Sidarai, A.H. One-step synthesis and characterizations of cerium oxide nanoparticles in an ambient temperature via Co-precipitation method. In *AIP Conference Proceedings*; AIP Publishing: Melville, NY, USA, 2018; p. 050026.
66. Farahmandjou, M.; Zarinkamar, M.; Firoozabadi, T.P. Synthesis of Cerium Oxide (CeO_2) nanoparticles using simple CO-precipitation method. *Rev. Mex. De Física* **2016**, *62*, 496–499.
67. Chen, P.-L.; Chen, I.W. Reactive Cerium(IV) Oxide Powders by the Homogeneous Precipitation Method. *J. Am. Ceram. Soc.* **1993**, *76*, 1577–1583.
68. Chen, H.-I.; Chang, H.-Y. Synthesis of nanocrystalline cerium oxide particles by the precipitation method. *Ceram. Int.* **2005**, *31*, 795–802. [CrossRef]
69. Ramachandran, M.; Subadevi, R.; Sivakumar, M. Role of pH on synthesis and characterization of cerium oxide (CeO_2) nano particles by modified co-precipitation method. *Vacuum* **2019**, *161*, 220–224. [CrossRef]
70. Nanda, H.S. Preparation and Biocompatible Surface Modification of Redox Altered Cerium Oxide Nanoparticle Promising for Nanobiology and Medicine. *Bioengineering* **2016**, *3*, 28. [CrossRef]
71. Nethi, S.K.; Nanda, H.S.; Steele, T.W.; Patra, C.R. Functionalized nanoceria exhibit improved angiogenic properties. *J. Mater. Chem. B* **2017**, *5*, 9371–9383. [CrossRef]
72. Corsi, F.; Caputo, F.; Traversa, E.; Ghibelli, L. Not Only Redox: The Multifaceted Activity of Cerium Oxide Nanoparticles in Cancer Prevention and Therapy. *Front. Oncol.* **2018**, *8*, 309. [CrossRef]
73. Gliga, A.R.; Edoff, K.; Caputo, F.; Källman, T.; Blom, H.; Karlsson, H.L.; Ghibelli, L.; Traversa, E.; Ceccatelli, S.; Fadeel, B. Cerium oxide nanoparticles inhibit differentiation of neural stem cells. *Sci. Rep.* **2017**, *7*, 9284. [CrossRef]
74. Celardo, I.; De Nicola, M.; Mandoli, C.; Pedersen, J.Z.; Traversa, E.; Ghibelli, L. Ce(3)+ ions determine redox-dependent anti-apoptotic effect of cerium oxide nanoparticles. *ACS Nano* **2011**, *5*, 4537–4549. [CrossRef]

75. Bumajdad, A.; Zaki, M.I.; Eastoe, J.; Pasupulety, L. Microemulsion-Based Synthesis of CeO2 Powders with High Surface Area and High-Temperature Stabilities. *Langmuir ACS J. Surf. Colloids* **2005**, *20*, 11223–11233. [CrossRef] [PubMed]
76. Kockrick, E. Synthesis and catalytic properties of microemulsion-derived cerium oxide nanoparticles. *J. Solid State Chem.* **2008**, *181*, 1614–1620. [CrossRef]
77. Richard, B.; Lemyre, J.L.; Ritcey, A.M. Nanoparticle Size Control in Microemulsion Synthesis. *Langmuir* **2017**, *33*, 4748–4757. [CrossRef] [PubMed]
78. Patil, S.; Sandberg, A.; Heckert, E.; Self, W.; Seal, S. Protein adsorption and cellular uptake of cerium oxide nanoparticles as a function of zeta potential. *Biomaterials* **2007**, *28*, 4600–4607. [CrossRef] [PubMed]
79. Arya, A.; Sethy, N.K.; Das, M.; Singh, S.K.; Das, A.; Ujjain, S.K.; Sharma, R.K.; Sharma, M.; Bhargava, K. Cerium oxide nanoparticles prevent apoptosis in primary cortical culture by stabilizing mitochondrial membrane potential. *Free Radic Res.* **2014**, *48*, 784–793. [CrossRef]
80. Masui, T. Characterization of Cerium(IV) Oxide Ultrafine Particles Prepared Using Reversed Micelles. *Chem. Mater.* **1997**, *9*, 2197–2204. [CrossRef]
81. Cimini, A.; D'Angelo, B.; Das, S.; Gentile, R.; Benedetti, E.; Singh, V.; Monaco, A.M.; Santucci, S.; Seal, S. Antibody-conjugated PEGylated cerium oxide nanoparticles for specific targeting of Aβ aggregates modulate neuronal survival pathway. *Acta Biomater* **2012**, *8*, 2056–2067. [CrossRef]
82. López, J.M.; Gilbank, A.L.; García, T.; Solsona, B.; Agouram, S.; Torrente-Murciano, L. The prevalence of surface oxygen vacancies over the mobility of bulk oxygen in nanostructured ceria for the total toluene oxidation. *Appl. Catal. B Environ.* **2015**, *174*, 403–412. [CrossRef]
83. Tok, A.I.Y. Hydrothermal synthesis of CeO_2 nano-particles. *J. Mater. Process. Technol.* **2007**, *190*, 217–222. [CrossRef]
84. Trenque, I.; Magnano, G.C.; Bolzinger, M.A.; Roiban, L.; Chaput, F.; Pitault, I.; Briançon, S.; Devers, T.; Masenelli-Varlot, K.; Bugnet, M.; et al. Shape-selective synthesis of nanoceria for degradation of paraoxon as a chemical warfare simulant. *Phys. Chem. Chem. Phys.* **2019**, *21*, 5455–5465. [CrossRef]
85. Zhang, Y.; Wu, X.; Hou, C.; Shang, K.; Yang, K.; Tian, Z.; Pei, Z.; Qu, Y.; Pei, Y. Dual-responsive dithio-polydopamine coated porous CeO_2 nanorods for targeted and synergistic drug delivery. *Int. J. Nanomed.* **2018**, *13*, 2161–2173. [CrossRef]
86. Singh, S.; Ly, A.; Das, S.; Sakthivel, T.S.; Barkam, S.; Seal, S. Cerium oxide nanoparticles at the nano-bio interface: Size-dependent cellular uptake. *Artif. Cells Nanomed. Biotechnol.* **2018**, *43* (Suppl. 3), S956–S963. [CrossRef] [PubMed]
87. Kannan, S.K.; Sundrarajan, A.M. Approach for the Synthesis of a Cerium Oxide Nanoparticle: Characterization and Antibacterial Activity. *Int. J. Nanosci.* **2014**, *13*, 1450018. [CrossRef]
88. Sangsefidi, F.S.; Nejati, M.; Verdi, J.; Salavati-Niasari, M. Green synthesis and characterization of cerium oxide nanostructures in the presence carbohydrate sugars as a capping agent and investigation of their cytotoxicity on the mesenchymal stem cell. *J. Clean. Prod.* **2017**, *156*, 741–749. [CrossRef]
89. Thovhogi, N.; Diallo, A.; Gurib-Fakim, A.; Maaza, M. Nanoparticles green synthesis by Hibiscus Sabdariffa flower extract: Main physical properties. *J. Alloy. Compd.* **2015**, *647*, 392–396. [CrossRef]
90. Kargar, H.; Ghazavi, H.; Darroudi, M. Size-controlled and bio-directed synthesis of ceria nanopowders and their in vitro cytotoxicity effects. *Ceram. Int.* **2015**, *41*, 4123–4128. [CrossRef]
91. Zhang, H.; He, X.; Zhang, Z.; Zhang, P.; Li, Y.; Ma, Y.; Kuang, Y.; Zhao, Y.; Chai, Z. Nano-CeO_2 exhibits adverse effects at environmental relevant concentrations. *Env. Sci. Technol.* **2011**, *45*, 3725–3730. [CrossRef]
92. Soren, S.; Jena, S.R.; Samanta, L.; Parhi, P. Antioxidant Potential and Toxicity Study of the Cerium Oxide Nanoparticles Synthesized by Microwave-Mediated Synthesis. *Appl. Biochem. Biotechnol.* **2015**, *177*, 148–161. [CrossRef]
93. Machmudah, S.; Winardi, S.; Kanda, H.; Goto, M. Synthesis of Ceria Zirconia Oxides using Solvothermal Treatment. In *MATEC Web of Conferences*; EDP Sciences: Les Ulis, France, 2018.
94. Kar, S.; Direct, S.; Chemistry, C.; Patel, C.; Temperature, S. Synthesis of Valence State Engineered UltraSmall Ceria Nanoparticles Investigation on the Role of Ethylenediamine as a Capping Agent. *J. Phys.* **2009**, *113*, 4862–4867.
95. Yu, T.; Joo, J.; Park, Y.I.; Hyeon, T. Large-scale nonhydrolytic sol-gel synthesis of uniform-sized ceria nanocrystals with spherical, wire, and tadpole shapes. *Angew. Chem. Int. Ed. Engl.* **2005**, *44*, 7411–7414. [CrossRef]

96. Darroudi, M. Green synthesis and evaluation of metabolic activity of starch mediated nanoceria. *Ceram. Int.* **2014**, *40*, 2041–2045. [CrossRef]
97. Elahi, B.; Mirzaee, M.; Darroudi, M.; Sadri, K.; Oskuee, R.K. Bio-based synthesis of Nano-Ceria and evaluation of its bio-distribution and biological properties. *Colloids Surf. B Biointerfaces* **2019**, *181*, 830–836. [CrossRef] [PubMed]
98. Gnanam, S.; Rajendran, V. Synthesis of CeO_2 or α–Mn_2O_3 nanoparticles via sol-gel process and their optical properties. *J. Solgel Sci. Technol.* **2011**, *58*, 62–69. [CrossRef]
99. Nourmohammadi, E. Cytotoxic activity of greener synthesis of cerium oxide nanoparticles using carrageenan towards a WEHI 164 cancer cell line. *Ceram. Int.* **2018**, *44*, 19570–19575. [CrossRef]
100. Yadav, T.P.; Srivastava, O.N. Synthesis of nanocrystalline cerium oxide by high energy ball milling. *Ceram. Int.* **2012**, *38*, 5783–5789. [CrossRef]
101. Li, X. Nanoscale Structural and Mechanical Characterization of a Natural Nanocomposite Material: The Shell of Red Abalone. *Nano Lett.* **2004**, *4*, 613–617. [CrossRef]
102. Lu, J.; Fang, Z.Z. Synthesis and Characterization of Nanoscaled Cerium (IV) Oxide via a Solid-State Mechanochemical Method. *J. Am. Ceram. Soc.* **2006**, *89*, 842–847. [CrossRef]
103. He, H. Preparation and Dispersion of Nanosize Ceria in High Electrolyte Slurry by Ball-Milling. *Integr. Ferroelectr.* **2015**, *161*, 36–44. [CrossRef]
104. Khorrami, M.B.; Sadeghnia, H.R.; Pasdar, A.; Ghayour-Mobarhan, M.; Riahi-Zanjani, B.; Hashemzadeh, A.; Zare, M.; Darroudi, M. Antioxidant and toxicity studies of biosynthesized cerium oxide nanoparticles in rats. *Int. J. Nanomed.* **2019**, *14*, 2915–2926. [CrossRef]
105. Vassie, J.A.; Whitelock John, M.; Lord Megan, S. Targeted Delivery and Redox Activity of Folic Acid-Functionalized Nanoceria in Tumor Cells. *Mol. Pharm.* **2018**, *15*, 994–1004. [CrossRef]
106. Grillone, A.; Li, T.; Battaglini, M.; Scarpellini, A.; Prato, M.; Takeoka, S.; Ciofani, G. Preparation, Characterization, and Preliminary In Vitro Testing of Nanoceria-Loaded Liposomes. *Nanomaterials* **2017**, *7*, 276. [CrossRef] [PubMed]
107. El Shaer, S.S.; Salaheldin, T.A.; Saied, N.M.; Abdelazim, S.M. In vivo ameliorative effect of cerium oxide nanoparticles in isoproterenol-induced cardiac toxicity. *Exp. Toxicol. Pathol.* **2017**, *69*, 435–441. [CrossRef] [PubMed]
108. Shanmugam, R.; Naik, P. Synthesis and biomedical applications of Cerium oxide nanoparticles – A Review. *Biotechnol. Rep.* **2017**, *17*. [CrossRef]
109. Reed, K. Exploring the properties and applications of nanoceria: Is there still plenty of room at the bottom? *Environ. Sci. Nano* **2014**, *1*, 390–405. [CrossRef]
110. Asati, A.; Santra, S.; Kaittanis, C.; Perez, J.M. Surface-charge-dependent cell localization and cytotoxicity of cerium oxide nanoparticles. *ACS Nano* **2010**, *4*, 5321–5331. [CrossRef]
111. Gagnon, J.; Fromm, K. Toxicity and Protective Effects of Cerium Oxide Nanoparticles (Nanoceria) Depending on Their Preparation Method, Particle Size, Cell Type, and Exposure Route. *Eur. J. Inorg. Chem.* **2015**, *2015*, 4510–4517. [CrossRef]
112. Kumar, A.; Das, S.; Munusamy, P.; Self, W.; Baer, D.R.; Sayle, D.C.; Seal, S. Behavior of nanoceria in biologically-relevant environments. *Environ. Sci. Nano* **2014**, *1*, 516–532. [CrossRef]

nanomaterials

Article

Polycaprolactone Composite Micro/Nanofibrous Material as an Alternative to Restricted Access Media for Direct Extraction and Separation of Non-Steroidal Anti-Inflammatory Drugs from Human Serum Using Column-Switching Chromatography

Hedvika Raabová [1], Lucie Chocholoušová Havlíková [1], Jakub Erben [2], Jiří Chvojka [2], František Švec [1] and Dalibor Šatínský [1,*]

1 The Department of Analytical Chemistry, Faculty of Pharmacy, Charles University, Akademika Heyrovského 1203, 50005 Hradec Králové, Czech Republic; raabovah@faf.cuni.cz (H.R.); havlikova@faf.cuni.cz (L.C.H.); svecfr@faf.cuni.cz (F.Š.)

2 The Department of Nonwovens and Nanofibrous Materials, Faculty of Textile Engineering, The Technical University of Liberec, Studentská 1402/2, 46001 Liberec, Czech Republic; erben.jaakub@gmail.com (J.E.); jiri.chvojka@tul.cz (J.C.)

* Correspondence: satinsky@faf.cuni.cz; Tel.: +420-495-067-228

Citation: Raabová, H.; Havlíková, L.C.; Erben, J.; Chvojka, J.; Švec, F.; Šatínský, D. Polycaprolactone Composite Micro/Nanofibrous Material as an Alternative to Restricted Access Media for Direct Extraction and Separation of Non-Steroidal Anti-Inflammatory Drugs from Human Serum Using Column-Switching Chromatography. *Nanomaterials* **2021**, *11*, 2669. https://doi.org/10.3390/nano11102669

Academic Editor: Angelo Ferraro

Received: 3 September 2021
Accepted: 6 October 2021
Published: 12 October 2021

Abstract: Application of the poly-ε-caprolactone composite sorbent consisting of the micro- and nanometer fibers for the on-line extraction of non-steroidal anti-inflammatory drugs from a biological matrix has been introduced. A 100 μL human serum sample spiked with ketoprofen, naproxen, sodium diclofenac, and indomethacin was directly injected in the extraction cartridge filled with the poly-ε-caprolactone composite sorbent. This cartridge was coupled with a chromatographic instrument via a six-port switching valve allowing the analyte extraction and separation within a single analytical run. The 1.5 min long extraction step isolated the analytes from the proteinaceous matrix was followed by their 13 min HPLC separation using Ascentis Express RP-Amide (100 × 4.6 mm, 5 μm) column. The recovery of all analytes from human serum tested at three concentration levels ranged from 70.1% to 118.7%. The matrix calibrations were carried out in the range 50 to 20,000 ng mL^{-1} with correlation coefficients exceeding 0.996. The detection limit was 15 ng mL^{-1}, and the limit of quantification corresponded to 50 ng mL^{-1}. The developed method was validated and successfully applied for the sodium diclofenac determination in real patient serum. Our study confirmed the ability of the poly-ε-caprolactone composite sorbent to remove the proteins from the biological matrix, thus serving as an alternative to the application of restricted-access media.

Keywords: restricted access media; nanofibers; microfibers; on-line extraction; biological samples; column-switching chromatography

1. Introduction

Non-steroidal anti-inflammatory drugs (NSAID) are widely used in pain, osteoarthritis, and rheumatoid arthritis treatment. A mechanism of their action lies in the inhibition of cyclo-oxygenase enzyme, an enzyme serving prostaglandin biosynthesis. These drugs play a crucial role in the suppression of the inflammatory response of an organism. The reduction in prostaglandin levels results in antipyretics, analgesics, and anti-inflammatory activity of NSAID. The most widely used NSAIDs in clinical practice are acetylsalicylic acid, paracetamol, indomethacin, ibuprofen, ketoprofen, sodium diclofenac, flurbiprofen, mefenamic acid, piroxicam, and nabumeton. Monitoring of NSAID in body fluid is essential for toxicological and pharmacokinetics studies. Their determination as a part of rational pharmacotherapy that aims to reduce risks and achieve disease treatment goals is less common. The levels are specifically monitored either in patients with an impaired renal function when the insufficient excretion occurs or when non-compliance is suspected [1,2]. The pharmaceutical industry is another sector where NSAID are determined in pharmaceutical formulations to ensure safety and effectiveness of drugs [3].

Numerous methods for NSAID determination in biological fluids including gas chromatography [4–6], capillary electrophoresis [7–9], and liquid chromatography [10,11] were developed over the past years. Although the analytical instruments were designed to provide increasingly faster and more sensitive analysis, the treatment of the biological matrixes prior to the instrumental analysis remains the bottleneck in the drug level monitoring [12]. The NSAID are usually determined in plasma, where the direct analysis of untreated matrix can lead to lower sensitivity, unprecise results, and in the worst case to irreversible damage of the chromatography column caused by protein precipitation. Thus, the sample pretreatment step cannot be omitted. This step is mainly based on time-consuming, laborious, and error-prone methods, such as protein precipitation, solid-phase extraction, and liquid–liquid extraction that can be potentially hazardous for laboratory staff [13]. Several approaches have been developed to overcome these shortcomings. The emphasis has been primarily placed on higher speed, automation, reduced consumption of organic solvents, smaller sample volumes, and increased selectivity [13,14]. Restricted access media (RAM) represent one of the applicable methods [15].

RAM sorbents separate the low molecular analytes from macromolecular interferences mostly via size exclusion effect. This analyte extraction and macromolecules removal occurs simultaneously after the injection of the untreated biological fluids [15]. The sample preparation is simplified and the analysis time is shortened while maintaining the desired extraction efficiency and avoiding column clogging [16]. The extraction mechanism of the original RAM is primarily based on their two surfaces. The outer surface is covered by hydrophilic groups avoiding access of macromolecules in the material, and the hydrophobic functional groups on the inner pore surface that are responsible for analyte retention. Additionally, the pores in the sorbent material contribute to the matrix removal by the size-exclusion mechanism [17].

The potential of RAM can be adequately utilized in on-line liquid chromatography extractions. The RAM cartridges are most often coupled with a column-switching chromatography system [16]. This setup was reported several times for NSAID analysis [18,19]. The main component of this system is a double position six-port selection valve that redirects the mobile phases between both extraction cartridge and analytical column. The extraction begins after the sample is injected in the cartridge. The undesired matrix components are removed from the analytes using the washing mobile phase. The column-switching valve then changes its position to complete the extraction step, and the analytes are eluted in the analytical column where they are separated using the separation mobile phase. The washing mobile is meanwhile transferred in the waste container. This process can be carried out either in a straight-flush or in a back-flush mode differing in the way the analytes are eluted from the extraction cartridge. They are injected and eluted from the extraction cartridge in the same direction as washing mobile phase flow in the former approach while the mobile phase elutes the analytes in the opposite direction to the sample injection in the latter [20].

Advanced approaches combine the RAM concept with the molecularly imprinted polymers, magnetism, or supramolecular solvents to enhance selectivity and specificity of the analyte extraction [21–23]. The use of nanomaterials is also included in these developments. For example, application of the restricted access nanotubes was reported [24]. We demonstrated first nanofibers as an alternative to RAM elsewhere [25]. We confirmed that the composite material produced from poly-ε-caprolactone exhibited satisfactory efficiency in extraction of parabens from human serum and bovine milk. The idea of using nanofibers as an alternative to RAM emerged as a result of their characteristics. The nanofibers have a large surface area to volume ratio enabling them to capture considerable quantities of low molecular weight analytes. In contrast, we speculated that the macromolecules are not retained because of the curvature of the fibers. The macromolecules appeared not to be flexible enough to be attached at multiple points to the highly curved nanofiber for their sufficient retention. This is not an issue with the low-molecular weight analytes [25]. Additionally, the large spaces between the fibers support the macromolecule passage through the sorbent to the waste.

We described the advantages of material combining poly-ε-caprolactone micro- and nanofibers (micro/nano PCL) for the on-line extraction of analytes from milk and serum matrixes in our previous study [25]. We confirmed the protein removal capability, good mechanical stability in the extraction/HPLC system, and re-usability of this sorbent for more than 300 analyses. The current study extends exploration of the potential of micro/nano PCL material in therapeutic drug monitoring and a method for NSAID determination in human serum samples was developed. This method was validated according to the International Council for Harmonization guideline and used for the determination of diclofenac content in a real patient serum.

2. Materials and Methods

2.1. Reagents and Materials

Ketoprofen (≥98.0%, KET), naproxen (≥98.0%, NAP), sodium diclofenac (≥98.5%, DCF), and indomethacin (≥98.5%, IND) were purchased from Sigma-Aldrich (Darmstadt, Germany) and used as model analytes. Phosphoric acid (Honeywell, Morris Plains, NJ, USA), acetonitrile (ACN), and methanol (MeOH) (VWR, Paris, France) served as mobile phase modifiers. The standard solutions were acidified by acetic acid from Penta (Prague, Czech Republic). The proteinaceous matrix was prepared by dissolution of lyophilized human serum Lyo hum N (Erba Lachema, Brno, Czech Republic). Water was purified by Millipore Milli-Q Direct Water Purification System from Merck (Darmstadt, Germany). The composite poly-ε-caprolactone sorbent was fabricated at the Technical University of Liberec from poly-ε-caprolactone (Mw 43.000) provided by Polysciences (Heddesheim, Germany), and chloroform (≥97%) and ethanol (99.97%) bought from Penta Chemicals (Pardubice, Czech Republic).

2.2. Instrumentation

2.2.1. Column-Switching Chromatography System

A Shimadzu Prominence instrument (Shimadzu Corporation, Kyoto, Japan) equipped with three LC-20AD Prominence pumps, a DGU-20A Prominence on-line degasser, a SIL 20AC Prominence autosampler, a CTO-20AC Prominence column oven, an SPD-M20A Prominence UV/VIS photodiode array detector, and a CBM-20A Prominence system controller was used for the separation and analysis. Switching between the analytical column and the extraction cartridge was provided via the six-port high-pressure flow line switching valve FCV-12AH. LC Solution software (version 5.97, Shimadzu Corporation, Kyoto, Japan) controlled the instrument. The separation of NSAID extracted from the human serum was carried out on the Ascentis Express RP-Amide (100 × 4.6 mm, 5 µm) analytical column combined with the RP-Amide (5 × 4.6 mm, 2.7 µm) guard column (Sigma-Aldrich, Darmstadt, Germany). The system configuration in extraction and separation mode is depicted in the Figure 1.

2.2.2. Meltblown and Electrospun Fibers

Poly-ε-caprolactone nanofibers and microfibers were fabricated using a process described in Supporting Information and was firstly reported elsewhere [26,27]. The meltblown equipment (Laboratory equipment J&M Laboratories, Ashland, OH, USA) in combination with DC electrospinning lab-made system based on a multi-needle spinner was used for the preparation of the fibers. The production equipment consisting of both electrospinning and meltblown systems is depicted in Figure S1 in Supporting Information section. A scanning electron microscope VEGA 3 (Tescan, Brno, Czech Republic) was used for the imaging of the fibrous sorbent. The morphology of micro/nanofibrous material is shown in Supporting Information section in Figure S2.

2.3. Preparation of Extraction Cartridge

A commercially available PEEK cartridge 4.6 × 10 mm (Merck, Darmstadt, Germany) was manually filled with 44 mg of the composite micro/nanofibrous PCL. The details of extraction cartridge preparation and schematic of filling are presented in Supporting Information section and in Figure S3. The cartridge in a plastic holder was then placed in the chromatographic system and washed with the mobile phase containing 10% ACN in 0.085% (*v*/*v*) aqueous phosphoric acid for 15 min and neat ACN was then injected six times. This washing was carried out to ensure complete removal of impurities that could remain in fibers after their production.

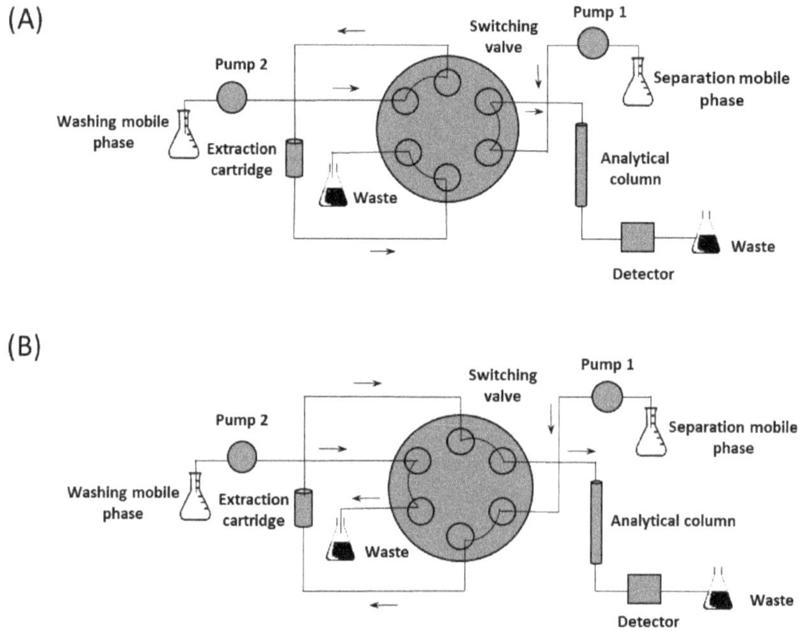

Figure 1. Column-switching chromatographic system in (**A**) on-line extraction and (**B**) on-line extraction-separation mode.

2.4. Preparation of Standard and Matrix Solutions

The stock KET, NAP, DCF, and IND solutions were prepared at a concentration of 1 mg mL^{-1} by dissolving the individual standard substances in MeOH. These solutions were stored in the dark at 4 °C. The mix stock solution containing 0.2 mg mL^{-1} NSAID was prepared by mixing the stock solutions. An appropriate volume of the mix stock solution was diluted with a water–acetonitrile–acetic acid solution (50/48/2, *v*/*v*/*v*) to obtain standards solutions at seven concentration levels of 50; 100; 500; 1000; 5000; 10,000; and 20,000 ng mL^{-1}. These solutions were used for the plotting the calibration curve. Commercially available lyophilized human serum was reconstituted according to manufacturer instructions and then diluted ten times with 20% aqueous ACN. This serum solution was spiked with the mix stock solution to obtain matrix solutions at concentrations of 50; 100; 500; 1000; 5000; 10,000; and 20,000 ng mL^{-1} and these solutions were centrifuged at 14,000× *g* rpm (21,578× *g*) for 15 min. The supernatant was then injected in the chromatographic system and the matrix calibration curve plotted. Matrix solutions were prepared fresh daily.

2.5. Real Sample

A real human serum was obtained from a patient administered with 250 mL continual intravenous infusion containing 75 mg of sodium diclofenac. Blood sampling was carried out in the Department of Clinical Biochemistry and Diagnostics of the University Hospital Hradec Králové. This real life serum was handled the same way as the matrix working solutions. First, it was diluted 10 times with 20% aqueous ACN and then centrifuged at 14,000× *g* rpm (21,578× *g*) for 15 min.

2.6. Analytical Method

The analytical run started with the extraction step when 100 µL sample, standard or matrix solution, was injected in the extraction cartridge filled with micro/nano PCL. The NSAID were captured by the sorbent while the proteins and other potentially interfering macromolecular substances were removed by the washing with the mobile phase composed of 10% ACN in 0.085% aqueous H_3PO_4. Simultaneously, the analytical column was conditioned using the 30% ACN in 0.085% aqueous H_3PO_4. The column-switching valve redirected this separation mobile phase after 1.5 min in the extraction cartridge. Hereby, the analytes were eluted in the analytical column and separated using the ACN gradient increasing the ACN percentage to 45% in 2.5 min and this mobile phase was pumped through the column for another 1.5 min. Then, the percentage was ramped to 55% in 4 min and to 75% in 0.5 min. Finally, the ACN percentage decreased to the initial concentration in 1 min at which the analytical column was re-equilibrated for 3 min. The flow rate of both mobile phases was held on 1 mL min^{-1}. The total analysis time was 15 min. The separated analytes were detected using the diode array detector. KET, DCF, and IND were monitored at 270 nm while NAP at 232 nm. The column was held at 20 °C since the micro/nano PCL dissolve in ACN at higher temperatures [26].

3. Results and Discussion

3.1. Optimization of Chromatographic Separation

We tested ACN and MeOH as organic components of the mobile phase with ACN produced peaks with a better symmetry. The NSAID drugs are weak acids. Therefore, 0.085% aqueous H_3PO_4 solution pH 2.6 was used as the aqueous part of the mobile phase to increase the retention in the column. We tested two analytical columns Ascentis Express RP-Amide (100 × 4.6 mm, 5 µm) and Ascentis Express F5 (100 × 4.6 mm, 5 µm) for the separations. Both columns use the core-shell particles technology improving the efficiency and reducing the flow resistance in column switching system. However, they are more susceptible to impurities from proteinaceous samples resulting in an increase in back pressure and eventually column clogging. Therefore, the Ascentis Express (5 × 4.6 mm, 2.7 µm) guard columns packed with the same stationary phase were inserted in the system to protect the analytical column. The Ascentis Express RP-Amide analytical column enabled a better separation of the analytes. Ascentis Express F5 stationary phase was not well suited because the extensive peak broadening and coelution of ketoprofen-naproxen and sodium diclofenac-indomethacin pairs as presented in Figure 2.

3.2. Optimization of On-Line Extraction

We reported elsewhere [25] that the removal of proteins from nanofibrous sorbent using an aqueous mobile phase occurs within the first minute. This finding was confirmed by monitoring elution at a wavelength of 280 nm that is the absorption maximum for proteins. Human serum contains in addition to proteins also more lipophilic ballast substances, such as, for example, lipoproteins and vitamins, that are difficult to remove from the system since aqueous washing is insufficient. In contrast, a higher concentration of organic solvents in the washing mobile phase and excessive washing lead to undesirable losses of desired analytes. Therefore, the composition of the washing mobile phase and the duration of extraction step had to be optimized to remove most of the lipophilic ballast molecules without any loss of the analytes. Removal of these compounds using the mobile phases containing ACN concentration of 5%, 10%, 15%, and 20% were tested. The recovery of adsorbed analytes on nanofibers decreased with the rising ACN concentration. However, the effect of this process on the removal of ballast substances was negligible. This is why we used a mixture comprising 10% ACN in 0.085% aqueous phosphoric acid. This washing mobile phase was the best compromise between analyte loss and ballast removal. The duration of the extraction step was studied applying the similar approach. The peak of matrix impurities was observed in the chromatogram at 270 nm after 1 min washing. Extension of the washing time by 30 s resulted in a reduction in the area of the matrix ballast peaks without reducing the peak areas of analytes. No further improvement was observed after further increasing the extraction time to 2 min. Thus, the extraction time was finally held at 1.5 min. Chromatogram of standard solution and spiked serum under the optimized conditions of extraction and separation is shown in Figure 3.

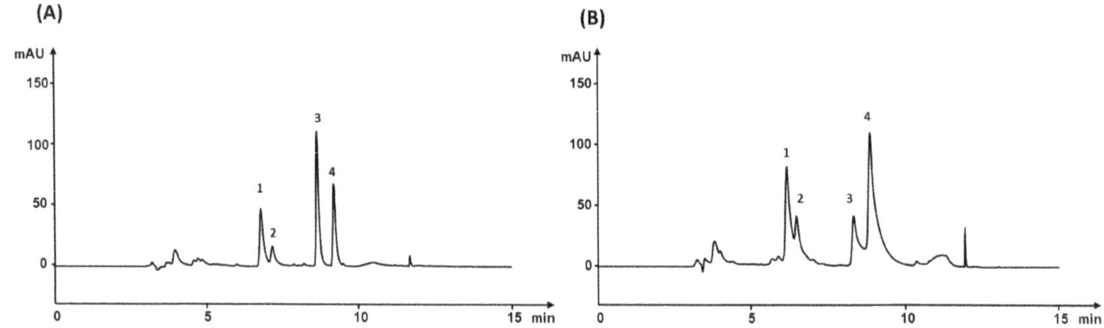

Figure 2. Chromatograms of standard solution spiked with 5000 ng mL^{-1} NSAID separated on (**A**) Ascentis Express RP-Amide column and (**B**) Ascentis Express F5 column at a wavelength 270 nm. 1—ketoprofen, 2—naproxen, 3—sodium diclofenac, 4—indomethacin.

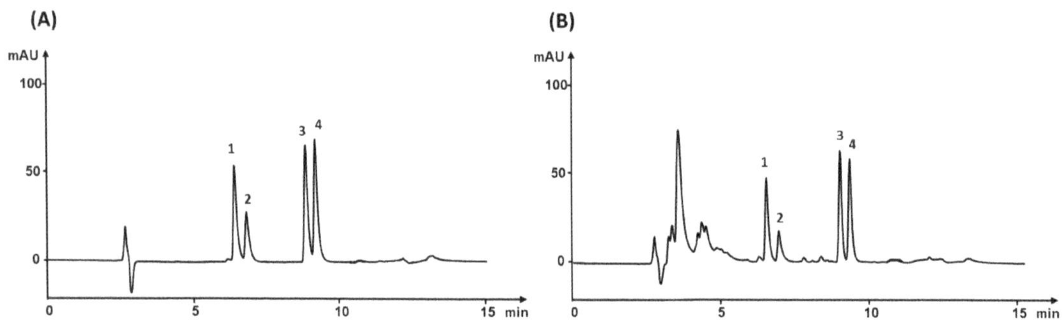

Figure 3. Chromatograms of (**A**) standard solution and (**B**) matrix solution spiked with 1000 ng mL^{-1} NSAID at a wavelength 270 nm. 1—ketoprofen, 2—naproxen, 3—sodium diclofenac, 4—indomethacin.

3.3. Extraction Efficiency

The analyte recovery from the human serum matrix was compared to the recovery obtained using extraction from standard solutions and expressed as a percentage value considering the standard solution being 100%. Figure 4 demonstrates that at levels up to 100 ng mL^{-1}, the analytes are extracted from the serum with an efficiency comparable to that found for the standard solutions. KET is an exception since its recovery exceeded 100% at almost all concentration levels. It can be caused by coelution with a matrix interference that increased the total peak area. This finding reduces the accuracy of the method for KET. However, the RSD values ranging from 1.16% to 0.017% and the precision is not affected.

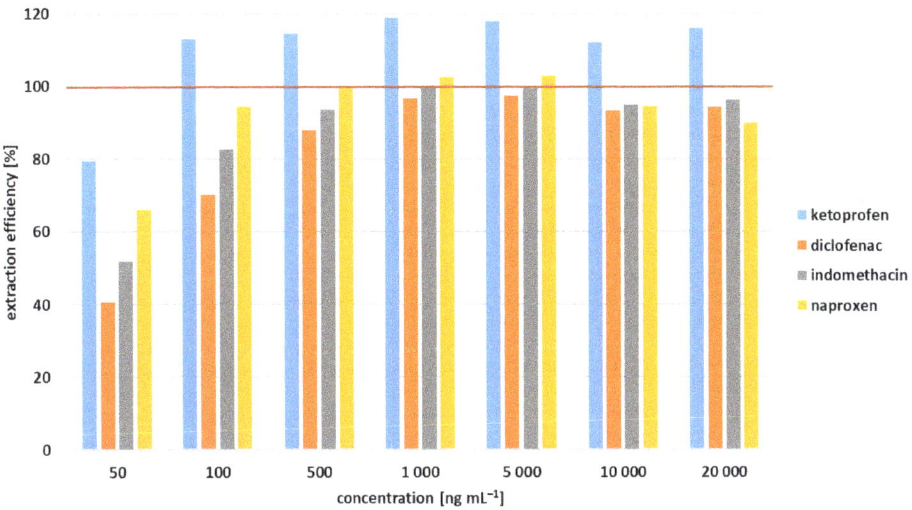

Figure 4. Extraction efficiency of micro/nano PCL sorbent expressed as a percentual recovery of NSAID in human serum.

3.4. Optimization of Injected Sample Volume

Different injection volumes were tested to achieve better sensitivity using UV detection and to simultaneously determine the protein removing capacity of the extraction cartridge. The proteins were completely removed in less than 1 min after the injection volumes 10, 25, and 50 μL. The cleaning of up of 100 μL centrifuged serum was achieved in 1.5 min. A further increase in the injection volume and simultaneous extension of the washing phase resulted in larger analyte loss. Therefore, 100 μL serum was designated as the maximal injection volume for the 10 × 4.6 mm i.d. extraction cartridge.

3.5. Validation

The method was validated with respect to the linearity, precision, accuracy, selectivity, and sensitivity following the ICH Q2 R1 guideline to evaluate the reliability of the results [28]. The validation parameters were chosen to primarily confirm micro/nanofibrous PCL applicability for the NSAID extraction since this sorbent is not standardized and commonly used. The ICH Q2 R1 protocol meets these requirements better than the M10 guideline that is usually used for the validation of bioanalytical methods. The chromatographic system suitability test was carried out confirm the suitability of the HPLC instrument for NSAIDs analysis. Mean values and standard deviations of the retention time, peak capacity, symmetry factor, resolution, and repeatability of the analytical run were calculated from results of six injections of standard solutions and evaluated according to the European Pharmacopoeia recommendations. The results are presented in the Table 1.

Linearity, Accuracy, Precision, and Limits of Detection and Quantification

All samples were measured in triplicate using the optimized conditions. The calibration curves were established for standard and serum matrix solutions at concentration levels of 50; 100; 500; 1000; 2000; 10,000; and 20,000 ng mL^{-1}. The linear relationship between the NSAID quantity and peak area was confirmed for KET, DCF, and IND in the concentration range 50 to 20,000 ng mL^{-1}. The calibration curve for NAP was linear for a range 50 to 10,000 ng mL^{-1}. The correlation coefficient for each drug was exceeded 0.996. Standard and matrix matched calibration curves including regression equations are demonstrated in Supporting Information section in Figure S4.

The accuracy was determined via the recovery study carried out for all the drugs in the human serum matrix. The recovery was calculated as a comparison of peak areas of

NSAID in standard and serum matrix solutions at three concentration levels of 100; 1000 and 10,000 ng mL^{-1} measured in triplicate. The results of recovery also determined the extraction efficiency for each drug.

Table 1. HPLC system suitability parameters. Sample injection was performed in six replicates.

	Ketoprofen	Naproxen	Sodium Diclofenac	Indomethacin
Retention time [min]	6.41	6.82	9.18	10.49
Retention time repeatability, RSD [%]	0.03	0.03	0.02	0.02
Repeatability of peak areas, RSD [%]	0.85	0.49	0.21	0.56
Peak resolution	1.82	10.32	5.99	
Peak symmetry	1.80	1.63	2.00	1.60
Peak capacity	14.03	13.18	21.65	18.80

The serum solutions were measured six times at three concentration levels of 100; 1000; and 10,000 ng mL^{-1} for intra-day precision determination. The results expressed as RSD (%) were determined in a range of 7.89% to 10.24% for a concentration level of 100 ng mL^{-1}, 0.17% to 0.90% for a concentration level of 1000 ng mL^{-1}, and 0.09% to 0.46% for a concentration level of 10,000 ng mL^{-1}. The inter-day precision was calculated for three consecutive measurements of matrix spiked with 1000 ng mL^{-1} NSAID in three days. The results expressed as RSD (%) were in a value range of 3.62 and 7.88%. The lowest concentration of the calibration curve equaled to 50 ng mL^{-1} (10σ) was established as a limit of quantification (LOQ). The limit of detection 15 ng mL^{-1} (LOD) was calculated from the LOQ value as a tree-folds (3σ) variation. The results of the validation are summarized in the Table 2. These results are comparable to those obtained using other methods described for diclofenac determination, including column-switching [18], protein precipitation [29], and SPE via commercial sorbent [30] applied as sample preparation procedures.

3.6. *Reusability of the Extraction Cartridge*

The manually packed extraction cartridge was used during the entire experiments corresponding to the analysis of more than 100 serum samples. Neither extraction efficiency nor the back pressure of PCL composite sorbent had changed during these experiments. Thus, our composite micro/nanofibrous sorbent is better suited for the extensive use in high-pressure liquid chromatography systems than nanofibers reported elsewhere [31]. The microfibers represent a stable scaffold resistant to the high pressure while, simultaneously, the large surface area to volume ratio of the nanofibers contributes to the high extraction efficiency. Moreover, the cotton-like texture of micro/nano composite material is easier to manually fill into the cartridge. The durability of our cartridge paralleled our previous study [20] where we applied 200 injections of 200 µL serum in commercial RAM LiChrospher RP-18 ADS column without observing a decrease in extraction efficiency.

3.7. *Analysis of Real Sample*

The real patient serum samples after continual intravenous infusion containing 75 mg sodium diclofenac were handled as a serum matrix and the content of diclofenac determined using our method. The diclofenac peak was recognized based on its retention time and UV spectrum. The calculated concentration of diclofenac was 32.57 ng mL^{-1} in ten times diluted serum corresponding to 325.7 ng mL^{-1} in the original patient serum. This result confirmed that the extraction using nanofibrous sorbent is enough sensitive to handle the real samples. The chromatogram of the patient serum is shown in Figure 5.

Table 2. HPLC method validation results.

		Ketoprofen	Naproxen	Sodium Diclofenac	Indomethacin
Standard calibration range [ng mL^{-1}] [1]		50–20,000	50–10,000	50–20,000	50–20,000
Correlation coefficient		0.9997	0.9996	0.9999	0.9998
Matrix calibration range [ng mL^{-1}] [2]		50–20,000 [1]	50–10,000 [2]	50–20,000 [1]	50–20,000 [1]
Correlation coefficient		0.9999	0.9960	0.9999	0.9999
Intra-day precision [%] [3]	c_1	8.54	7.89	10.24	8.37
	c_2	0.60	0.90	0.38	0.17
	c_3	0.46	0.29	0.09	0.21
Inter-day precision [%] [4]		6.79	3.62	7.44	7.88
Accuracy [%] [5]	c_1	112.91 ± 4.50	94.31 ± 4.09	70.07 ± 6.20	82.76 ± 6.96
	c_2	118.66 ± 0.84	102.28 ± 0.46	96.42 ± 0.15	100.06 ± 0.22
	c_3	111.87 ± 0.70	94.24 ± 0.23	93.12 ± 0.83	94.81 ± 0.20
LOD [ng mL^{-1}] [6]		2.49	24.40	3.58	26.57
LOQ [ng mL^{-1}] [7]		8.29	81.34	11.92	88.57

[1] Each concentration of calibration standard was measured in triplicate; [2] Each concentration of matrix calibration standard was measured in triplicate. Calibration curves were linear for (1) ketoprofen, diclofenac sodium, and indomethacin at seven concentration levels, (2) naproxen at six concentration levels; [3] Relative standard deviation (RSD) was calculated from six injections of the matrix solutions spiked with analytes at concentration levels c_1 = 100 ng mL^{-1}, c_2 = 1000 ng mL^{-1} and c_3 = 10,000 ng mL^{-1}; [4] Relative standard deviation (RSD) was calculated from the average of three injection of matrix solutions spiked with analytes at concentration level c = 1000 ng mL^{-1} for three days; [5] Accuracy was determined as a method recovery of matrix solutions at concentration levels c_1 = 100 ng mL^{-1}, c_2 = 1000 ng mL^{-1} and c_3 = 10 000 ng mL^{-1} measured in triplicate.; [6] Limit of detection (LOD) was calculated from the signal-to-noise ratio in a 3-fold (3σ) variation; [7] Limit of quantification (LOQ) was calculated from the signal-to-noise ratio in a 10-fold (10σ) variation.

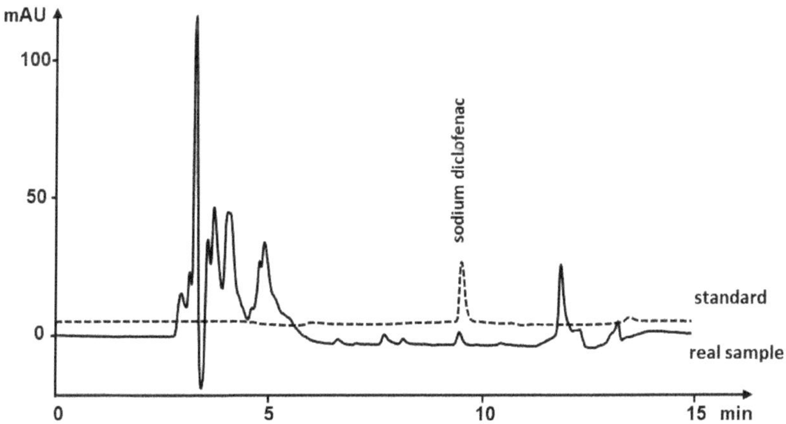

Figure 5. Standard solution spiked with 50 ng mL^{-1} diclofenac and real patient human serum containing the same drug. The concentration of diclofenac in real sample was equal to 32.57 ng mL^{-1} in ten times diluted serum.

4. Conclusions

Our study built on previous extraction experiments with micro/nano PCL that had promising properties enabling the direct extraction of analytes from proteinaceous matrix. We tested the micro/nano PCL fibers as the extraction sorbent for on-line solid-phase extraction of NSAID. After the injection of 100 μL human serum matrix, the sorbent enabled removal of the proteins and the majority of other macromolecular interferences within 1.5 min. The analyte recovery from the matrix was comparable to that obtained with the extraction from solutions of standards. This demonstrated that micro/nano PCL is a promising sorbent for KET, NAP, DCF, and IND even from the proteinaceous matrixes. The applicability of micro/nano PCL sorbent for the therapeutic drug monitoring was explored using a real sample containing diclofenac. The extraction provided good sensitivity even at low concentrations typical of biological samples and purified them sufficiently. Based on our results, we expect that the nano/micro fibrous PCL sorbent has a potential to improve the bioanalytical methods for sample pretreatment in terms of higher effectiveness at a lower costs compared to commercial RAM. This makes it a valuable tool for the on-line extraction/chromatography methods developed for the therapeutic drug monitoring enabling fast, safe, and precise analysis.

Supplementary Materials: The following are available online at https://www.mdpi.com/article/10.3390/nano11102669/s1. Figure S1. The combination of meltblown and electrospinning: 1—drum collector, 2—composite micro-nanofiber layer, 3—airstream with fibers, 4—multi needle spinner, 5—needle counter electrode, 6—meltblown die, 7—pumps, 8—extruder, 9—transmission, 10—high voltage sources, 11—engine; Figure S2. The morphology of micro/nanofibrous PCL material before and after using in high pressure chromatography system; Figure S3. The filling process of the extraction cartridge; Figure S4. Calibration curves—matrix (green), standard (orange).

Author Contributions: Conceptualization, H.R. and D.Š.; methodology, H.R., L.C.H. and D.Š.; software, H.R.; validation, H.R., L.C.H., D.Š. and F.Š.; formal analysis, J.E. and J.C.; investigation, D.Š. and H.R.; resources, J.E. and J.C.; data curation, H.R.; writing—original draft preparation, H.R. and L.C.H.; writing—review and editing, H.R., F.Š. and D.Š.; visualization, H.R.; supervision, D.Š.; project administration, J.C. and D.Š.; funding acquisition, D.Š. and F.Š. All authors have read and agreed to the published version of the manuscript.

Funding: The authors gratefully acknowledge the financial support of the GAČR project No. 20-19297S, STARSS project (Reg. No. CZ.02.1.01/0.0/0.0/15_003/0000465) co-funded by the ERDF, and GAUK project No. 1134119 and SVV 260 548.

Acknowledgments: The authors gratefully acknowledge the cooperation with Department of Clinical Biochemistry and Diagnostics at the Faculty Hospital in Hradec Králové for providing the patient serum samples.

Conflicts of Interest: The authors declare no conflict of interest.

Abbreviations

ACN	Acetonitrile
DCF	Sodium diclofenac
H_3PO_4	Phosphoric acid
IND	Indomethacin
KET	Ketoprofen
LOD	Limit of detection
LOQ	Limit of quantification
MeOH	Methanol
NAP	Naproxen
NSAID	Non-steroidal anti-inflammatory drugs
micro/nano PCL	Composite sorbent containing poly-ε-caprolactone micro- and nanofibers
RAM	Restricted access media

References

1. Sun, Y.; Takaba, K.; Kido, H.; Nakashima, M.N.; Nakashima, K. Simultaneous determination of arylpropionic acidic nonsteroidal anti-inflammatory drugs in pharmaceutical formulations and human plasma by HPLC with UV detection. *J. Pharm. Biomed. Anal.* **2003**, *30*, 1611–1619. [CrossRef]
2. Castilhos, T.S.; Barreto, F.; Meneghini, L.; Bergold, A.M. Development of high-throughput multi-residue method for non-steroidal anti-inflammatory drugs monitoring in swine muscle by LC-MS/MS. *Food Addit. Contam. Part A* **2016**, *33*, 1166–1174. [CrossRef]
3. Martin, M.J.; Pablos, F.; Gonzalez, A.G. Simultaneous determination of caffeine and non-steroidal anti-inflammatory drugs in pharmaceutical formulations and blood plasma by reversed-phase HPLC from linear gradient elution. *Talanta* **1999**, *49*, 453–459. [CrossRef]

4. El Haj, B.M.; Al Ainri, A.M.; Hassan, M.H.; Bin Khadem, R.K.; Marzouq, M.S. The GC/MS analysis of some commonly used non-steriodal anti-inflammatory drugs (NSAIDs) in pharmaceutical dosage forms and in urine. *Forensic Sci. Int.* **1999**, *105*, 141–153. [CrossRef]
5. Kim, K.R.; Yoon, H.R. Rapid screening for acidic non-steroidal anti-inflammatory drugs in urine by gas chromatography mass spectrometry in the selected-ion monitoring mode. *J. Chromatogr. B Biomed. Appl.* **1996**, *682*, 55–66. [CrossRef]
6. Arroyo, D.; Ortiz, M.C.; Sarabia, L.A. Optimization of the derivatization reaction and the solid-phase microextraction conditions using a D-optimal design and three-way calibration in the determination of non-steroidal anti-inflammatory drugs in bovine milk by gas chromatography-mass spectrometry. *J. Chromatogr. A* **2011**, *1218*, 4487–4497. [CrossRef]
7. Heitmeier, S.; Blaschke, G. Direct assay of nonopioid analgesics and their metabolites in human urine by capillary electrophoresis and capillary electrophoresis-mass spectrometry. *J. Chromatogr. B Biomed. Sci. Appl.* **1999**, *721*, 109–125. [CrossRef]
8. Bechet, I.; Fillet, M.; Hubert, P.; Crommen, J. Quantitative-analysis of nonsteroidal antiinflammatory drugs by capillary zone electrophoresis. *J. Pharm. Biomed. Anal.* **1995**, *13*, 497–503. [CrossRef]
9. Maboundou, C.W.; Paintaud, G.; Berard, M.; Bechtel, P.R. Separation of 15 nonsteroidal antiinflammatory drugs using micellar electrokinetic capillary chromatography. *J. Chromatogr. B Biomed. Appl.* **1994**, *657*, 173–183. [CrossRef]
10. Navas, N.; Urena, R.; Capitan-Vallvey, L.-F. Determination of celecoxib, rofecoxib, sodium diclofenac and niflumic acid in human serum samples by HPLC with DAD detection. *Chromatographia* **2008**, *67*, 55–61. [CrossRef]
11. Dadashzadeh, S.; Vali, A.M.; Rezagholi, N. LC determination of piroxicam in human plasma. *J. Pharm. Biomed. Anal.* **2002**, *28*, 1201–1204. [CrossRef]
12. Kataoka, H.; Saito, K. Recent advances in column switching sample preparation in bioanalysis. *Bioanalysis* **2012**, *4*, 809–832. [CrossRef]
13. Kataoka, H. New trends in sample preparation for clinical and pharmaceutical analysis. *Trends Anal. Chem.* **2003**, *22*, 232–244. [CrossRef]
14. Nazario, C.E.D.; Fumes, B.H.; da Silva, M.R.; Lancas, F.M. New materials for sample preparation techniques in bioanalysis. *J. Chromatogr. B Anal. Technol. Biomed. Life Sci.* **2017**, *1043*, 81–95. [CrossRef] [PubMed]
15. Cassiano, N.M.; Barreiro, J.C.; Moraes, M.C.; Oliveira, R.V.; Cass, Q.B. Restricted access media supports for direct high-throughput analysis of biological fluid samples: Review of recent applications. *Bioanalysis* **2009**, *1*, 577–594. [CrossRef] [PubMed]
16. Rudolphi, A.; Boos, K.S.; Grimm, C.H. Column switching and restricted access materials for sample preparation. *Abstr. Pap. Am. Chem. Soc.* **1997**, *214*, 186.
17. Yang, S.H.; Fan, H.; Classon, R.J.; Schug, K.A. Restricted access media as a streamlined approach toward on-line sample preparation: Recent advancements and applications. *J. Sep. Sci.* **2013**, *36*, 2922–2938. [CrossRef] [PubMed]
18. Lee, H.S.; Jeong, C.K.; Choi, S.J.; Kim, S.B.; Lee, M.H.; Ko, G.I.; Sohn, D.H. Simultaneous determination of aceclofenac and diclofenac in human plasma by narrowbore HPLC using column-switching. *J. Pharm. Biomed. Anal.* **2000**, *23*, 775–781. [CrossRef]
19. Suenami, K.; Lim, L.W.; Takeuchi, T.; Sasajima, Y.; Sato, K.; Takekoshi, Y.; Kanno, S. Direct determination of non-steroidal anti-inflammatory drugs by column-switching LC-MS. *J. Sep. Sci.* **2006**, *29*, 2725–2732. [CrossRef]
20. Brabcová, I.; Šatínský, D.; Solich, P. A fast on-line micro-column sample pretreatment with a HPLC column switching technique and large volume sample injection for determination of non-steroidal anti-inflammatory drugs in human serum. *Anal. Methods* **2013**, *5*, 6153–6159. [CrossRef]
21. Puoci, F.; Iemma, F.; Cirillo, G.; Curcio, M.; Parisi, O.I.; Spizzirri, U.G.; Picci, N. New restricted access materials combined to molecularly imprinted polymers for selective recognition/release in water media. *Eur. Polym. J.* **2009**, *45*, 1634–1640. [CrossRef]
22. Xiao, D.L.; Liu, S.B.; Liang, L.Y.; Bi, Y.P. Magnetic restricted-access microspheres for extraction of adrenaline, dopamine and noradrenaline from biological samples. *Mikrochim. Acta* **2016**, *183*, 1417–1423. [CrossRef]
23. de Faria, H.D.; Abrao, L.C.D.; Santos, M.G.; Barbosa, A.F.; Figueiredo, E.C. New advances in restricted access materials for sample preparation: A review. *Anal. Chim. Acta* **2017**, *959*, 43–65. [CrossRef] [PubMed]
24. Barbosa, A.F.; Barbosa, V.M.P.; Bettini, J.; Luccas, P.O.; Figueiredo, E.C. Restricted access carbon nanotubes for direct extraction of cadmium from human serum samples followed by atomic absorption spectrometry analysis. *Talanta* **2015**, *131*, 213–220. [CrossRef]
25. Raabová, H.; Háková, M.; Chocholoušová-Havliková, L.; Erben, J.; Chvojka, J.; Solich, P.; Švec, F.; Šatínský, D. Poly-epsilon-caprolactone Nanofibrous Polymers: A Simple Alternative to Restricted Access Media for Extraction of Small Molecules from Biological Matrixes. *Anal. Chem.* **2020**, *92*, 6801–6805. [CrossRef]
26. Háková, M.; Chocholoušová-Havliková, L.; Chvojka, J.; Erben, J.; Solich, P.; Švec, F.; Šatínský, D. A comparison study of nanofiber, microfiber, and new composite nano/microfiber polymers used as sorbents for on-line solid phase extraction in chromatography system. *Anal. Chim. Acta* **2018**, *1023*, 44–52. [CrossRef]
27. Erben, J.; Pilařová, K.; Sanetrnik, F.; Chvojka, J.; Jenčová, V.; Blažková, L.; Havlíčeka, J.; Novák, O.; Mikeš, P.; Prosecká, E.; et al. The combination of meltblown and electrospinning for bone tissue engineering. *Mater. Lett.* **2015**, *143*, 172–176. [CrossRef]
28. European Medicines Association. *Note for Guidance on Validation of Analytical Procedures: Text and Methodology*; European Medical Agency: Amsterdam, The Netherlands, 1995.
29. Emara, L.H.; Taha, N.F.; El-Ashmawy, A.A.; Raslan, H.M.; Mursi, N.M. A rapid and sensitive bioanalytical hplc method for determining diclofenac sodium in human plasma for bioequivalence studies. *J. Liq. Chromatogr. Relat. Technol.* **2012**, *35*, 2203–2216. [CrossRef]

30. Sahoo, N.K.; Sahu, M.; Rao, P.S.; Ghosh, G. Solid phase extraction and quantification of diclofenac sodium in human plasma by liquid chromatography-tandem mass spectrometry. *J. Anal. Chem.* **2015**, *70*, 424–430. [CrossRef]
31. Háková, M.; Chocholoušová-Havliková, L.; Chvojka, J.; Solich, P.; Šatínský, D. An on-line coupling of nanofibrous extraction with column-switching high performance liquid chromatography—A case study on the determination of bisphenol A in environmental water samples. *Talanta* **2018**, *178*, 141–146. [CrossRef]

nanomaterials

Article

Optimization of Microalga *Chlorella vulgaris* Magnetic Harvesting

Maria G. Savvidou [1], Maria Myrto Dardavila [2,*], Ioulia Georgiopoulou [2], Vasiliki Louli [2], Haralambos Stamatis [3], Dimitris Kekos [1] and Epaminondas Voutsas [2]

1 Biotechnology Laboratory, School of Chemical Engineering, National Technical University of Athens, 9 Iroon Polytechniou Str, Zografou Campus, 15780 Athens, Greece; msavvid@central.ntua.gr (M.G.S.); kekos@chemeng.ntua.gr (D.K.)
2 Laboratory of Thermodynamics and Transport Phenomena, School of Chemical Engineering, National Technical University of Athens, 9 Iroon Polytechniou Str, Zografou Campus, 15780 Athens, Greece; ioulia.p.geo@gmail.com (I.G.); svlouli@chemeng.ntua.gr (V.L.); evoutsas@chemeng.ntua.gr (E.V.)
3 Laboratory of Biotechnology, Department of Biological Applications and Technologies, University of Ioannina, 45110 Ioannina, Greece; hstamati@uoi.gr
* Correspondence: mirtodar@mail.ntua.gr; Tel.: +30-210-7723230

Abstract: Harvesting of microalgae is a crucial step in microalgae-based mass production of different high value-added products. In the present work, magnetic harvesting of *Chlorella vulgaris* was investigated using microwave-synthesized naked magnetite (Fe_3O_4) particles with an average crystallite diameter of 20 nm. Optimization of the most important parameters of the magnetic harvesting process, namely pH, mass ratio (m_r) of magnetite particles to biomass (*g/g*), and agitation speed (rpm) of the *C. vulgaris* biomass–Fe_3O_4 particles mixture, was performed using the response surface methodology (RSM) statistical tool. Harvesting efficiencies higher than 99% were obtained for pH 3.0 and mixing speed greater or equal to 350 rpm. Recovery of magnetic particles via detachment was shown to be feasible and the recovery particles could be reused at least five times with high harvesting efficiency. Consequently, the described harvesting approach of *C. vulgaris* cells leads to an efficient, simple, and quick process, that does not impair the quality of the harvested biomass.

Keywords: *Chlorella vulgaris*; microwave-synthesized magnetite particles; response surface methodology; harvesting process optimization

Citation: Savvidou, M.G.; Dardavila, M.M.; Georgiopoulou, I.; Louli, V.; Stamatis, H.; Kekos, D.; Voutsas, E. Optimization of Microalga *Chlorella vulgaris* Magnetic Harvesting. *Nanomaterials* **2021**, *11*, 1614. https://doi.org/10.3390/nano11061614

Academic Editor: Jose L. Arias

Received: 22 May 2021
Accepted: 17 June 2021
Published: 20 June 2021

1. Introduction

Microalgae have been extensively investigated over the past decades as a source for biofuel production due to their high lipid and carbohydrate yields, as well as being a natural source of high value-added bioactive compounds such as polyphenols, carotenoids, fatty acids, and antibiotics [1–3]. Bioactive chemicals derived from natural sources present higher biological activity and acceptance by consumers compared to the synthetic alternatives [4].

Nowadays, microalgae are of significant importance in the fields of human health, cosmetics, food, and animal feed. In comparison to terrestrial crop plants, microalgae can provide higher productivity and photosynthetic performance, and since they can be cultivated on infertile land, they do not compete with existing food production methods [5,6]. Of the several major production steps of microalgae components, harvesting is both energy and time demanding. It is estimated that microalgae biomass harvesting is responsible for 20–30% of the total biomass production cost [7,8]. Moreover, the harvesting step is crucial for the downstream process, since it leads to a slurry of highly concentrated solid matter [9].

Various harvesting techniques have been introduced over the years such as centrifugation, sedimentation, flocculation, filtration, flotation, and their various combinations. These methods provide (variable) sufficient harvesting efficiencies, however, they are relatively energy and time consuming. Furthermore, none of them can be used as an efficient method for all microalgal cell types, and their application depends on the desired end-product of microalgae cells [10,11]. The reduced lipid levels regarding flocculation, as well as clogging and fouling in filtration, are also considerable disadvantages [12,13]. Moreover, the added cost of the chemicals used in chemical-flocculation or the microorganisms in bio-flocculation, along with the potential contamination of the harvested biomass, are further drawbacks [13–16].

Magnetic separation, which has been extensively used in the food industry, wastewater treatment, and the steel industry, Refs. [3,17,18] has gathered popularity throughout the past decade as a method for harvesting microalgal cells. Magnetic biomass harvesting, which uses micro- or nano-sized magnetic particles and an applied external magnetic field, is a highly efficient process for cell separation; the method is fast and efficient [17,19–22]. The process is based on the interactions between the magnetic particles and the surface of the microalgae cells. The magnetic particles can attach to the cells via electrostatic forces, hydrogen bonds, acid-base interactions or van der Waals forces, forming magnetic particle–cell bonding [23]. Key factors affecting the magnetic separation efficiency include cell type and their growth stage, magnetic particle dosage, pH, ions in the cultivation medium, temperature, and gradient of the magnetic field [3]. Both naked and surface-functionalized (using polyacrylamides, chitosan, poly diallyldimethylammonium chloride (PDDA), aminoclay, polyethylenimine (PEI), 3-aminopropyl triethoxysilane (APTES)) magnetic nanoparticles have been applied for the harvesting of *C. reinhardtii*, *C. vulgaris*, *N. salina*, *N. maritime*, *S. dimorfus*, and *S. ovalternus* microalgae, among others [2,3,24–29]. Nonetheless, successful upscaling of the magnetic harvesting method depends on some important characteristics of the magnetic particles i.e., their biocompatibility, reusability, and ease of use, as well as the optimization of the parameters affecting the harvesting efficiency [30].

Chlorella sp. is a microalga rich in vitamins, polysaccharides, minerals, lipids, and other high-value products [31]. Lutein, a compound with proven anti-cataract properties, is also found in *Chlorella sp.*, while its extracts exhibit anti-tumor, anti-inflammatory, antioxidant, and anti-microbial properties, and reduce blood pressure and cholesterol levels [32–34]. As a result, *Chlorella sp.* is one of the most extensively used microalga in cosmetics, pharmaceuticals, food, and animal feed industries [31,35].

Although *Chlorella sp.* holds second place in the global "algal products" market [36], its harvesting process is still a major challenge. That is why the harvesting of freshwater microalgae *Chlorella vulgaris* (using microwave-synthesized Fe_3O_4 magnetic particles) was examined in this work. The purpose of this study is to develop an efficient *C. vulgaris* harvesting method by examining and optimizing the most important parameters. Specifically, three harvesting parameters have been investigated, namely pH, mass ratio of magnetite particles to biomass, and agitation speed; these were characterized using response surface methodology (RSM). Reusability of the magnetic particles, as well as integrity of the microalgae cells, were also examined. Scanning electron microscopy (SEM) was used to explore the interactions between the microalgae cells and the magnetic particles, while the adsorption mechanism of the *C. vulgaris* cells on the magnetite particles was deduced by measuring the adsorption isotherm at ambient conditions. Finally, zeta potential measurements were performed in order to examine the separation mechanism of *C. vulgaris* microalgae with the aid of the synthesized iron oxide magnetic particles.

2. Materials and Methods

2.1. Chemical Reagents

For the synthesis of the magnetite (Fe_3O_4) particles, iron (II) sulfate ($FeSO_4 \cdot 7H_2O$) salt (Chem-Lab NV, Zedelgem, Belgium) and sodium hydroxide pellets (NaOH) (Panreac Quimica, SA) were employed. The cultivation medium of the *Chlorella vulgaris* strain was composed of sodium nitrate ($NaNO_3$), calcium chloride ($CaCl_2 \cdot 2H_2O$), magnesium sulfate ($MgSO_4 \cdot 7H_2O$), potassium hydrogen phosphate ($K_2HPO_4 \cdot 3H_2O$), sodium chloride (NaCl), EDTA disodium (Na_2EDTA), iron (III) chloride ($FeCl_3$), manganese (II) chloride ($MnCl_2$) and zinc chloride ($ZnCl_2$) and cobalt chloride ($CoCl_2$) (Sigma–Aldrich, St. Louis, MO, USA). Potassium dihydrogen phosphate (KH_2PO_4), vitamin B1, and vitamin B12 provided from Merck (KGaA, Darmstadt, Germany). Deionized water was used for the solubilization of the chemical reagents. Magnetite nano powder (50–100 nm) was purchased from Sigma-Aldrich (St. Louis, MO, USA) for comparison with the microwave-synthesized Fe_3O_4 particles used in this work. A trypan blue assay was performed for examining cell membrane integrity (Merck, KGaA, Darmstadt, Germany). All the reagents used were of analytical grade.

2.2. *Microwave Synthesis of Magnetite (Fe_3O_4) Particles*

Magnetic particles were prepared using a simple, rapid, and low-cost precipitation method that has been previously reported [20,27]. Briefly, 1 g of $FeSO_4 \cdot 7H_2O$ was diluted in 100 mL of deionized water at ambient temperature. Gradual addition of 1 M NaOH (with continuous magnetic stirring) was used to adjust the pH of the solution to 12. As a result, a black precipitate of $Fe(OH)_2$ was formed. By adding deionized water, the final volume of the solution was fixed at 200 mL. The solution was then placed in a common microwave oven where it was radiated for 10 min at 700 W. As a consequence, the $Fe(OH)_2$ precipitate was oxidized to Fe_3O_4. Subsequently, the magnetite particles were collected by employing a NdFeB magnet (50.8 mm × 50.8 mm × 25.4 mm, Supermagnete, Gottmadingen, Germany) with 12,600–12,900 G (N40) magnetic induction intensity. Thereafter, the Fe_3O_4 particles were washed several times with deionized water in order to remove any residual ions, dried in an oven at 60 °C under vacuum, pulverized to a fine powder using a mortar and pestle, and stored in dry conditions for further use.

2.3. *Microalgal Strain and Cultivation*

Chlorella vulgaris strain (UTEX 1809) was purchased from Culture Collection of Algae at the University of Texas (Austin, TX, USA). The strain was grown in a medium with the following composition (per liter): $NaNO_3$, 250 mg; $CaCl_2 \cdot 2H_2O$, 25 mg; $MgSO4 \cdot 7H_2O$, 75 mg; $K_2HPO_4 \cdot 3H_2O$; 75 mg; KH_2PO_4, 175 mg; NaCl, 25 mg; trace element solution, 6 mL from stock solution; vitamin B1, 0.12 µg; vitamin B12, 0.10 µg. One liter of trace element solution contained Na_2EDTA, 750 mg; $FeCl_3$, 97.0 mg; $MnCl_2$, 41.0 mg; $ZnCl_2$, 5.0 mg; $CoCl_2$, 2.0 mg; and $Na_2MoO_4 \cdot 2H_2O$, 4.0 mg. The pH was adjusted to 6.8–7.0 and the temperature was controlled at 24 °C. Cell cultures of *C. vulgaris* were grown in 1000 mL Erlenmeyer flasks containing a culture volume of 600 mL at 150 rpm. The injected concentration was adapted to an optical density of 0.1 at 600 nm (OD 600nm). Periodic purity assessment was performed by microscopic examination. The photobioreactor was illuminated at 50 µmol photon m^{-2} s^{-1} with 24/7 white LED lamps at the cell culture surface. *C. vulgaris* growth was monitored by measuring the OD at 600 nm with a UV-VIS S-22 BOECO (Shimadzu, Kyoto, Japan) for up to 15 days. All experiments were conducted in duplicate.

2.4. *Characterization of Fe_3O_4 Particles and Microalgae*

2.4.1. X-ray Diffraction of Magnetite Particles and Magnetic Properties

In order to determine the crystallographic structure and phase composition of the synthesized magnetic particles, X-ray diffraction was employed by using a Brooker X-ray D8 advance diffractometer equipped with a Cu-Kα radiation source. The radiation of the specimen was generated at 40 kV and 40 mA at room temperature, with a scanning rate of 0.1° per minute, from 2θ 28° up to 89°. The phenomenon mean diameter size of the magnetite crystallites was calculated via the Debye–Scherrer formula method [37] by exploiting the obtained XRD raw data using the Diffrac. Suite Eva software. Furthermore, the same XRD study was performed for the high purity Fe3O4 nanoparticles purchased from Sigma-Aldrich, in order to compare the two spectra and deduce on the purity and microstructural characteristics of the microwave-synthesized magnetic particles. The magnetic properties of Fe_3O_4 were assessed at room temperature using a vibrating sample magnetometer.

2.4.2. Observation with Scanning Electron Microscopy Coupled with Energy Dispersive X-ray Spectroscopy (SEM-EDS)

The morphological characteristics of treated and untreated *C. vulgaris* algal cells, were studied via scanning electron microscopy (SEM) using an FEI Quanta 200 scanning electron microscope.

Using the same technique, the *C. vulgaris*—Fe_3O_4 interaction was also evaluated. For this purpose, magnetically harvested *C. vulgaris* cells were fixed in 2.5% (*v*/*v*) paraformaldehyde at 4 °C for 1 h, washed in 0.05 M phosphate buffer (pH 7.4), and consecutively dehydrated in 20%, 35%, 50%, 75%, 90%, and 100% (*v*/*v*) ethanol solution [22].

All the aforementioned samples were sputter-coated with Au prior to SEM examination. Moreover, EDS technique was employed (using an EDAX analyzer) in order to obtain the elemental compositional analysis of the microalgal cells and the magnetite particles.

2.4.3. Microalgae Analysis

Algae concentration (g L^{-1}) was calculated using a calibration curve of known optical density at 600 nm and respective dry weights that were determined gravimetrically after drying the algae cells at 60 °C. The optical density was measured with a UV-VIS S-22 BOECO spectrophotometer.

2.4.4. Zeta Potential Measurements

The zeta potential values of the prepared Fe_3O_4 particles and the *C. vulgaris* cells were measured with the Zetasizer Nano-ZS (Malvern, UK) at ambient temperature and calculated according to the Smoluchowski equation using the Zetasizer software. In order to perform these measurements, Milli-Q water was used as the dispersion medium for the magnetite particles (3 mg L^{-1}) and the microalgae cells (50 mg L^{-1}). The pH of the mixtures was fixed to 3.0, 5.0, and 7.0 by adding small quantities of aqueous NaOH and HCl. In each case, at least three zeta potential measurements were recorded, and the reported values correspond to their arithmetic mean.

2.5. C. vulgaris Magnetic Harvesting Using Fe_3O_4 Particles

2.5.1. Magnetic Harvesting Experimental Procedure

The magnetic harvesting experiments of *C. vulgaris* biomass were carried out under ambient conditions in batch mode. The experimental procedure included the following steps: (a) The pH of the microalgal cultivation was fixed at the desired value through the addition of aqueous NaOH or HCl solutions, (b) in order to facilitate the flocculation between the algal cells and the particles, 10 mL of the microalgae cultivation was mixed with a given amount of magnetic particles for 10 min using a mechanical stirrer operating at desired rpm, and (c) the flocs were separated from the cultivation medium with the application of a strong magnetic field, imposed using the same permanent magnet which was employed during the synthesis of the magnetite particles, for 3 min.

Finally, the harvesting efficiency (*HE*%) of the process was calculated from Equation (1):

$$HE\% = \frac{OD_0 - OD_1}{OD_0} \cdot 100 \quad (1)$$

where, OD_0 is the initial absorbance of the microalgae cultivation at a wavelength of 600 nm, and OD_1 is the absorbance at the same wavelength of the supernatant liquid that separates from the microalgae-particles flocs after the application of the magnetic field. The OD_0 and OD_1 values were measured with the UV-VIS S-22 BOECO spectrophotometer.

2.5.2. Effect of Harvesting Process Parameters on Microalgae Magnetic Harvesting Using Fe_3O_4 Particles

In order to maximize the harvesting efficiency, the effect of the pH of the mixture, the mass ratio (m_r) of magnetite particles (g)/dry biomass (g) in the algal broth, and the agitation speed (rpm) of the *C. vulgaris* cultivation-Fe_3O_4 particles mixture, were examined. Three values of pH (3.0, 5.0, and 7.0), three m_r ratios (10:1, 12:1, and 14:1) and three agitation speeds (250, 350, and 450 rpm) were compared. Each separate magnetic harvesting experiment was performed three times, and the reported *HE*% results are the mean values.

Preliminary harvesting experiments aiming to evaluate the effect of the duration of flocs' magnetic separation were also performed. According to the results, the strong attachment of Fe_3O_4 on the cell membrane, along with the application of a powerful magnetic field, results in the rapid harvesting of the flocculated biomass, therefore the study of this parameter was excluded.

2.5.3. Experimental Design for Optimization of *C. vulgaris* Harvesting Using the Response Surface Methodology

The optimum conditions for magnetic harvesting with Fe_3O_4 particles were established using central composite design (CCD) under the application of response surface methodology (RSM). The RSM is a combination of mathematical and statistical techniques applicable when several interactive parameters are examined and is considered useful in order to evaluate the relative significance of the operating factors on magnetic harvesting. In the present work, based on preliminary experiments, the independent parameters affecting magnetic harvesting were found to be pH, mass ratio, and agitation speed. Thus, three-level full factorial design (3^k) was used, considering these operational parameters. The experimental design was devised based on the central level (0) between the minimum (−1) and the maximum levels (+1) of the normalized values. Therefore, twenty experiments were carried out using different values of the examined factors (Table 1): pH (3.0–7.0), mass ratio (10:1–14:1), and agitation speed (250–450 rpm). For a complex system, such as the one under consideration, the response cannot be mathematically expressed through a first-order or a second-order polynomial equation. Therefore, the results of the harvesting experiments were fitted to a modified model, and the quality of the fitted model was quantitatively assessed by the ANOVA method to characterize the interaction effect between independent factors and the microalgae-harvesting efficiency.

Table 1. Independent variables and the distribution of their level.

Variables	Level −1	Level 0	Level +1
pH	3	5	7
mass ratio ($g_{magn.}/g_{biomass}$)	10	12	14
agitation speed (rpm)	250	350	450

Regression analysis and estimation of the coefficients were performed using Design Expert® trial version 12 software.

2.5.4. Adsorption Isotherms

The elucidation of the adsorption capacity and mechanism of the microwave-synthesized magnetite particles on the *C. vulgaris* cells is provided by the adsorption isotherms. In the present study, two different models—Langmuir and Freundlich—were employed in order to analyze the experimental data. These models are frequently reported in related literature as sufficiently accurate to describe the adsorption of magnetic particles on microalgae cells [19,21].

According to the Langmuir model, the amount of dry algae cells adsorbed per unit weight of magnetic particles (Q_e [g g^{-1}]), is given by Equation (2):

$$Q_e = Q_m \cdot \frac{C_e \cdot K_l}{1 + C_e \cdot K_l} \tag{2}$$

which can be transformed into the following linear form:

$$\frac{C_e}{Q_e} = \frac{1}{Q_m \cdot K_l} + \frac{C_e}{Q_m} \tag{3}$$

According to the Freundlich model, Q_e is given by Equation (4):

$$Q_e = K_f \cdot C_e^{\frac{1}{n_f}} \tag{4}$$

which can be transformed into the following linear form:

$$lnQ_e = lnK_f + \frac{1}{n_f} \cdot lnC_e \tag{5}$$

In Equations (2)–(5), Q_m [g g^{-1}] is the maximum adsorption capacity, C_e [g L^{-1}] is the concentration of the microalgae cells in the supernatant after completion of the harvesting process, K_l [L g^{-1}] is the Langmuir adsorption constant, K_f [g g^{-1}] is the Freundlich constant (which is related to the adsorption capacity), and n_f [dimensionless] is the Freundlich factor of the heterogeneity of the adsorption sites.

The adsorption experiments were conducted at ambient temperature in a batch mode in duplicates. For each experiment, an amount of Fe_3O_4 was added to 10 mL of algae cultivation of known initial concentration (C_0 [g L^{-1}]) in order to realize the magnetic harvesting procedure that is described in Section 2.5.1.

Consequently, the amount of dry algae adsorbed per unit weight of magnetite particles (Q_e [g g^{-1}]) was calculated according to Equation (6).

$$Q_e = \frac{C_0 - C_e}{m} \cdot V \tag{6}$$

where, m [g] is the mass of the magnetic particles used and V [L] is the volume of microalgae culture used, i.e., 0.01 L.

2.5.5. Reusability of Magnetic Particles

The study on the reusability of the magnetic particles was realized by carrying out tests with the same (used) particles for cycles, in accordance with the procedure outlined by Markeb et al. [38]. The magnetic particles and microalgae cells obtained from magnetic harvesting were mixed with 10 mL of NaOH (0.5 M) and agitated at 200 rpm for 10 min. The suspension was then sonicated for 10 min. Afterwards, 3 mL of methanol and 3 mL of chloroform were added, and the solution was sonicated for another 10 min. Following sonication, the magnetite particles were collected with a permanent magnet, washed three times, and dried overnight. The detached magnetic particles were used to evaluate the harvesting efficiency tests under the same conditions in order to investigate their reusability.

2.6. Statistical Analysis

Tukey's method, based on one factor ANOVA at the 5% confidence level, was used for the statistical analysis, which was performed with SPSS 15.0.1 software (SPSS Inc., Chicago, IL, USA). Statistically significant differences were reported when the probability of the results (p) value is less than 0.05 assuming the null hypothesis.

2.7. Cell Membrane Integrity of *C. vulgaris* Cells

The cell integrity of *C. vulgaris* was determined by the trypan blue staining method. Total of 10 uL of cells were harvested and after addition of 10 uL 1% trypan blue solution the cells were incubated for 10 min at room temperature. The intact cells (viable) remained green (no penetration of the trypan blue solution) while the broken cells appeared blue (stain diffused in the protoplasm).

3. Results and Discussion

3.1. Characterization of Fe_3O_4 Particles

Figure 1 demonstrates the X-ray diffraction spectra of the magnetite particles purchased from Sigma-Aldrich (Figure 1a) and the magnetic particles synthesized in this work (Figure 1b). According to the analysis conducted via XRD software Diffrac. Suite Eva, one predominant phase was identified, that of magnetite (PDF 02-1035 Fe_3O_4 Magnetite), for both materials.

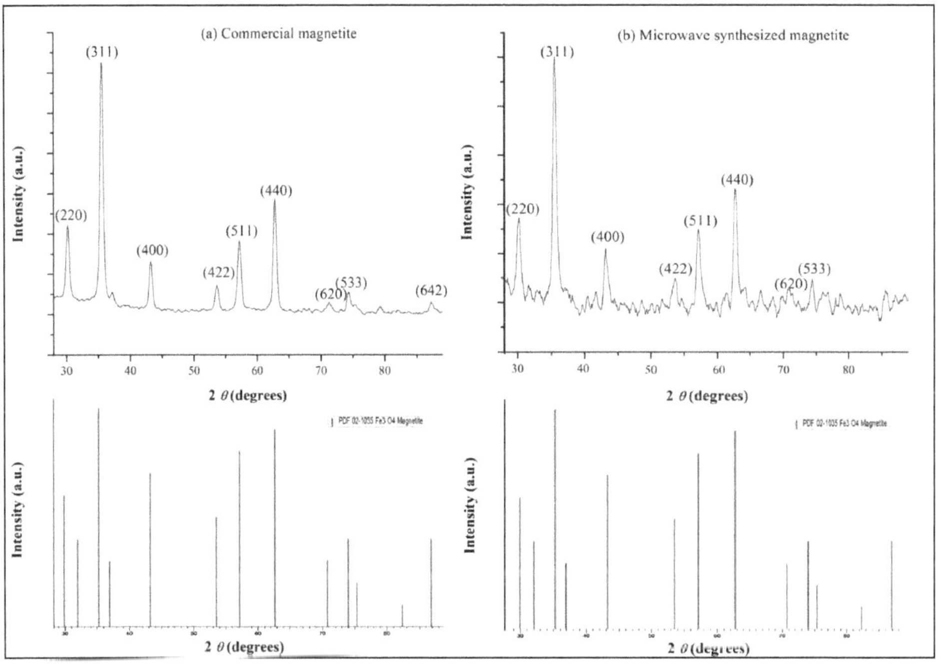

Figure 1. X-ray diffraction spectra of (**a**) commercially available magnetite particles and (**b**) the microwave-synthesized magnetite particles.

The noteworthy similarity between the two spectra suggests that they have close microstructural characteristics. Particularly, regarding the microwave-synthesized magnetite (Figure 1b), eight different relatively strong diffraction peaks are displayed in the XRD spectrum at 2 θ 30.10°, 35.48°, 42.96°, 53.71°, 57,17°, 62.77°, 70.80°, and 74.27° which correspond to the crystallographic planes (220) (311), (400), (422), (511), (440), (620), and (533) respectively, indicating a cubic inverse spinel structure. Likewise, the same crystallographic planes, can be observed for the commercial magnetite powder (Figure 1a) at 2 θ 30.18°, 35.55°, 43.20°, 53.64°, 57,20°, 62.82°, 71.20°, and 74.34°. One more crystallographic plane is present in the spectrum of commercial magnetite at 2 θ 87.36°, namely the (642). This last plane cannot be identified with certainty in the XRD pattern of the synthesized magnetite, due to the relatively higher noise encountered. In both cases, it is clearly observed that five diffraction peaks, namely (220), (311), (400), (511), (440) are the main ones. The aforementioned analysis of the Bragg reflections for both magnetite particles was realized by identifying each diffraction peak from the raw XRD data using the Diffrac. Suite Eva software.

Using the tools provided by the same software, the phenomenon mean diameter size of magnetite crystallites was calculated according to the Debye–Scherrer equation [37] for each diffraction peak that was identified. So, an average crystallite diameter size equal to

20 ± 4 nm and 20 ± 3 nm was calculated for the synthesized magnetite and the commercial one respectively.

For the determination of the magnetic properties of the microwave-synthesized magnetite particles, the saturated mass magnetization was determined by vibrating the sample in a magnetometer. The magnetization loop is illustrated in Figure 2. From these results, it is evident that the particles follow a unhysteretic loop, illustrating superparamagnetic behavior at room temperature, thus indicating that their size is below 35 nm [39], which is verified in our case as the particles demonstrate an average diameter of 20 nm. The synthesized magnetic particles (Fe_3O_4) exhibit saturation magnetization of 60 emu g^{-1}.

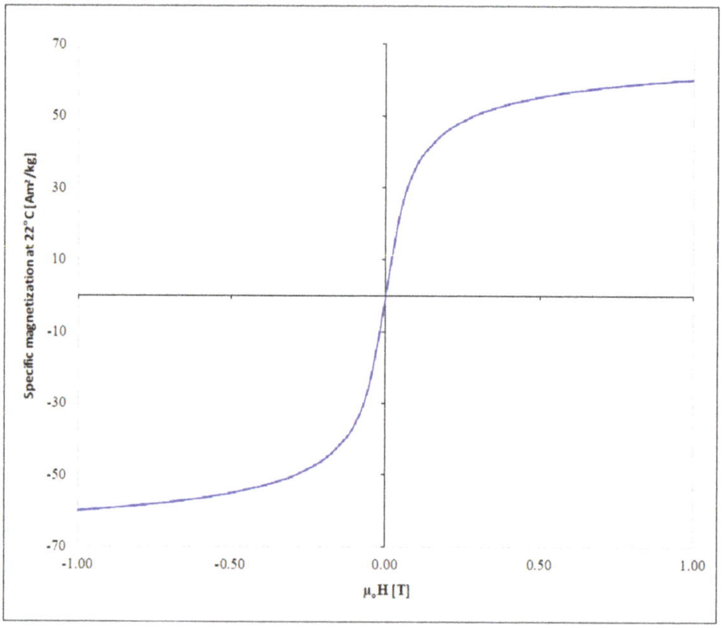

Figure 2. Magnetization curve for the synthesized Fe_3O_4.

Thus, it is concluded that the simple, rapid, and energy-efficient synthesis route followed in this work leads to the formation of high purity nanocrystalline magnetite, which is similar in microstructure to the commercially available that was employed as a means of comparison.

3.2. *Algae Harvesting Efficiency Optimization*

3.2.1. Characterization of *C. vulgaris* and Magnetic Particles Interaction Using SEM

Evaluation of the interactions between the microalgae and the (synthesized) magnetic particles was tested by characterizing *C. vulgaris* before (control) and after mixing with the magnetic particles using the scanning electron microscopy (SEM). The morphology of *C. vulgaris* cells alone and with magnetic particles, are illustrated in Figure 3. As it can be seen in Figure 3a, the surface of the untreated cells is smooth and spherical, whereas after the magnetic harvesting at pH 3.0 (Figure 3c), it can be observed that all the cell surfaces are fully covered with magnetic particles. The cell surfaces appear rougher and the patterns that are visible in Figure 3a are not discerned in Figure 3c. This confirms that the amount of magnetic particles we chose to use is enough to fully cover every single cell. Moreover, energy dispersive X-ray spectroscopy (EDX) reports the iron peak in Figure 3d, verifying the presence of magnetite on the cells in contrast to the untreated cells (Figure 3b).

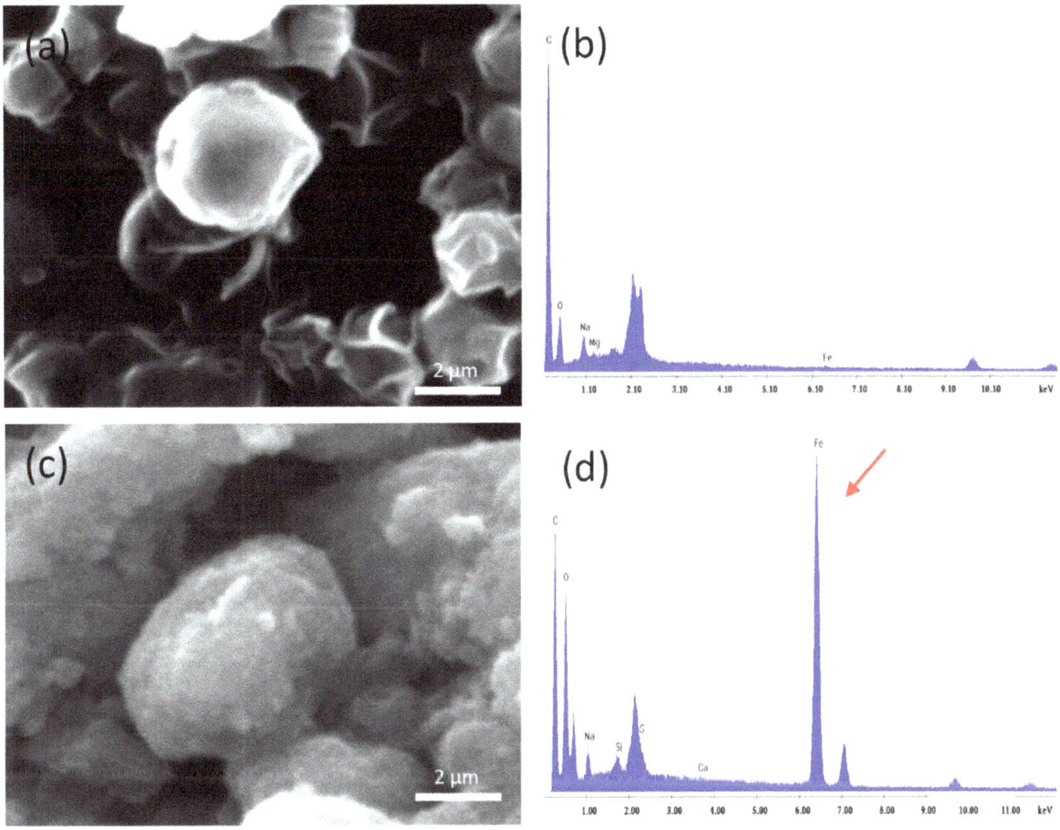

Figure 3. Scanning electron microscopy (SEM) images performed on (**a**) *Chlorella vulgaris* cells before magnetic harvesting, (**b**) EDX analysis for the elemental composition of the cells where the absence of iron peak is reported, (**c**) harvested *Chlorella vulgaris* cells at pH 3.0, (**d**) EDX analysis for the elemental composition of the harvested *Chlorella vulgaris* cells where the iron peak (arrow) is shown.

3.2.2. Magnetic Harvesting of *C. vulgaris*

A face-centered (distance of each axial point from the center: alpha = 1) composite design was developed for modelling and optimizing (statistically) the magnetic harvesting of *C. vulgaris*. Twenty experiments were carried out examining the effect of pH, mass ratio (m_r), and agitation speed on harvesting efficiency. The range of the examined parameters was chosen based on preliminary experiments and corresponding literature. Specifically, studies on *Chlorella vulgaris* and related species [2,12,40,41] have been conducted in a pH range between 2 and 12, while maximum harvesting efficiency was achieved in most cases at acidic or neutral pH. The *Chlorella vulgaris* cell walls are very robust and remain resistant at acidic pH, in comparison to other microalgal species that do not attain such a resilient cell wall and might be affected. Although harsh conditions can affect the viability of the cells, the harvesting process followed in this study is very quick, without significantly affecting the biomass composition. Regarding stirring, low agitation speed values [19,42] of around 250 rpm [43], or even higher values at 800 rpm [44] have been used for harvesting *Chlorella* species. This data led us to choose a pH range from 3 to 7 and an agitation speed range between 250 and 450 rpm.

The experimental results are presented in Table 2. It is observed that the decrease of pH, and the increase of the other two parameters, improve the harvesting efficiency.

Thus, the best experimental harvesting efficiencies higher than 99% were obtained for pH 3.0 and mixing speed greater or equal to 350 rpm. However, even at a pH similar to that of the cell culture (pH 7.0), the harvesting efficiency is proved to be satisfactory (>85%) for m_r = 14:1 and an agitation speed of 450 rpm. These results are in agreement with those reported by Zhu et al. [9] and Bharte and Desai [12], for different *Chlorella* species that demonstrated maximum harvesting efficiency in acidic pH, thus verifying our findings. Analogous studies on *C. vulgaris* harvesting have shown that harvesting based on a flocculation method can lead to high harvesting efficiency, but under different optimum conditions compared to this study. For example, Tork et al. [42] reported a 90% efficiency at basic pH (pH = 11.7) values using cationic starch nanoparticles. Similarly, in the study of Leite et al. [40] for *Chlorella Sorokiniana*, a harvesting efficiency greater than 97% (combined with the appropriate velocity gradient and agitation time) was achieved under highly alkaline conditions (pH = 12). Moreover, a lower optimal agitation rate of 150 rpm was reported by Almomani [19] and Razack et al. [45] (using as a flocculant iron oxide nanoparticles and seed powder of clearing nut respectively). However, the above differences concerning optimal harvesting conditions may be due to the different biomass and magnetic materials, absence or presence of surface modification etc.

Table 2. Design table formed by RSM—CCD, presenting the experimental conditions and the experimental values of harvesting efficiency.

Run	pH	Mass Ratio	Agitation Speed (rpm)	Measured Harvesting Efficiency (%)
1	7	10:1	450	64.7
2	3	14:1	450	99.6
3	3	10:1	250	83.3
4	7	14:1	250	72.3
5	7	14:1	450	87.2
6	7	10:1	250	51.7
7	3	10:1	450	99.4
8	3	14:1	250	87.8
9	5	12:1	350	76.9
10	3	12:1	350	99.2
11	5	12:1	450	88.8
12	5	12:1	250	66.5
13	7	12:1	350	65.0
14	5	12:1	350	76.7
15	5	10:1	350	70.3
16	5	12:1	350	77.9
17	5	14:1	350	92.5
18	5	12:1	350	80.8
19	5	12:1	350	77.6
20	5	12:1	350	77.1

Analysis of variance (ANOVA) was used to determine the relationship between the dependent (harvesting efficiency %) and independent variables (pH, mass ratio, and agitation). Transformation of the response was considered necessary in order to bound harvest efficiency within reasonable limits. The transformation chosen is presented below:

$$y' = \ln\left(\frac{y - lower}{upper - y}\right) \quad (7)$$

where, y stands for the harvesting efficiency (%), lower and upper for the boundaries (40 and 100 respectively), and y' for the transformed harvesting efficiency.

Table 3 shows the statistical results of harvesting efficiency for RSM using ANOVA, while a linear model with respect to the examined factors and their interactions is obtained as follows:

$$y' = -2.473 - 2.022 * A + 0.299 * B + 0.027 * C - 0.003 * A * C + 0.229 * A^2 \quad (8)$$

where, y' stands for the transformed harvesting efficiency, A for pH, B for the mass ratio ($g_{magn.\ par.}/g_{biomass}$), and C for the agitation speed (rpm). It is clarified that through Equation (8) the relation between the independent variables and the transformed dependent variable (y') is expressed. The calculated harvest efficiency (y) is finally determined by solving Equation (7) with respect to y:

$$y = \frac{100 * e^{y'} + 40}{1 + e^{y'}} \quad (9)$$

Table 3. Analysis of variance results for RSM model regarding harvesting process, where A is pH, B is mass ratio, and C is agitation speed.

Source	Sum of Squares	df	Mean Square	F-Value	p-Value Prob > F
Model	51.99	5	10.40	47.58	<0.0001
A-pH	28.47	1	28.47	130.29	<0.0001
B-mass ratio	3.58	1	3.58	16.37	0.0012
C-agitation speed	12.54	1	12.54	57.39	<0.0001
AC	3.21	1	3.21	14.69	0.0018
A^2	4.19	1	4.19	19.15	0.0006
Residual	3.06	14	0.22		
Std. Dev.	0.47		**R^2**	0.94	
Mean	1.1		**Adj R^2**	0.93	
C.V. %	42.52		**Pred R^2**	0.86	
PRESS	7.96		**Adeq Precision**	26.60	

The model was evaluated based on the F- and p-values. For an F-value > 1 and p-value < 0.01, the model under consideration is deemed valid. In the present work, both the high F-value (47.58) and low p-value (<0.0001) of the model prove its significance and accuracy. All the factors considered were significant (high F-value and p-value < 0.05). Specifically, the most important variant is shown to be pH (higher F-value). The obtained value of R^2 (0.94) implies the high correlation between the dependent and independent variables, while the adjusted R^2 is only slightly lower (0.93) and in reasonable agreement with the predicted R^2 (0.86). Finally, the adequate precision (26.60) is higher than 4 indicating that the model can be used to navigate the design space.

The juxtaposition of the predicted against the experimental values of harvesting efficiency (shown in Figure 4) proves that most of the data points are close to the 45-degree line, proving the reliability of the model.

3.3. Separation Mechanism

In order to study the possible electrostatic interactions between the algal biomass and the microwave-synthesized magnetic particles, their zeta potential (ζ) values were measured within the pH range of the magnetic harvesting experiments (pH = 3 to 7). It was found that both the magnetic particles and the biomass cells attain a negative charge. In particular, measurements show ζ values in the range of -17.4 to -33.2, and of -16.6 to -24.9, for the Fe_3O_4 particles and the *C. vulgaris* cells respectively, in the specific dispersing medium (Milli-Q water).

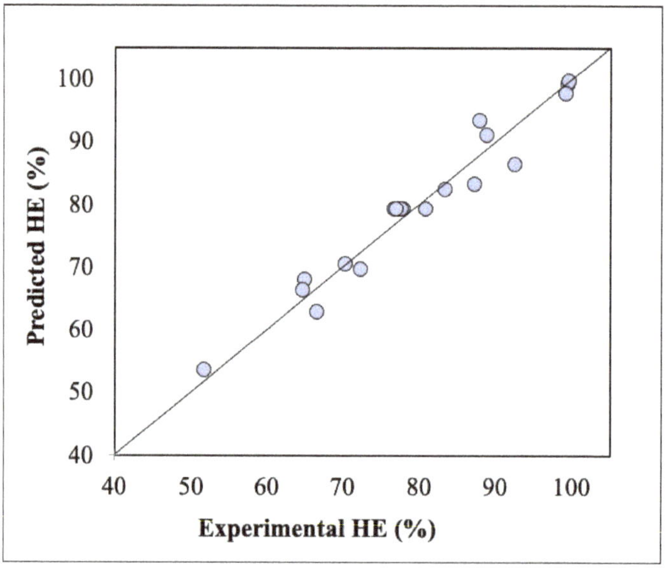

Figure 4. Predicted versus experimental values of harvesting efficiency.

As shown in Figure 5, the zeta potential of the magnetite particles increases with decreasing pH, whereas the *C. vulgaris* cell ζ values remain practically the same as the pH decreases from 7 to 5, and then increases with a further decrease of the pH. Negative surface charge of microalgae biomass is expected in acidic solutions due to the presence of negatively charged groups on the surface of the cells. Concerning the bare Fe_3O_4 particles, a net negative surface charge has been previously reported [20,46].

Since in the present study the magnetic particles and the *C. vulgaris* cells are both negatively charged, it could be assumed that the interactions between them are not dependent on a charge neutralization mechanism. Nevertheless, in any aqueous solution, metal oxide (MeOx) particle surfaces can undergo the following protonation/deprotonation reaction:

$$\mathrm{MeOxO^-} + \mathrm{H_2O} \overset{\mathrm{H_2O}}{\leftrightarrow} \mathrm{MeOxOH} \overset{\mathrm{H_2O}}{\leftrightarrow} \mathrm{MeOxOH_2}^+ \quad (10)$$

Consequently, under acidic conditions and as the pH decreases, an increase of the positively charged sites on the surface of the metal oxide particles is expected, resulting in an increase in the zeta potential values [28] without necessarily altering the net charge of the particles. This is in agreement with the ζ values obtained here for the microwave-synthesized Fe_3O_4 particles. The maximum harvesting efficiency values measured appear at the lowest applied pH values, i.e., at pH = 3. Hence, this behavior may be attributed to local electrostatic interactions between positively charged sites on the surface of the magnetite particles and the negatively charged surfaces of the *C. vulgaris* cells. This assumption is further confirmed by the fact that the harvesting efficiency increases with decreasing pH.

Nevertheless, electrostatic interactions may not be the primary binding mechanism of the synthesized magnetite nanoparticles with *C. vulgaris* cells. Particularly, Fe_3O_4 particles can be attached to the microalgal cells through hydrogen bonding; due to the protonation of the magnetite particles under acidic conditions, hydrogen bond donor chemical species OH_2^+ formed on Fe_3O_4 can interact with hydrogen bond acceptor groups present on *C. vulgaris* cells, such as amino or carboxy groups [47,48]. Furthermore, due to the nano-size of the Fe_3O_4 particles, a large specific surface area and high surface energy is expected, characteristics that imply strong adsorption of the particles to the cells [49,50].

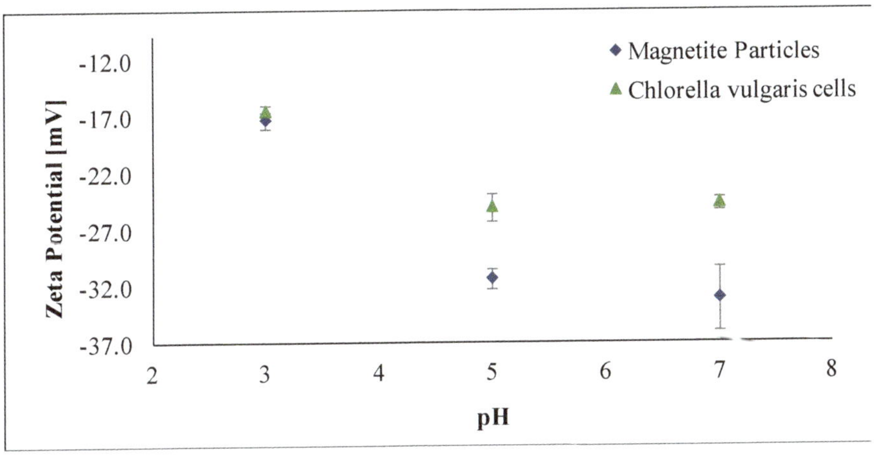

Figure 5. Zeta Potential values of microwave-synthesized Fe_3O_4 particles and *C. vulgaris* cells at different pH values of the dispersing medium.

3.4. Adsorption Isotherms

The adsorption isotherms of the synthesized magnetite particles on the *C. vulgaris* cells are illustrated in Figure 6a,b along with the fitted curves using the Langmuir and Freundlich models respectively. The parameters of both models were obtained using the least squares linear fitting and are presented in Table 4.

Table 4. Parameters estimated using the Langmuir and Freundlich models.

Langmuir Model			Freundlich Model		
Q_m [g g^{-1}]	K_l [L g^{-1}]	R^2	K_f [g g^{-1}]	n_f	R^2
22.95	95.10	0.99	41.30	3.81	0.94

The results indicate that the adsorption isotherm of Fe_3O_4 particles on the *C. vulgaris* cells is better described by the Langmuir model of adsorption, according to which, a full coverage of the microalgae cells by the magnetic particles occurs. The maximum adsorption capacity, Q_m, is predicted to be equal to 22.95 g of dry biomass per gram of magnetic particles. Moreover, the high K_l value given by the Langmuir model denotes a strong attraction between the *C. vulgaris* cells and the microwave-synthesized magnetic particles [28]. The adsorption mechanism of Fe_3O_4 particles on *C. vulgaris* presented here has been previously reported for the same microalgae species both on naked and surface functionalized magnetite particles [19,21].

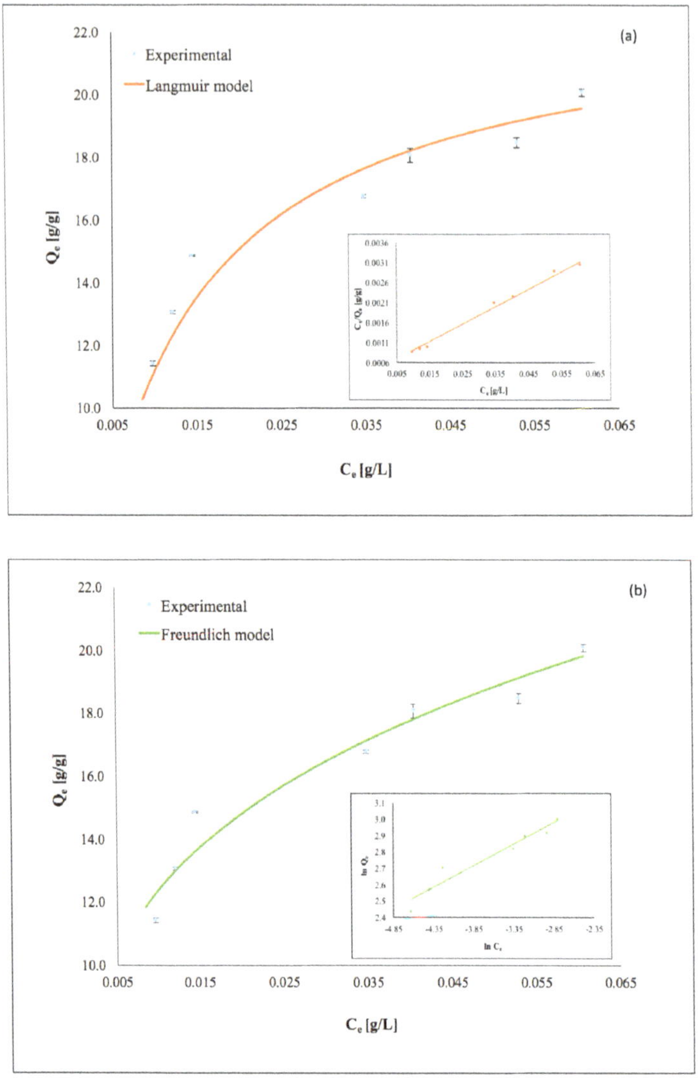

Figure 6. Experimental adsorption isotherm of microwave-synthesized magnetite particles on the *C. vulgaris* cells at 25 °C, fitted with (**a**) Langmuir model and (**b**) Freundlich model; pH = 7, rpm = 450, *C. vulgaris* cell concentration 0.22 g L^{-1}.

3.5. Regeneration and Reusability of Fe_3O_4 Particles

Both microalgae harvesting and the downstream processing of harvested biomass (e.g., to obtain biofuels or bio-products) call for recovery and regeneration of the magnetic particles, alongside testing the cell integrity to ensure the intracellular abidance of bioactive compounds and thereby their efficient and easy extraction. Our data indicate that the magnetic particles can be reused for at least five cycles, with less than 20% decrease of the algae harvesting efficiency—even during the 5th cycle, (Figure 7). This is similar to the results reported by Almomani [19]. These findings indicate that the regeneration procedure followed is an acceptable method, since an efficient reusability of the magnetic particles is observed.

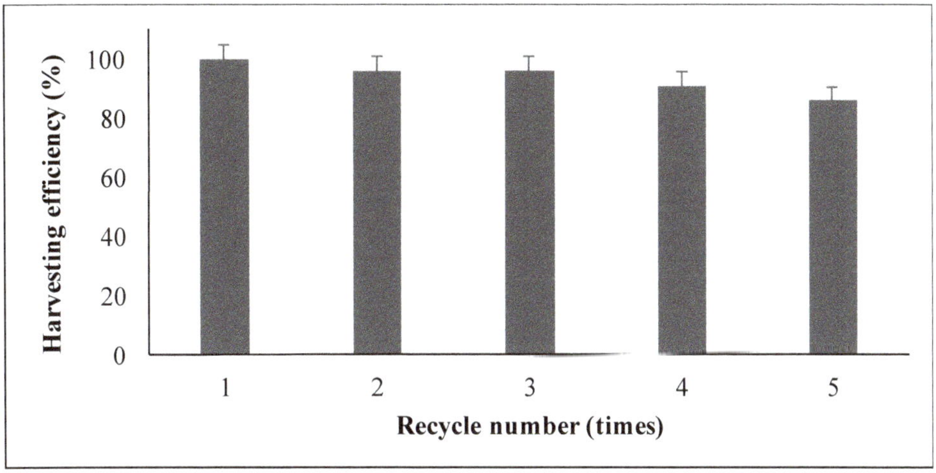

Figure 7. *Chlorella vulgaris* harvesting efficiency (initial concentration of 0.22 g L^{-1}) at five different consecutive cycles.

Analogous studies demonstrated a reusability efficiency of 80% after three cycles with chloroform: methanol treatments and ultrasonication [51], and at least 85% after five cycles of acid–base treatment combined with ultrasonication [52]. Markeb et al. [38], using NaOH, methanol, and chloroform as organic solvents, and ultrasonication, reported a minimum 80% recovery of the magnetic nanoparticles. Treatment with NaOH at pH = 12 resulted in 95% recovery for polypropylene/iron oxide nanoparticles, with the recovered magnetic particles retained almost the same microalgae biomass harvesting efficiency as per the newly synthesized ones [53]. An 80% reusability efficiency of cationic surfactant-decorated iron oxide nanoparticles after microalgae detachment using SDS and sonication has also been reported [54].

Staining with trypan blue (confirming compromised plasma membranes) verified the integrity of the bounded cells on the magnetic particles (Figure 8) and confirmed that no precious intracellular macromolecules were released in the harvesting medium. The trypan blue staining was performed at the most efficient pH for harvesting efficiency (pH = 3). Thus, the cells maintain their ability to produce high value-added products suitable for further use in various biotechnological applications and/or biofuels production.

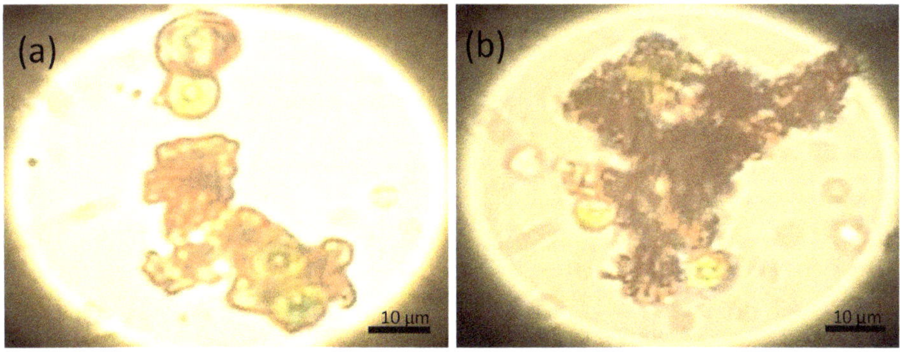

Figure 8. Trypan blue staining (**a**,**b**) on *Chlorella vulgaris* cells bound with magnetic particles (at pH 3.0) indicates the integrity of the cell membrane. Intact cells (viable) remained green, without the penetration of the trypan blue solution, while the broken cells appeared blue as stain diffused in the protoplasm.

4. Conclusions

Chlorella vulgaris harvesting was assessed using iron oxide magnetic particles synthesized using a Fe (II) precursor with the aid of microwave irradiation. The magnetic particles were characterized by X-ray diffraction, identifying one predominant phase, namely that of magnetite, with a mean diameter size of 20 nm. The simple and rapid production route followed in this work leads to the formation of high purity nanocrystalline magnetite with a superparamagnetic behavior at room temperature. The adsorption isotherm of the Fe_3O_4 particles on the *C. vulgaris* cells at ambient temperature, demonstrated a full coverage of the microalgae cells by the magnetic particles in accordance with the Langmuir model. Furthermore, the magnetite particles do not impose loss of microalgae cells' integrity. Response surface methodology verified the experimental optimum operational harvesting conditions, namely pH = 3, mass ratio = 14:1, and agitation speed = 450 rpm, under which a harvesting efficiency equal to 99.6% was achieved. pH was proved as the most crucial variant. The magnetic particles were successfully detached from the microalgal cells and reused for five cycles maintaining at least 80% of their initial harvesting efficiency. The separation mechanism was primarily attributed to the formation of hydrogen bonds between the magnetite particles and the microalgae cells under acidic conditions and to the expression of nano-size effects related to the high surface energy and large specific surface area of the particles. The high magnetic harvesting efficiencies (greater than 99%) obtained at pH 3.0 and mixing speed greater or equal to 350 rpm using microwave-synthesized naked magnetite (Fe_3O_4) particles, the cell integrity after the harvesting procedure and the ability to reuse the synthesized magnetic particles for at least five cycles of harvesting, contribute to the novelty of the present work and indicate that the proposed process is efficient and very promising.

Author Contributions: Conceptualization, E.V.; methodology, M.G.S., M.M.D.; validation, M.G.S., M.M.D., V.L., E.V.; software, I.G.; investigation, M.G.S., M.M.D.; resources, M.G.S., M.M.D., E.V.; data curation, M.G.S., M.M.D., V.L., H.S., D.K., E.V.; writing—original draft preparation, M.G.S., M.M.D.; writing—review and editing, V.L., E.V.; supervision, H.S., E.V.; project administration, V.L., E.V. All authors have read and agreed to the published version of the manuscript.

Funding: This research has been co-financed by the European Union and Greek national funds through the Operational Program Competitiveness, Entrepreneurship and Innovation, under the call, "Special Actions, Aquaculture-Industrial Materials-Open Innovation in Culture" (project code: T6YBP-00033).

Institutional Review Board Statement: Not applicable.

Informed Consent Statement: Not applicable.

Data Availability Statement: Additional data for this study are not available on public database, the corresponding author can provide them upon request.

Acknowledgments: The authors would like to thank Petros Schinas and Nikolaos Panagiotou for the assistance in SEM analysis and XRD, as well as Evangelos Hristoforou for the assistance in VSM analysis.

Conflicts of Interest: The authors declare no conflict of interest.

References

1. Khan, M.I.; Shin, J.H.; Kim, J.D. The promising future of microalgae: Current status, challenges and optimization of a sustainable and renewable industry for biofuels, feed, and other products. *Microb. Cell Fact.* **2018**, *17*, 36. [CrossRef] [PubMed]
2. Sharma, R.K.; Saharia, M.; Srivstava, R.; Kumar, S.; Sahoo, L. Tailoring microalgae for efficient biofuel production. *Front. Mar. Sci.* **2018**, *5*, 382. [CrossRef]
3. Hossain, N.; Mahlia, T.M.I.; Saidur, R. Latest development in microalgae-biofuel production with nano-additives. *Biotechnol. Biofuels* **2019**, *12*, 125. [CrossRef] [PubMed]
4. Novoveská, L.; Ross, M.E.; Stanley, M.S.; Pradelles, R.; Wasiolek, V.; Sassi, J.-F. Microalgal Carotenoids: A Review of Production, Current Markets, Regulations, and Future Direction. *Mar. Drugs* **2019**, *17*, 640. [CrossRef] [PubMed]

5. Gouveia, L.; Oliveira, A.C. Microalgae as a raw material for biofuels production. *J. Ind. Microbiol. Biotechnol.* **2009**, *36*, 269–274. [CrossRef]
6. Clarens, A.F.; Resurreccion, E.P.; White, M.A.; Colost, L.M. Environmental life cycle comparison of algae to other bioenergy feedstocks. *Environ. Sci. Technol.* **2010**, *44*, 1813–1819. [CrossRef]
7. Coons, J.E.; Kalb, D.M.; Dale, T.; Marrone, B.L. Getting to low-cost algal biofuels: A monograph on conventional and cutting-edge harvesting and extraction technologies. *Algal Res.* **2014**, *6 Pt B*, 250–270. [CrossRef]
8. Sharma, K.K.; Garg, S.; Li, Y.; Malekizadeh, A.; Schenk, P.M. Critical analysis of current microalgae dewatering techniques. *Biofuels* **2013**, *4*, 397–407. [CrossRef]
9. Zhu, L.-D.; Hiltunen, E.; Li, Z. Using magnetic materials to harvest microalgal biomass: Evaluation of harvesting and detachment efficiency. *Environ. Technol.* **2019**, *40*, 1006–1012. [CrossRef]
10. Milledge, J.J.; Heaven, S. A review of the harvesting of micro-algae for biofuel production. *Rev. Environ. Sci. Biotechnol.* **2013**, *12*, 165–178. [CrossRef]
11. Molina Grima, E.; Belarbi, E.H.; Acién Fernández, F.G.; Robles Medina, A.; Chisti, Y. Recovery of microalgal biomass and metabolites: Process options and economics. *Biotechnol. Adv.* **2003**, *20*, 491–515. [CrossRef]
12. Bharte, S.; Desai, K. Harvesting Chlorella species using magnetic iron oxide nanoparticles. *Psychol. Res.* **2019**, *67*, 128–133. [CrossRef]
13. Papazi, A.; Makridis, P.; Divanach, P. Harvesting Chlorella minutissima using cell coagulants. *J. Appl. Phycol.* **2010**, *22*, 349–355. [CrossRef]
14. Chen, L.; Wang, C.; Wang, W.; Wei, J. Optimal conditions of different flocculation methods for harvesting *Scenedesmus* sp. cultivated in an open-pond system. *Bioresour. Technol.* **2013**, *133*, 9–15. [CrossRef] [PubMed]
15. Xu, Y.; Purton, S.; Baganz, F. Chitosan flocculation to aid the harvesting of the microalga Chlorella sorokiniana. *Bioresour. Technol.* **2013**, *129*, 296–301. [CrossRef] [PubMed]
16. Matter, A.I.; Bui, V.K.H.; Jung, M.; Seo, J.Y.; Kim, Y.-E.; Lee, Y.-C.; Oh, Y.-K. Flocculation Harvesting Techniques for Microalgae: A Review. *Appl. Sci.* **2019**, *9*, 3069. [CrossRef]
17. Borlido, L.; Azevedo, A.M.; Roque, A.C.A.; Aires-Barros, M.R. Magnetic separations in biotechnology. *Biotechnol. Adv.* **2013**, *31*, 1374–1385. [CrossRef]
18. Yavuz, C.T.; Prakash, A.; Mayo, J.T.; Colvin, V.L. Magnetic separations: From steel plants to biotechnology. *Chem. Eng. Sci.* **2009**, *64*, 2510–2521. [CrossRef]
19. Almomani, F. Algal cells harvesting using cost-effective magnetic nano-particles. *Sci. Total Environ.* **2020**, *720*, 137621. [CrossRef]
20. Boli, E.; Savvidou, M.; Logothetis, D.; Louli, V.; Pappa, G.; Voutsas, E.; Kolisis, F.; Magoulas, K. Magnetic harvesting of marine algae *Nannochloropsis oceanica*. *Sep. Sci. Technol.* **2017**, *6395*, 1–8. [CrossRef]
21. Gerulová, K.; Bartošová, A.; Blinová, L.; Bártová, K.; Dománková, M.; Garaiová, Z.; Palcut, M. Magnetic Fe_3O_4-polyethyleneimine nanocomposites for efficient harvesting of *Chlorella zofingiensis*, *Chlorella vulgaris*, *Chlorella sorokiniana*, *Chlorella ellipsoidea* and *Botryococcus braunii*. *Algal Res.* **2018**, *33*, 165–172. [CrossRef]
22. Savvidou, M.G.; Boli, E.; Logothetis, D.; Lymperopoulou, T.; Ferraro, A.; Louli, V.; Mamma, D.; Kekos, D.; Magoulas, K.; Kolisis, F.N. A Study on the Effect of Macro-and Micro-Nutrients on Nannochloropsis oceanica Growth, Fatty Acid Composition and Magnetic Harvesting Efficiency. *Plants* **2020**, *9*, 660. [CrossRef] [PubMed]
23. Taghizadeh, S.M.; Berenjian, A.; Chew, K.W.; Show, P.L.; Mohd Zaid, H.F.; Ramezani, H.; Ghasemi, Y.; Raee, M.J.; Ebrahiminezhad, A. Impact of magnetic immobilization on the cell physiology of green unicellular algae *Chlorella vulgaris*. *Bioengineered* **2020**, *11*, 141–153. [CrossRef]
24. Eroglu, E.; Agarwal, V.; Bradshaw, M.; Chen, X.; Smith, S.M.; Raston, C.L.; Swaminathan Iyer, K. Nitrate removal from liquid effluents using microalgae immobilized on chitosan nanofiber mats. *Green Chem.* **2012**, *14*, 2682–2685. [CrossRef]
25. Hu, Y.R.; Guo, C.; Wang, F.; Wang, S.K.; Pan, F.; Liu, C.Z. Improvement of microalgae harvesting by magnetic nanocomposites coated with polyethylenimine. *Chem. Eng. J.* **2014**, *242*, 341–347. [CrossRef]
26. Prochazkova, G.; Podolova, N.; Safarik, I.; Zachleder, V.; Branyik, T. Physicochemical approach to freshwater microalgae harvesting with magnetic particles. *Colloids Surf. B Biointerfaces* **2013**, *112*, 213–218. [CrossRef] [PubMed]
27. Prochazkova, G.; Safarik, I.; Branyik, T. Harvesting microalgae with microwave synthesized magnetic microparticles. *Bioresour. Technol.* **2013**, *130*, 472–477. [CrossRef]
28. Xu, L.; Guo, C.; Wang, F.; Zheng, S.; Liu, C.Z. A simple and rapid harvesting method for microalgae by in situ magnetic separation. *Bioresour. Technol.* **2011**, *102*, 10047–10051. [CrossRef]
29. Fraga-Garcia, P.; Kubbutat, P.; Brammen, M.; Schwaminger, S.; Berensmeier, S. Bare iron oxide nanoparticles for magnetic harvesting of microlagae: From interaction behavior to process realization. *Nanomaterials* **2018**, *8*, 292. [CrossRef]
30. Franzreb, M.; Siemann-Herzberg, M.; Hobley, T.J.; Thomas, O.R.T. Protein purification using magnetic adsorbent particles. *Appl. Microbiol. Biotechnol.* **2006**, *70*, 505–516. [CrossRef]
31. Sharma, R. Effects of Culture Conditions on Growth and Biochemical Profile of Chlorella Vulgaris. *J. Plant Pathol. Microbiol.* **2012**, *3*, 1000131. [CrossRef]
32. Zhang, R.; Chen, J.; Mao, X.; Qi, P.; Zhang, X. Anti-inflammatory and anti-aging evaluation of pigment-protein complex extracted from *Chlorella Pyrenoidosa*. *Mar. Drugs* **2019**, *17*, 586. [CrossRef]

33. Shibata, S.; Natori, Y.; Nishihara, T.; Tomisaka, K.; Matsumoto, K.; Sansawa, H. Antioxidant and Anti-Cataract Effects of Chlorella on Rats with Streptozotocin-Induced Diabetes. *J. Nutr. Sci. Vitaminol.* **2003**, *49*, 334–339. [CrossRef]
34. Ishiguro, S.; Robben, N.; Burghart, R.; Cote, P.; Greenway, S.; Thakkar, R.; Upreti, D.; Nakashima, A.; Suzuki, K.; Comer, J.; et al. Cell wall membrane fraction of *Chlorella sorokiniana* enhances hot antitumor immunity and inhbits colon carcinoma growth in mice. *Integr. Cell Ther.* **2020**, *19*, 1–10.
35. Belasco, W. Algae burgers for a hungry world? The rise and fall of chlorella cuisine. *Technol. Cult.* **1997**, *38*, 608–634. [CrossRef]
36. Rani, K.; Sandal, N.; Sahoo, P.K. A comprehensive review on chlorella-its composition, health benefits, market and regulatory scenario. *J. Pharm. Innov.* **2018**, *7*, 584–589.
37. Dardavila, M.M.; Hamilakis, S.; Loizos, Z.; Kollia, C. Ni/ZrO_2 composite electrodeposition in the presence of coumarin: Textural modifications and properties. *J. Appl. Electrochem.* **2015**, *45*, 503–514. [CrossRef]
38. Markeb, A.A.; Llimós-Turet, J.; Ferrer, I.; Blánquez, P.; Alonso, A.; Sánchez, A.; Moral-Vico, J.; Font, X. The use of magnetic iron oxide based nanoparticles to improve microalgae harvesting in real wastewater. *Water Res.* **2019**, *159*, 490–500. [CrossRef]
39. Aivazoglou, E.; Metaxa, E.; Hristoforou, E. Microwave-assisted synthesis of iron oxide nanoparticles in biocompatible organic environment. *AIP Adv.* **2018**, *8*, 048201. [CrossRef]
40. Leite, L.D.S.; Daniel, L.A. Optimization of microalgae harvesting by sedimentation induced by high pH. *Water Sci. Technol.* **2020**, *82*, 1227–1236. [CrossRef]
41. Zhao, Y.; Liang, W.; Liu, L.; Li, F.; Fan, Q.; Sun, X. Harvesting Chlorella vulgaris by magnetic flocculation using Fe_3O_4 coating with polyaluminium chloride and polyacrylamide. *Bioresour. Technol.* **2015**, *198*, 789–796. [CrossRef]
42. Tork, M.B.; Khalilzadeh, R.; Kouchakzadeh, H. Efficient harvesting of marine Chlorella vulgaris microalgae utilizing cationic starch nanoparticles by response surface methodology. *Bioresour. Technol.* **2017**, *243*, 583–588. [CrossRef] [PubMed]
43. Fayad, N.; Yehya, T.; Audonnet, F.; Vial, C. Harvesting of microalgae Chlorella vulgaris using electro-coagulation-flocculation in the batch mode. *Algal Res.* **2017**, *25*, 1–11. [CrossRef]
44. Liu, Y.; Jin, W.; Zhou, X.; Han, S.F.; Tu, R.; Feng, X.; Wang, Q. Efficient harvesting of Chlorella pyrenoidosa and Scenedesmus obliquus cultivated in urban sewage by magnetic flocculation using nano-Fe_3O_4 coated with polyethyleneimine. *Bioresour. Technol.* **2019**, *290*, 121771. [CrossRef]
45. Razack, S.A.; Duraiarasan, S.; Shellomith, A.S.; Muralikrishnan, K. Statistical optimization of harvesting Chlorella vulgaris using a novel bio-source, Strychnos potatorum. *Biotechnol. Rep.* **2015**, *7*, 150–156. [CrossRef]
46. Zhao, Y.; Fan, Q.; Wang, X.; Jiang, X.; Jiao, L.; Liang, W. Application of Fe_3O_4 coated with modified plant polyphenol to harvest oleaginous microalgae. *Algal Res.* **2019**, *38*, 101417. [CrossRef]
47. Wang, T.; Yang, W.L.; Hong, Y.; Hou, Y.L. Magnetic nanoparticles grafted with amino-riched dendrimer as magnetic flocculant for efficient harvesting of oleaginous microalgae. *Chem. Eng. J.* **2016**, *297*, 304–314. [CrossRef]
48. Wang, X.; Zhao, Y.; Jiang, X.; Liu, L.; Li, X.; Li, H.; Liang, W. In-situ self-assembly of plant polyphenol-coated Fe_3O_4 particles for oleaginous microalgae harvesting. *J. Environ. Manag.* **2018**, *214*, 335–345. [CrossRef]
49. Hu, Y.R.; Wang, F.; Wang, S.K.; Liu, C.Z.; Guo, C. Efficient harvesting of marine microalgae Nannochloropsis maritima using magnetic nanoparticles. *Bioresour. Technol.* **2013**, *138*, 387–390. [CrossRef]
50. Li, Y.G.; Gao, H.S.; Li, W.L.; Xing, J.M.; Liu, H.Z. In situ magnetic separation and immobilization of dibenzothiophene-desulfurizing bacteria. *Bioresour. Technol.* **2009**, *100*, 5092–5096. [CrossRef]
51. Lin, Z.; Xu, Y.; Zhen, Z.; Fu, Y.; Liu, Y.; Li, W.; Zhang, D. Application and reactivation of magnetic nanoparticles in Microcystis aeruginosa harvesting. *Bioresour. Technol.* **2015**, *190*, 82–88. [CrossRef] [PubMed]
52. Xu, Y.; Wang, X.; Fu, Y.; Hu, F.; Qian, G.; Liu, Q.; Sun, Y. Interaction energy and detachment of magnetic nanoparticles-algae. *Environ. Sci. Technol.* **2020**, *41*, 2618–2624. [CrossRef] [PubMed]
53. Hena, S.; Fatihah, N.; Tabassum, S.; Lalung, J.; Jing, S.Y. Magnetophoretic harvesting of freshwater microalgae using polypyrrole/Fe_3O_4 nanocomposite and its reusability. *J. Appl. Phycol.* **2016**, *28*, 1597–1609. [CrossRef]
54. Seo, J.Y.; Praveenkumar, R.; Kim, B.; Seo, J.-C.; Na, J.-G.; Seo, J.C.; Park, J.-Y.; Na, J.-G.; Jeon, S.G.; Park, S.B.; et al. Downstream integration of microalgae harvesting and cell disruption by means of cationic surfactant-decorated Fe_3O_4 nanoparticles. *Green Chem.* **2016**, *18*, 3981–3989. [CrossRef]

nanomaterials

MDPI

Article

Papain Covalently Immobilized on Chitosan–Clay Nanocomposite Films: Application in Synthetic and Real White Wine

Ilaria Benucci [1], Claudio Lombardelli [1], Ilaria Cacciotti [2,*] and Marco Esti [1]

[1] Department of Agriculture and Forestry Science (DAFNE), Tuscia University, via S. Camillo de Lellis snc, 01100 Viterbo, Italy; ilaria.be@unitus.it (I.B.); claudiolombardelli@libero.it (C.L.); esti@unitus.it (M.E.)

[2] Department of Engineering, University of Rome "Niccolò Cusano", INSTM RU, Via Don Carlo Gnocchi, 3, 00166 Rome, Italy

* Correspondence: ilaria.cacciotti@unicusano.it

Received: 12 July 2020; Accepted: 17 August 2020; Published: 19 August 2020

Abstract: Increasing attention has been recently paid to the development of nanocomposite materials for food application as new tool to enhance the mechanical and thermal properties of polymers. In this study, novel chitosan–clay nanocomposite films were produced as carriers for the covalent immobilization of papain, by using a fixed amount of chitosan (1% *w*/*v*) and a food-grade activated montmorillonite (Optigel, OPT) or a high-purity unmodified montmorillonite (SMP), in four different weight percentages with respect to chitosan (i.e., 20, 30, 50, 70% *w*/*w*). Both nanoclays (OPT and SMP) improved the mechanical properties of the obtained nanocomposites, and the OPT films showed the highest Young modulus and mechanical resistance (σ_{max}). The nanocomposites were used as carriers for the covalent immobilization of papain, which was preliminarily characterized in model wine towards a synthetic substrate, showing the highest efficiency in the release of the reaction product when it was bound on OPT-30 and OPT-50 films. Finally, the latter biocatalyst (papain on OPT-50 film) was applied for the protein stabilization of two different unfined white wines, and it efficiently reduced both the haze potential and the protein content.

Keywords: chitosan; clay; nanocomposite films; papain; covalent immobilization; wine

1. Introduction

The application of enzymes as powerful catalysts in a large number of industrial fields, including pharmaceuticals and food, allows the development of efficient processes with low environmental impact [1]. In the wine-making industry, some enzymes have been traditionally used whereas others have been only recently introduced. Proteases, able to catalyze the hydrolysis of proteins, have been proposed as a selective alternative to the conventional bentonite fining for removing the haze-forming proteins, which are responsible for haze formation in white wine under post-bottling conditions.

In recent decades, immobilization techniques have been successfully applied to enhance enzyme properties, in particular stability and reusability [2]. The material employed for the support preparation is of fundamental importance in the immobilization process, affecting the catalytic properties of the produced biocatalyst. A wide range of inorganic and organic materials, as well as composites, may be applied as carriers for enzyme immobilization [3,4] in different shapes (e.g., films [5], beads [6,7] and fibers [8]). Recently, increasing attention has been paid to organic–inorganic composite materials for food application. Regarding composite films as supports for enzyme immobilization, Yang et al. [9] reported that the chemical and thermal stability of organic membranes could be enhanced by an inorganic phase. The latter usually allows the stabilization of enzyme–support interactions [10].

Among biopolymers, chitosan (CS, a linear polysaccharide made of *N*-acetyl-D-glucosamine and D-glucosamine units [11]) is the most commonly used for the production of immobilization supports as it is, as well as in composite/nanocomposite forms [2], to be applied in the food and pharmaceutical industries.

Several studies have been aimed at improving the mechanical and thermal stability of neat CS films [12–14] by seeking CS-based nanocomposites, with the addition of specific nanoparticles to enhance the aforementioned properties [3,13,15]. The nanomaterials dispersed within the chitosan matrix, e.g., nanoclays, carbon structures, and metal/metal oxides, via physical or chemical interaction [16], allow not only to enhance the physical, mechanical, and thermal stability but also to endow the composite with their intrinsic features, e.g., high specific surface area [11].

Among the nanoparticles, clay is a fine-grained soil material that contains metal oxides or hydroxides with traces of organic matters. Due to its smallest dimension, excellent colloidal properties [11], and ability to interact with CS by electrostatic force, clay has been used for the preparation of CS–clay nanocomposites through adsorption, gelation, or intercalation [13,17].

Recently, CS–clay nanocomposite films have been prepared by using different nanoclays (i.e., bentonite, sepiolite and montmorillonite (MMT)) and have been applied as supports for the covalent immobilization of stem bromelain in wine production [13,15]. In our previous works, we proposed nanocomposites based on CS and nanoclays in very low concentrations (1–5% *w*/*w*) [12] and systems based on CS and nanoclays in high concentrations (70–80% *w*/*w*) [14], maintaining, in both cases, the total weight of the composite constant. Papain from Carica papaya L. latex (EC 3.4.22.2), a cysteine protease applied in brewing for the removal of chill haze [18] and for the clarification treatment of pomegranate juice [19], has been covalently immobilized on CS. Moreover, Benucci et al. [20] immobilized papain on commercial chitosan beads by direct linkage for the protein stabilization of white wines.

This study was focused on the production of novel CS–clay nanocomposite films, using a fixed amount of CS (1% *w*/*v*) and a food-grade activated MMT (Optigel, OPT) or a high-purity unmodified MMT (SMP), in four different weight percentages with respect to CS (i.e., 20, 30, 50, 70% *w*/*w*), as carriers for the covalent immobilization of papain.

The physical properties, in terms of the morphology, thermal stability and the mechanical behavior, of the produced nanocomposites were investigated. Finally, the catalytic properties of immobilized papain were tested in synthetic wine and its effectiveness in protein stabilization was investigated in real white wines.

2. Materials and Methods

2.1. Enzyme, Chemicals and Wines

Papain from Carica Papaya latex (EC 3.4.22.2), shellfish-derived chitosan (CS) powder (low molecular weight 50–190 kDa; percentage of deacetylation 75%), glutaraldehyde (GDH, 25% v/v), and glycerol (≥99.5%) were obtained from Sigma-Aldrich (Milan, Italy). Two different nanoclays, i.e., activated food-grade montmorillonite (Optigel, OPT) and high-purity unmodified montmorillonite (SMP), were kindly provided by BYK Additives GmbH (Wesel, Germany) and Zhejiang Fenghong New Material Co., Ltd. (Huzhou, China), respectively. The selected synthetic tripeptide chromogenic substrate (Bz–Phe–Val–Arg–p-nitroaniline (pNA)), applied for the kinetic characterization of immobilized papain, was purchased from Bachem (Bubendorf, Switzerland). All the other chemicals were of analytical grade (Sigma Aldrich, Milan, Italy). Manzoni and Sauvignon Blanc unfined white wines (vintage 2018) were kindly provided by Casale del Giglio winery (Le Ferriere, LT, Italy) and their oenological parameters are summarized in Table 1.

Table 1. Oenological parameters of the unfined Manzoni and Sauvignon Blanc wines.

	Manzoni	Sauvignon Blanc
pH	3.37 ± 0.01	3.34 ± 0.01
Total acidity (g/L tartaric acid)	6.00 ± 0.05	6.00 ± 0.05
Alcohol level (% *v*/*v*)	11.2 ± 0.3	13.2 ± 0.2
Free SO_2 (mg/L)	6 ± 1	10 ± 1
Total SO_2 (mg/L)	30 ± 2	48 ± 3
Total phenols (mg/L catechin)	217 ± 1	219 ± 4
Total protein (mg/L)	465 ± 43	120 ± 7
ΔNTU $Index_0$	600 ± 11	116 ± 8

Δ Nephelometric Turbidity Units (NTU) $Index_0$, difference between the turbidity of the initial wine and that of the wine after the heat test.

2.2. Preparation of CS–Clay Nanocomposite Films by Solvent Casting

CS–clay supports were obtained by solvent casting technique, using low molecular weight CS (1% *w*/*v*) blended with glycerol (CS:glycerol 75:25, in %*w*/*w*) [5], and adding two different nanoclay types, i.e., OPT and SMP, in four different weight concentrations (i.e., 20, 30, 50, 70% *w*/*w*). The set-up procedure was reported elsewhere [5]. Briefly, nanoclays aqueous suspensions were prepared by ultrasonication using Sonics Vibracell CV33 (Sonics & Materials Inc., Newtown, CT, USA) in the following conditions: power 750 W, frequency 20 kHz, amplitude 30%, time 30 min. Afterwards, acetic acid (2% *v*/*v*), glycerol and CS were added and the mixtures were maintained under continuous magnetic stirring overnight. The final suspensions were solvent cast onto plastic Petri dishes, in a fume hood, and dried at room temperature for 48 h. Furthermore, as a reference, a system composed of only CS and glycerol was produced. The obtained samples were designed as Clay-free for the CS/glycerol-based film and OPT-*x* and SMP-*x* in the case of clay-loaded composites (where *x* is the w/w percentage of the used nanoclay with respect to the CS, i.e., 20, 30, 50, and 70% *w*/*w*).

2.3. Physical Characterization of CS–Clay Nanocomposite Films

The produced films were morphologically, thermally and mechanically analyzed by Field-Emission Gun Scanning Electron Microscope (FEG-SEM, Leo Supra 35, Carl Zeiss SMT Ltd., Cambridge, UK), differential scanning calorimetry (DSC, TAInstruments Q2000, New Castle, DE, USA) and uniaxial tensile tests (Lloyd LRX, Lloyd Instruments Ltd., West Sussex, UK), respectively. In detail, the SEM micrographs were acquired on gold-coated samples, applying a voltage of 3–5 kV. DSC measurements were carried out between 25 and 400 °C (heating rate 10 °C·min^{-1}), in a nitrogen atmosphere (N_2 flow rate 50 cc·min^{-1}), using a sample weight of ~5 mg. For the mechanical investigation, the ASTM D1708 and ASTM D882 standards were followed, and the nominal specimen cross-section was considered for the measurement of all mechanical features. Dog-bone specimens (width 4.8 mm, length 22.25 mm) were obtained from the produced films. The tests were performed at 1.2 mm·min^{-1}, using a 50 N load cell. All measurements were taken in triplicate.

2.4. Papain Immobilization on CS–Clay Nanocomposite Films

Before enzyme immobilization, the CS–film samples were neutralized by overnight shaking in 26% (*v*/*v*) ethanol–NaOH 2 M solution and then cut into squares (10 mm × 10 mm) with a razor blade. The immobilization procedure was performed as briefly reported in the following [13]. The film surface was activated by adding 1 mL of 3% (*v*/*v*) GDH as a cross-linker, and keeping at room temperature under constant agitation (120 rpm) for 2 h. The activated films were thoroughly washed with distilled water; thereafter, 1 mL of papain preparation (0.12 $mg_{protein}$/mL solubilized in the tartaric buffer) was added. After the overnight incubation (150 rpm at 20 °C), the obtained biocatalysts were carefully washed with tartaric buffer, and then left to stand for 20 min in 0.1 M glycine solution. At the end of the immobilization procedure, the biocatalysts were washed three times with 2 M NaCl solution to remove all non-covalently bound proteins. The immobilization yield (IY, %) was calculated as the difference between the protein concentration in the enzyme solution, before and after immobilization. The protein concentration was evaluated by the Bradford method [21] using bovine serum albumin (BSA) as a standard protein.

2.5. *Proteolytic Activity Assay*

Proteolytic activity toward the tripeptide chromogenic substrate (Bz–Phe–Val–Arg–pNA) was tested in model wine (0.03 M, tartaric acid/sodium tartrate solution pH 3.2, containing 12% *v*/*v* of ethanol) at 20 °C, as reported by Benucci et al. [13].

2.6. *Kinetic Characterization of Immobilized Papain*

A kinetic study of papain immobilized on CS–clay nanocomposite films was carried out in model wine, fortified with Bz–Phe–Val–Arg–pNA substrate (0–750 μM). Kinetic parameters (V_{max} and K_M (Michaelis–Menten constant)) were determined according to the Michaelis–Menten equation using a non-linear regression procedure (GraphPad Prism 5.01, GraphPad Software, Inc., San Diego, CA, USA). Moreover, k_{cat} (turnover number) and K_a (affinity constant) were calculated as described by Benucci et al. [13].

2.7. *Wine Stabilization Treatment in the Batch-Scale Stirred Reactor*

Ten milliliters of each white wine were treated using the immobilized biocatalyst in a laboratory-scale stirred reactor (120 rpm, 20 °C) for 24 h. Trials were performed in triplicate, in three different stirred reactors, in order to obtain three independent replicates. To investigate the potential absorption effect of the CS–clay nanocomposite films, a blank correction was made using the supports covered with protease firstly immobilized and then deactivated with NaOH 16 N.

2.8. *Wine Protein Content Determination*

The total protein content of both white wines, before and after the enzymatic treatment in the laboratory-scale stirred reactor, was determined using the potassium dodecyl sulphate method, according to Gaspar et al. [22]. All measurements were made in triplicate.

2.9. *Heat Test*

The protein stability of both white wines, before and after the enzymatic treatment in laboratory-scale stirred reactor, was investigated by heat test, incubating the wines at 80 °C for 6 h, and then keeping them at 4 °C for 16 h [23]. Haze formation was measured using HD 25.2 turbidimeter (Delta Ohm, Padua, Italy), and the turbidity was expressed in nephelometric turbidity units (NTU). Wine stability was calculated as the difference in the wine NTU after and before the heat test (haze potential). The turbidity removal yield (TRY (%)) was expressed as the percentage of haze potential decrease due to the proteolytic treatment. All measurements were made in triplicate.

2.10. *Statistical Analysis*

One-way completely randomized analysis of variance (ANOVA), with an EXCEL® Add-in macro, was employed to analyze the data, obtained from the average of three replicate measurements. Then, a Tukey honestly significant difference (Tukey HSD) post-hoc test ($p = 0.05$) for the multiple comparisons of samples was applied.

3. Results and Discussion

3.1. *Physical Properties of CS–Clay Nanocomposite Films*

All the samples were morphologically analyzed. As an example, the SEM micrographs of SMP70 and OPT70 are compared in Figure 1. Two very different filler distribution behaviors were observed.

OPT clays were better dispersed, promoting a more uniform and homogeneous surface. On the contrary, the SMP addition induced the development of several agglomerates, leading to a very rough surface, characterized by several defects (Figure 1).

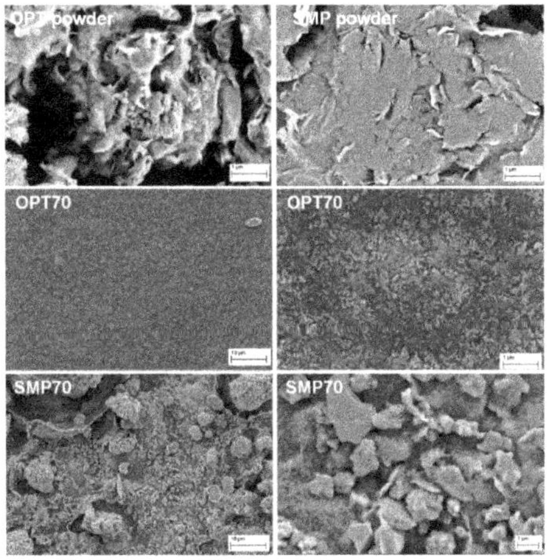

Figure 1. SEM micrographs of the food-grade activated montmorillonite (Optigel, OPT) and high-purity unmodified montmorillonite (SMP) powders, and the OPT- and SMP-based films.

The influence of OPT and SMP on the thermal properties was explored by DSC measurements. In Table 2, the acquired thermal properties are collected and the DSC thermograms of OPT20 and SMP20, as an example, are compared to that of the clay-free sample in Figure 2.

Table 2. Thermal properties of the chitosan and nanoclay-based supports.

	Clay-Free	OPT			SMP		
		20	30	50	20	30	50
T_{IHB} (°C)	121 ± 0.5	118 ± 0.4	124 ± 0.7	111 ± 0.6	118 ± 0.4	115 ± 0.4	105 ± 0.3
Td (°C)	287 ± 1.2	274 ± 1.3	272 ± 0.9	260 ± 1.0	279 ± 1.1	280 ± 1.4	284 ± 1.5
ΔH_{mI} (J/g)	259 ± 1.5	259 ± 1.4	238 ± 1.2	142 ± 0.9	274 ± 1.1	296 ± 1.3	335 ± 2.0

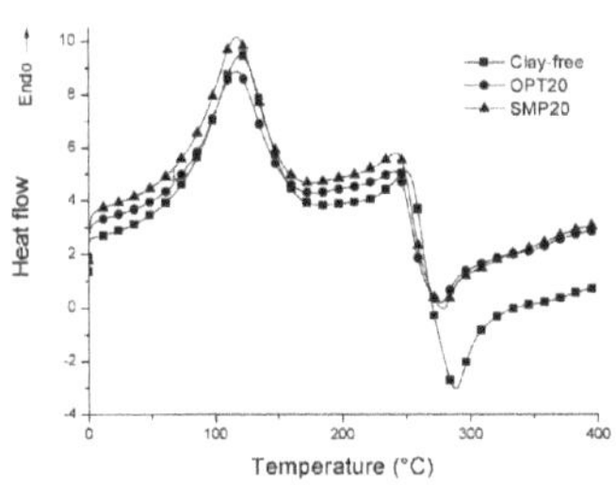

Figure 2. Differential scanning calorimetry (DSC) thermograms of Clay-free, OPT20 and SMP20 samples.

In all cases, two main peaks, one endothermic and the other one exothermic, were detected at approximately 105–120 °C and 235–335 °C (Table 2), respectively. The first one was ascribed to the dissociation process of interchain hydrogen bonding of CS (T_{IHB}) [24,25], whereas the second one to CS decomposition (T_d) [26–28]. In all the composite systems, a decrement of T_{IHB} temperature was revealed, compared to Clay-free (121 °C). Particularly, a progressive decrease increasing the nanoclay amount was evidenced in the case of SMP-based films. For the related enthalpies, different trends were recorded for OPT- and SMP-based films. For the former, a decrement of the enthalpy was evidenced increasing OPT amount, whereas, an opposite trend was revealed for SMP-based composites. This experimental evidence indicates that higher energy is necessary in order to induce the dissociation process of interchain hydrogen bonding of CS in the case of SMP-based systems. Concerning the chitosan degradation temperature, a decrease was evidenced in the case of composite systems with respect to Clay-free film, suggesting a not good filler dispersion within the polymeric matrix, with the consequent development of many agglomerates.

The effect of nanoclays on the mechanical behavior was analyzed performing uniaxial tensile tests (Figure 3, Table 3).

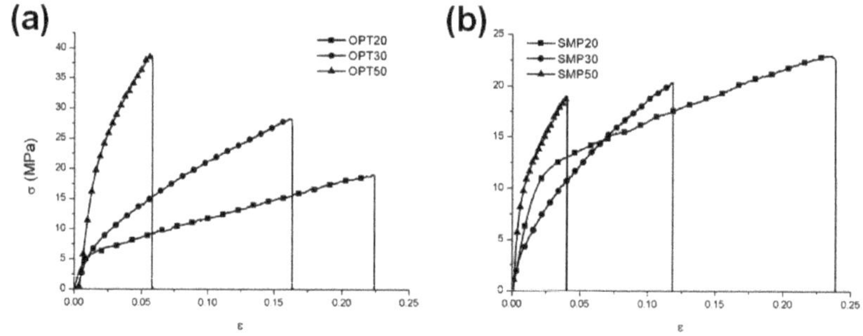

Figure 3. σ–ε curves of OPT (**a**) and SMP (**b**) based films.

Table 3. Mechanical properties of the OPT- and SMP-based films, compared to the neat one (Clay-free).

	E (MPa)	σ_{max} (MPa)	ε_{max}
Clay-Free	579 ± 23	16 ± 1	0.30 ± 0.01
OPT-20	738 ± 20	21± 6	0.22 ± 0.08
OPT-30	653 ± 71	23 ± 4	0.13 ± 0.05
OPT-50	1670 ± 202	36 ± 7	0.06 ± 0.03
SMP-20	580 ± 55	24 ± 4	0.28 ± 0.05
SMP-30	528 ± 4	22 ± 3	0.12 ± 0.03
SMP-50	1582 ± 210	19 ± 4	0.04 ± 0.01

An improvement in the mechanical responsiveness, in terms of the Young modulus and maximum stress value (σ_{max}), was evidenced in the case of all composite systems, particularly those based on OPT, due to its better distribution within the polymeric matrix, as testified by the reported SEM micrographs (Figure 1) and DSC data (Table 2). The OPT and SMP-based films presented different mechanical behavior. In detail, a progressive increment of the ultimate tensile strength (σ_{max}) was evidenced for OPT-based samples with the OPT amount, whereas an opposite trend was observed in the case of SMP-based ones (Table 3), probably due to the presence of several agglomerates whose amount tended to increase with filler percentage (Figure 1), as expected.

Additionally, a remarkable increase in the Young Modulus value was observed for a nanoclay amount of 50% w/w, independently of the considered nanoclay. On the contrary, up to 30% w/w), comparable values were recorded (Table 3). As expected, by increasing the nanoclay amount, a progressive decrease in the elongation at break (ε_{max}) was revealed.

Taking into account all the collected morphological, thermal and mechanical properties, it is evident that OPT presented a higher affinity towards CS, as testified by its better distribution within the polymeric matrix, promoting a good filler-polymeric chains interaction.

Indeed, SMP and OPT led to distinct behaviour due to their different chemical nature and microstructure. In details, even if both SMP and OPT powders were composed of platelet-like particles (Figure 1), a higher tendency to agglomeration was evidenced in the case of SMP powder, whereas

OPT nanoclays were more homogeneously dispersed within the polymeric matrix. These experimental evidences could be ascribed to OPT good wettability and compatibility, able to promote a significant improvement of the mechanical behavior (Table 3).

3.2. *Kinetic Properties of Papain Immobilized on CS–Clay Nanocomposite Films*

CS–clay films, produced by the addition of two nanoclays (OPT and SMP) in different amounts (20, 30, 50, 70% *w/w* with respect to CS), were applied as supports for the papain covalent immobilization.

Data in Table 4 show that the IY was significantly higher for SMP-70 (60%) with respect to Clay-free sample (45%), thus proving that the addition of such nanoclay remarkably increased the amount of covalently bound protein, probably due to its higher surface roughness and better affinity between SMP clay and the papain. Similarly, Basak et al. [29] demonstrated that the addition of bentonite into chitosan beads (ratio 50:50% *w/w*) significantly increased the catalase immobilization yield. Furthermore, the addition of OPT nanoclay significantly decreased the IY, with the only exception of OPT 70. These data appeared consistent with those described by Cacciotti et al. [15], who found that the IY slightly decreased when the bromelain from pineapple stem was immobilized on nanocomposites obtained combining MMT (i.e., SMP and OPT) with low amounts of CS in different relative ratios (MMT:CS 70:30, 75:25 and 80:20, % *w/w*), compared to Clay-free support. Moreover, the lowest IY was found for OPT-30 and OPT-50 films (26 and 29%, respectively), whereas no relevant differences were revealed comparing the other samples.

Table 4. Immobilization yield and kinetic properties of papain immobilized on the CS–clay nanocomposite systems.

Sample	IY (%)	V_{max} (mIU $mg^{-1}{}_{PI}$)	K_M (μM)	k_{cat} (min^{-1})	K_a (min^{-1} μM^{-1})	R^2
Clay-free	$45 \pm 6^{b,c}$	6.9 ± 0.2^{e}	88 ± 9^{a}	3183 ± 0^{e}	36 ± 1^{d}	0.99
OPT-20	$41 \pm 4^{c,d}$	15.3 ± 0.9^{c}	$75 \pm 16^{a,b}$	6535 ± 1^{c}	87 ± 4^{c}	0.97
OPT-30	26 ± 5^{e}	21.4 ± 1.0^{b}	106 ± 23^{a}	12507 ± 1^{b}	118 ± 6^{b}	0.98
OPT-50	29 ± 2^{de}	38.9 ± 2.0^{a}	102 ± 14^{a}	20417 ± 2^{a}	199 ± 3^{a}	0.99
OPT-70	55 ± 3^{b}	11.6 ± 0.5^{d}	$50 \pm 8^{b,c}$	4225 ± 0^{d}	85 ± 3^{c}	0.98
SMP-20	$54 \pm 7^{a,b,c}$	2.6 ± 0.1^{f}	34 ± 8^{c}	1017 ± 0^{h}	$30 \pm 2^{d,e}$	0.97
SMP-30	$41 \pm 5^{c,d}$	3.5 ± 0.2^{f}	$46 \pm 11^{b,c}$	642 ± 0^{i}	14 ± 1^{f}	0.97
SMP-50	56 ± 2^{b}	4.5 ± 0.1^{f}	$49 \pm 2^{b,c}$	1721 ± 0^{g}	25 ± 0^{e}	1.00
SMP-70	60 ± 4^{a}	7.3 ± 0.1^{e}	29 ± 3^{c}	2503 ± 0^{f}	86 ± 1^{c}	1.00

IY = immobilization yield, V_{max} = maximum velocity, K_M = Michaelis–Menten constant, k_{cat} = turnover number, K_a = affinity constant. For each parameter, values with different roman letters (a–i) are significantly different according to Tukey's test ($p < 0.05$).

The kinetic characterization of the biocatalysts was performed in order to explore the effect of the nanoclays' inclusion within the polymeric matrix on the behavior of immobilized papain. All the biocatalysts followed the hyperbolic trend described by the Michaelis–Menten equation (Figure 4) and the related catalytic properties are reported in Table 4. Papain on OPT films showed higher V_{max} and k_{cat} values with respect to the enzyme linked to all the other supports (Clay-free and SMP-based films). Notably, the highest V_{max} and k_{cat} values were found for papain immobilized on OPT-50 (38.9 mIU/mg_{IP} and 20,417 min^{-1}, respectively), thus indicating a greater product release velocity. However, both parameters significantly increased as the percentage of clay in the nanocomposite increased, except for the ratio 30:70% *w/w* (Table 4). Basak et al. [29] described similar results, finding an increased V_{max} value (about +20%) for lipase immobilized on CS–clay (hydrophilic bentonite) with respect to the neat CS support. As widely described in the literature for α-amylase and α-amylase immobilized on clay-CS carriers, an increased V_{max} value may arise from the conformational changes of biocatalysts, which usually occur after the linkage to the support [30].

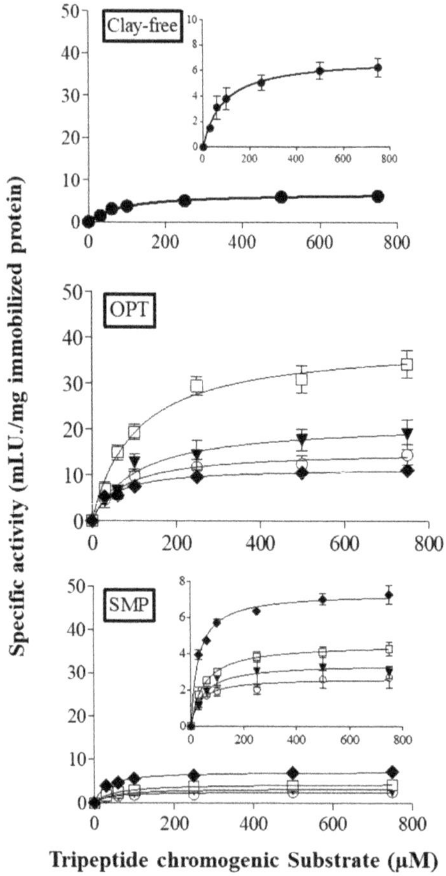

Figure 4. Kinetic curves of papain immobilized on CS–clay composite systems loaded with OPT or SMP at various weight percentages: 20 (○), 30 (▼), 50 (□) and 70 (♦) % *w/w*.

In contrast, papain immobilized on SMP-20, SMP-30 and SMP-50 films showed significantly lower V_{max} and k_{cat} values with respect to the enzyme linked to Clay-free support, whereas no significant differences were revealed for SMP-70 (Table 4). These results are in agreement with what Cacciotti et al. [14] reported for another protease belonging to the papain-superfamily (bromelain from pineapple stem).

All SMP biocatalysts and OPT-70 showed a significant decrease in K_M values (on average −55% and −43% for SMP-based samples and OPT-70, respectively) as compared with Clay-free. No remarkable differences were found for OPT-20, OPT-30 and OPT-50 (Table 4), thus indicating no changes in the enzyme–substrate complex formation in comparison to the reference biocatalyst. Similar results were described by Basak et al. [29], who found comparable K_M values for the lipase immobilized on CS and on CS–clay. Moreover, the K_a values revealed for papain immobilized on OPT films were 2.4 (OPT-20 and OPT-70) to 5.5-fold (OPT-50) higher as compared with the enzyme immobilized on Clay-free, suggesting a greater apparent affinity of the protease on the aforementioned nanocomposites for the synthetic substrate. Concerning the biocatalysts on the SMP-based supports, their K_a values always appeared lower or not statistically different (i.e., SMP-20) with respect to Clay-free film.

Overall, in spite of the lowest IY observed for papain immobilized on OPT-based films, OPT-30 and OPT-50 exhibited the best catalytic performances, both in terms of V_{max}, k_{cat} and K_a, thus indicating a higher efficiency in the release of the reaction product. As known, the decrease in the biocatalyst activity could be ascribed to a great amount of enzyme molecules immobilized on the carrier, since the latter could limit the accessibility of the substrate to the active sites, due to diffusion limitation [31]. Furthermore, as described in the literature [13], the improved catalytic properties observed for papain bound on OPT films could be likely due to the hydrophilic nature of OPT clay, which contributes to create an environment conducive to the interaction between substrate and enzyme. Otherwise, the hydrophobic clay which constitutes SMP-based films, does not facilitate the aforementioned interaction [13].

Thus, among all the produced biocatalysts, papain immobilized on the OPT-50 film was selected as the most suitable to be applied for the protein stabilization treatment of white wines.

3.3. Wine Stabilization Treatment in Batch-Scale Stirred Reactor

The partial stabilization of white wines via proteolytic treatment has been already described by other authors [1]. Benucci et al. [20] described the efficacy of a laboratory bench-scale packed-bed reactor, containing papain bound on chitosan beads, in stabilizing white wines.

The effectiveness of papain immobilized on OPT-50 film was tested in a laboratory-scale stirred reactor with the aim of reducing the protein instability of two real white wines (i.e., Manzoni and Sauvignon Blanc), characterized by a different haze potential (601 ± 11 ΔNTU and 116 ± 8 ΔNTU, respectively), as well as by a different initial protein content (465 ± 63 mg_{BSAeq}/L and 124 ± 1 mg_{BSAeq}/L, respectively). Papain covalently immobilized on OPT-50 nanocomposite film significantly reduced both the haze potential and the protein content in the two wine samples (Table 5). The biocatalyst exhibited the greatest stabilization efficiency in Manzoni wine, with a TRY reduction rate of about 83% and a decrease in total protein amount of approximately 73%. The lowest stabilization efficiency revealed in the Sauvignon Blanc wine (TRY reduction: 31%; total protein reduction: 12%) could be ascribed to the higher content of potential inhibitors found in Sauvignon Blanc with respect to Manzoni wine, especially SO_2 free (about 2-fold higher in the former compared to the latter wine wine) and the highest alcohol content (about 2% v/v). Indeed, it has been proven that these substances exhibit a strong inhibiting action against the catalytic activity of papain [20].

Table 5. Protein stability data of Manzoni and Sauvignon blanc white wines, treated in a laboratory-scale stirred reactor with papain immobilized on OPT-50 carrier at 20 °C (120 rpm, 24 h). Reported data are mean of triplicate measurements.

Treatment	Net Haze after Heat Test		Residual Protein Content	
	ΔNTU	Reduction (TRY, %)	mg $_{BSAeq}$/L	Reduction (%)
		Manzoni		
Untreated wine	601 ± 11 ^{a}	-	465 ± 63 ^{a}	-
OPT-50	102 ± 41 ^{b}	83	124 ± 35 ^{b}	73
		Sauvignon Blanc		
Untreated wine	116 ± 8 ^{a}	-	124 ± 1 ^{a}	-
OPT-50	80 ± 1 ^{b}	31	109 ± 1 ^{b}	12

TRY (%) = Turbidity removal yield expressed as the percentage of haze potential decrease due to the proteolytic treatment. For each parameter, values with different roman letters (a,b) are significantly different according to Tukey's test ($p < 0.05$).

4. Conclusions

The CS-based nanocomposite films produced by the addition of two nanoclays (OPT and SMP) showed improved mechanical properties. The better distribution of OPT clay within the polymeric matrix, as proved by the SEM micrographs and DSC data, imparted a higher Young modulus and σ_{max} to the OPT nanocomposites.

Papain was successfully bound on such carriers, showing the highest efficiency in the release of the reaction product after the covalent immobilization on OPT-30 and OPT-50. The latter biocatalyst reliably reduced both the haze potential and the protein content in the two tested white wine samples, namely Manzoni and Sauvignon Blanc, even if the stabilization efficiency was affected by the content of the potential inhibitors in the wine.

Author Contributions: Conceptualization, I.B. and I.C.; methodology, I.B. and I.C.; validation, I.B., I.C. and M.E.; formal analysis, I.B. and I.C.; investigation, I.B., C.L. and I.C.; resources, I.C. and M.E.; data curation, I.B. and I.C.; writing—original draft preparation, I.B. and I.C.; writing—review and editing, M.E.; supervision, I.C. and M.E.; project administration, I.C. and M.E.; funding acquisition, I.C. and M.E. All authors have read and agreed to the published version of the manuscript.

Funding: This research was supported by BioEnBi project "Biotecnologie enzimatiche innovative per processi di chiarifica sostenibili nel settore birrario" (Grant 85-2017-15362), funded by Lazio Innova Spa, Lazio Region (Italy) in the context of Progetti Gruppi di Ricerca, Lazio Innova 2018–2020.

Conflicts of Interest: The authors declare no conflict of interest.

References

1. Ottone, C.; Romero, O.; Aburto, C.; Illanes, A.; Wilson, L. Biocatalysis in the winemaking industry: Challenges and opportunities for immobilized enzymes. *Compr. Rev. Food Sci. Food Saf.* **2020**, *19*, 595–621. [CrossRef]
2. Verma, M.L.; Kumar, S.; Das, A.; Randhawa, J.S.; Chamundeeswari, M. Chitin and chitosan-based support materials for enzyme immobilization and biotechnological applications. *Environ. Chem. Lett.* **2020**, *18*, 315–323. [CrossRef]
3. Zdarta, J.; Meyer, A.S.; Jesionowski, T.; Pinelo, M. A general overview of support materials for enzyme immobilization: Characteristics, properties, practical utility. *Catalysts* **2018**, *8*, 92. [CrossRef]
4. Liburdi, K.; Straniero, R.; Benucci, I.; Garzillo, A.M.V.; Esti, M. Lysozyme immobilized on micro-sized magnetic particles: Kinetic parameters at wine pH. *Appl. Biochem. Biotechnol.* **2012**, *166*, 1736–1746. [CrossRef]
5. Zappino, M.; Cacciotti, I.; Benucci, I.; Nanni, F.; Liburdi, K.; Valentini, F.; Esti, M. Bromelain immobilization on microbial and animal source chitosan films, plasticized with glycerol, for application in wine-like medium: Microstructural, mechanical and catalytic characterizations. *Food Hydrocoll.* **2015**, *45*, 41–47. [CrossRef]
6. Benucci, I.; Lombardelli, C.; Cacciotti, I.; Liburdi, K.; Nanni, F.; Esti, M. Chitosan beads from microbial and animal sources as enzyme supports for wine application. *Food Hydrocoll.* **2016**, *61*, 191–200. [CrossRef]
7. Benucci, I.; Mazzocchi, C.; Lombardelli, C.; Cacciotti, I.; Esti, M. Multi-enzymatic systems immobilized on chitosan beads for pomegranate juice treatment in fluidized bed reactor: Effect on haze-active molecules and chromatic properties. *Food Bioprocess Tech.* **2019**, *12*, 1559–1572. [CrossRef]
8. Cacciotti, I.; Pallotto, F.; Scognamiglio, V.; Moscone, D.; Arduini, F. Reusable optical multi-plate sensing system for pesticide detection by using electrospun membranes as smart support for acetylcholinesterase immobilisation. *Mater. Sci. Eng. C* **2020**, *111*, 110744. [CrossRef]
9. Yang, Y.; Yang, H.; Yang, M.; Shen, G.; Yu, R. Amperometric glucose biosensor based on a surface treated nanoporous ZrO_2/chitosan composite film as immobilization matrix. *Anal. Chim. Acta* **2004**, *525*, 213–220. [CrossRef]
10. Yang, H.C.; Hou, J.; Chen, V.; Xu, Z.K. Surface and interface engineering for organic—Inorganic composite membranes. *J. Mater. Chem. A* **2016**, *4*, 9716–9729. [CrossRef]
11. Kumar, S.; Ye, F.; Dobretsov, S.; Dutta, J. Chitosan nanocomposite coatings for food, paints, and water treatment applications. *Appl. Sci.* **2019**, *9*, 2409. [CrossRef]
12. Wu, Z.; Huang, X.; Li, Y.C.; Xiao, H.; Wang, X. Novel chitosan films with laponite immobilized Ag nanoparticles for active food packaging. *Carbohydr. Polym.* **2018**, *199*, 210–218. [CrossRef] [PubMed]

13. Benucci, I.; Liburdi, K.; Cacciotti, I.; Lombardelli, C.; Zappino, M.; Nanni, F.; Esti, M. Chitosan/clay nanocomposite films as supports for enzyme immobilization: An innovative green approach for winemaking applications. *Food Hydrocoll.* **2018**, *74*, 124–131. [CrossRef]
14. Jamróz, E.; Kulawik, P.; Kopel, P. The effect of nanofillers on the functional properties of biopolymer-based films: A review. *Polymers* **2019**, *11*, 675. [CrossRef] [PubMed]
15. Cacciotti, I.; Lombardelli, C.; Benucci, I.; Esti, M. Clay/chitosan biocomposite systems as novel green carriers for covalent immobilization of food enzymes. *J. Mater. Res. Technol.* **2019**, *8*, 3644–3652. [CrossRef]
16. Saleh, T.A. Nanomaterials: Classification, properties, and environmental toxicities. *Environ. Technol. Inno.* **2020**, 101067. [CrossRef]
17. Islam, M.A.; Tan, Y.L.; Islam, M.A.; Romić, M.; Hameed, B.H. Chitosan–bleaching earth clay composite as an efficient adsorbent for carbon dioxide adsorption: Process optimization. *Colloids Surf. A Physicochem. Eng. Asp.* **2018**, *554*, 9–15. [CrossRef]
18. Mastanjević, K.; Krstanović, V.; Lukinac, J.; Jukić, M.; Vulin, Z.; Mastanjević, K. Beer—The importance of colloidal stability (non-biological haze). *Fermentation* **2018**, *4*, 91. [CrossRef]
19. Cerreti, M.; Liburdi, K.; Benucci, I.; Spinelli, S.E.; Lombardelli, C.; Esti, M. Optimization of pectinase and protease clarification treatment of pomegranate juice. *LWT Food Sci. Technol.* **2017**, *82*, 58–65. [CrossRef]
20. Benucci, I.; Lombardelli, C.; Liburdi, K.; Acciaro, G.; Zappino, M.; Esti, M. Immobilised native plant cysteine proteases: Packed-bed reactor for white wine protein stabilisation. *J. Food Sci. Tech.* **2016**, *53*, 1130–1139. [CrossRef]
21. Bradford, M.M. A rapid and sensitive method for the quantification of microgram quantities of protein utilizing the principle of protein-dye binging. *Anal. Biochem.* **1976**, *72*, 248–254. [CrossRef]
22. Gaspar, L.M.; Machado, A.; Coutinho, R.; Sousa, S.; Santos, R.; Xavier, A.; Simões, J.S. Development of potential yeast protein extracts for red wine clarification and stabilization. *Front. Microbial.* **2019**, *10*, 2310. [CrossRef] [PubMed]
23. Colangelo, D.; Torchio, F.; De Faveri, D.M.; Lambri, M. The use of chitosan as alternative to bentonite for wine fining: Effects on heat-stability, proteins, organic acids, colour, and volatile compounds in an aromatic white wine. *Food Chem.* **2018**, *264*, 301–309. [CrossRef] [PubMed]
24. Chuang, W.Y.; Young, T.H.; Yao, C.H.; Chiu, W.Y. Properties of the poly (vinyl alcohol)/chitosan blend and its effect on the culture of fibroblast in vitro. *Biomaterials* **1999**, *20*, 1479–1487. [CrossRef]
25. Kittur, F.S.; Prashanth, K.H.; Sankar, K.U.; Tharanathan, R.N. Characterization of chitin, chitosan and their carboxymethyl derivatives by differential scanning calorimetry. *Carbohydr. Polym.* **2002**, *49*, 185–193. [CrossRef]
26. Sakurai, K.; Maegawa, T.; Takahashi, T. Glass transition temperature of chitosan and miscibility of chitosan/poly (N-vinyl pyrrolidone) blends. *Polymer* **2000**, *41*, 7051–7056. [CrossRef]
27. Zeng, M.; Fang, Z.; Xu, C. Effect of compatibility on the structure of the microporous membrane prepared by selective dissolution of chitosan/synthetic polymer blend membrane. *J. Membr. Sci.* **2004**, *230*, 175–181. [CrossRef]
28. Garavand, F.; Rouhi, M.; Razavi, S.H.; Cacciotti, I.; Mohammadi, R. Improving the integrity of natural biopolymer films used in food packaging by crosslinking approach: A review. *Int. J. Biol. Macromol.* **2017**, *104*, 687–707. [CrossRef]
29. Basak, E.; Aydemir, T.; Dinçer, A.; Becerik, S.Ç. Comperative study of catalase immobilization on chitosan, magnetic chitosan and chitosan-clay composite beads. *Artif. Cells Nanomed. Biotechnol.* **2013**, *41*, 408–413. [CrossRef]
30. Chang, M.Y.; Juang, R.S. Activities, stabilities, and reaction kinetics of three free and chitosan–clay composite immobilized enzymes. *Enzyme Microb. Technol.* **2005**, *36*, 75–82. [CrossRef]
31. Jiang, D.S.; Long, S.Y.; Huang, J.; Xiao, H.Y.; Zhou, J.Y. Immobilization of pycnoporus sanguineus laccase on magnetic chitosan microspheres. *Biochem. Eng. J.* **2005**, *25*, 15–23. [CrossRef]

MDPI
St. Alban-Anlage 66
4052 Basel
Switzerland
Tel. +41 61 683 77 34
Fax +41 61 302 89 18
www.mdpi.com

Nanomaterials Editorial Office
E-mail: nanomaterials@mdpi.com
www.mdpi.com/journal/nanomaterials

www.ingramcontent.com/pod-product-compliance
Ingram Content Group UK Ltd.
Pitfield, Milton Keynes, MK11 3LW, UK
UKHW060458050826
14089UKWH00044B/131

9 783036 547008